Biology and Integrated Management of Turfgrass Diseases

---

by

Gary W. Beehag

Dr Nathan R. Walker

Dr Percy T.W. Wong

and

Jyri Kaapro

CABI is a trading name of CAB International

CABI
Nosworthy Way
Wallingford
Oxfordshire OX10 8DE
UK

CABI
200 Portland Street
Boston
MA 02114
USA

Tel: +44 (0)1491 832111
E-mail: info@cabi.org
Website: www.cabi.org

Tel: +1 (617)682-9015
E-mail: cabi-nao@cabi.org

© Gary W. Beehag, Nathan R. Walker, Percy T.W. Wong, Jyri Kaapro 2024. All rights reserved. No part of this publication may be reproduced in any form or by any means, electronically, mechanically, by photocopying, recording or otherwise, without the prior permission of the copyright owners.

The views expressed in this publication are those of the author(s) and do not necessarily represent those of, and should not be attributed to, CAB International (CABI). Any images, figures and tables not otherwise attributed are the author(s)' own. References to internet websites (URLs) were accurate at the time of writing.

CAB International and, where different, the copyright owner shall not be liable for technical or other errors or omissions contained herein. The information is supplied without obligation and on the understanding that any person who acts upon it, or otherwise changes their position in reliance thereon, does so entirely at their own risk. Information supplied is neither intended nor implied to be a substitute for professional advice. The reader/user accepts all risks and responsibility for losses, damages, costs and other consequences resulting directly or indirectly from using this information.

CABI's Terms and Conditions, including its full disclaimer, may be found at https://www.cabi.org/terms-and-conditions/.

A catalogue record for this book is available from the British Library, London, UK.

ISBN-13: 9781789246216 (hardback)
9781789246223 (ePDF)
9781789246230 (ePub)

DOI: 10.1079/9781789246230.0000

Commissioning Editor: Rebecca Stubbs
Editorial Assistant: Lauren Davies/Emma McCann
Production Editor: Shankari Wilford

Typeset by Straive, Pondicherry, India
Printed and bound in the UK by CPI Group (UK) Ltd, Croydon, CR0 4YY

Mark
Best
[signature]

Biology and Integrated Management of Turfgrass Diseases

Contents

About the Authors	xi
Foreword	xiii
Preface	xv
Acknowledgements	xvii

1 Turfgrass Ecosystems — 1
  Turfgrass Plants — 1
    Primary grasses — 2
    Secondary grasses — 3
    Experimental grasses — 3
    Mown forbs — 4
  Physiological Classification, Geographic Origin and Climatic Adaptation — 5
  Turfgrass Microbiome — 6
  Turfgrass Pathosystem — 6
  Diseases in Monoculture versus Polyculture Turfgrass Systems — 7
    Factors influencing monoculture diseases — 7
  New-Encounter Turfgrass Pathogens and Diseases — 9
  Economic Significance of Turfgrass Diseases — 10
    Disease management versus control — 12
  Layout of This Book — 12

2 Environment, Host and Pathogen: Disease Epidemiology — 16
  Favourable Environment — 16
    Phyllosphere — 16
    Thatch-mat region — 17
    Rhizosphere — 17
  Susceptible Host — 18
    Primary, secondary and reservoir hosts — 18
    Inherited plant factors — 18
    Management-induced factors — 19
    Turfgrass stress as a disease factor — 20

		Page
	Host defence mechanisms: how turfgrasses defend against pathogens	21
	Plant disease resistance	22
	Phytopathogenic Microbes	23
	Host–pathogen relationship	23
	Disease pressure	23
	Pathogen attack mechanisms: how pathogens infect turfgrass hosts	23
	How invading pathogens disrupt their host	25
	Probable Impacts of Climate Change on Turfgrass Diseases	26
3	**Ecological Groups and the Environment of Turfgrass Microorganisms**	**29**
	Microflora and Microfauna Groups	29
	Microbial Lifestyles: Parasitic, Pathogenic and Saprophytic	30
	Symbiotic Relationships: Endophyte, Mycorrhizae and Rhizobacteria	30
	The Environment of Turfgrass Microbes	31
	Temperature	32
	Moisture	32
	Rootzone oxygen	35
	Organic matter accumulation	35
	Rootzone pH	35
	Host nutrition	35
	Hydrophobicity	35
	Disease-suppressive soils	35
	Solar radiation	36
	Wind	36
	Turfgrass Disease Cycles	36
	Overwintering/oversummering of pathogens	38
	Disease Epidemics, Complexes and Succession	38
4	**Monitoring, Forecasting, Symptomology, Sampling and Diagnosis**	**42**
	Surveillance and Biosecurity	42
	Monitoring of Turfgrass Quality and Key Weather Factors	42
	Disease Prediction and Forecasting	43
	Disease Symptomology	45
	Recognition of early, advancing and late-stage symptoms	47
	Detection of disease signs	49
	Turfgrass Disease Diagnosis	50
	Sample collection, preservation, packaging and submission	52
	Laboratory diagnostic methodologies	53
	Koch's postulates	55
	Understanding and Interpretation of Disease Diagnostic Reports	56
	Understanding Turfgrass Disease Trial Data	57
5	**Genetic, Cultural and Biological Management of Turfgrass Diseases**	**59**
	Disease-Resistant Cultivars	59
	Sanitation	61
	Primary Cultural Practices	61
	Mowing regime	61
	Irrigation scheduling	62
	Fertilization regime	63
	Rolling scheduling	65
	Secondary Cultural Practices	65
	Consumable Chemical Products	68
	Biological Turfgrass Disease Management Options	70

	Microbial inoculants	70
	Organic amendments and suppressive composts	72
	Biofumigation	72
	Putting It All Together – Compiling an Integrated Disease Management Plan	73
6	**Cool-Weather Fungal Diseases of Mown Grasses**	**78**
	Foliage-infecting fungal diseases	78
	Downy Mildew	79
	Limonomyces-Incited Diseases	80
	Pink patch	80
	Cream leaf blight	82
	Powdery Mildew	82
	Pythium-Incited Diseases	83
	Pythium nomenclature	84
	Cool-weather *Pythium* diseases	86
	Damping-off	86
	Pythium crown and root rot	87
	Pythium spring dead spot	87
	Red Thread	88
	Rust and Smut Diseases	90
	Rusts	91
	Smuts	91
	Stem- and root-infecting fungal diseases	92
	Fairy Ring Formations	93
	Edaphic fairy rings	95
	Lectophilic fairy rings	97
	Thatch collapse	98
	Gaeumannomyces-Incited Diseases	101
	Take-all patch of cool-season grasses	101
	Microdochium Patch	107
	Ophiosphaerella-Incited Diseases	114
	Spring dead spot	114
	Necrotic ring spot	121
	Dead spot	123
	Rhizoctonia- and *Waitea*-Incited Diseases	125
	Rhizoctonia and *Waitea* nomenclature	125
	Cool-weather *Rhizoctonia* diseases	127
	Large patch	127
	Yellow patch	129
	Root Decline of Warm-Season Grasses	130
	Characterization of the causal agents	132
	Global distribution and host range	133
	Specific root decline diseases of warm-season grasses	134
	Bermudagrass decline	134
	Fairway patch	134
	Summer decline	135
	Lesser-known root decline diseases	135
	Diagnosis of root decline diseases of warm-season grasses	135
	Snow Mould Diseases	143
	Snow moulds and their causal agents	143
	Lesser-Known, Cool-Weather Fungal Diseases	148

7	**Warm-Weather Fungal Diseases of Mown Grasses**	158
	Foliage-borne fungal diseases	158
	Colletotrichum-Incited Diseases	159
	Anthracnose on cool-season and warm-season grasses	159
	Copper Spot	164
	Dollar Spot	166
	Grey Leaf Spot	171
	Helminthosporium Group Diseases	174
	Bipolaris and *Exsorohilum*	176
	Pyrenophora (*Drechslera*)	176
	Curvularia-incited diseases	176
	Signs and symptoms of leaf spot diseases	177
	Leptosphaerulina Leaf Blight	180
	Nigrospora Leaf Blight	182
	Lesser-Known, Warm-Weather, Foliage Fungal Diseases	183
	Stem- and root-infecting fungal diseases	184
	Kikuyu Yellows	184
	Magnaporthiopsis-Incited Diseases	186
	Summer patch	187
	Warm-Weather *Pythium* Diseases	190
	Pythium blight	191
	Pythium root dysfunction	193
	Warm-Weather *Rhizoctonia*- and *Waitea*-Incited Diseases	194
	Brown patch	194
	Brown ring patch	199
	Basal leaf blight	200
	Leaf and sheath spot	201
	Southern Blight	204
8	**Fungal Diseases of Mown Forbs**	215
	Mown Forbs	215
	Chamomile	215
	Clover	216
	Colobanthus	216
	Cotula	216
	Dichondra	216
	Pinto peanut	216
	Starweed	216
	Fungal Diseases of Mown Forbs	216
	Alternaria leaf spot	218
	Cercospora leaf spot	218
	Fairy ring	218
	Gold bracelet disease	218
	Leaf rust	220
	Phytophthora root rot	220
	Rhizoctonia brown patch	221
	Rolf's disease (southern blight)	221
	Sclerotinia patch	222
	Winter Pythium patch	224
	White patch disease	225

9	**Plant-Pathogenic Bacteria, Primitive Microbes and Viruses**	227
	Bacterial and Phytoplasma Diseases	228
	Bacterial wilt	228
	Bacterial decline	230
	Bermudagrass white leaf	230
	Cyanobacteria and yellow spot disease	231
	Primitive Foliar and Root-Infecting Microbes	234
	Polymyxa root rot	235
	Rapid blight	235
	Slime moulds	237
	Viruses and Viral Diseases	237
	St. Augustine decline (SAD)	239
	Mosaic disease of St. Augustinegrass	240
	Centipede grass mosaic disease (CGMD)	240
	Mosaic disease of Queensland blue couch	240
	Zoysia dwarf virus/mosaic virus	241
	Management of Plant-Pathogenic Bacterial, Phytoplasma and Virus Diseases	241
10	**Plant-Parasitic and Beneficial Nematodes**	245
	Significance of Turfgrass-Parasitic Nematodes	245
	Nematode Feeding Mechanisms	246
	Genera of Turfgrass-Parasitic Nematodes	247
	Parasitic Nematode Feeding Groups	247
	Economically Important Turfgrass-Parasitic Nematodes	248
	Ectoparasitic	248
	Endoparasitic	250
	Foliar nematode parasites	251
	Basic Nematode Biology	252
	Nematode population dynamics	252
	Nematode damage thresholds	254
	Relative pathogenicity	255
	Root-Feeding Nematode Signs and Damage Symptoms	256
	Aboveground symptoms	256
	Belowground symptoms	256
	Confirmation of root-feeding nematodes	258
	Systems Approach to Parasitic Nematode Management	259
	Surveillance and biosecurity programmes	260
	Nematode population monitoring and sampling	260
	Nematode-resistant varieties	261
	Cultural practices	261
	Nematicides and bionematicides	263
11	**Understanding Turfgrass Protectant Pesticides**	271
	Registration of Turfgrass Protectants	271
	Turfgrass protectant label nomenclature	271
	Turfgrass Protectant Formulations	272
	Formulation components	272
	How Fungicides and Biofungicides Work	272
	Biochemical mode of action	273
	Fungicide phytomobility	274
	Preventive versus curative strategy	274

Product Application and Placement	274
Foliar application and rootzone targeting	275
Managing foliar application	276
Managing rootzone placement	277
Maximizing fumigant efficacy	278
Fungicide Tank Mixtures, Compatibility and Synergism	278
Chemistry compatibility testing	279
Generalized tank-mixing guidelines	280
Spray tank water considerations	280
Spray nozzle selection	281
Spray system cleaning	281
Environmental Behaviour and Fates of Turfgrass Protectants	282
Aboveground behaviour and fates	282
Belowground behaviour and fates	283
Minimizing fungicide non-target impacts	285
Fungicide Resistance in Turfgrass Systems	287
Determination of fungicide resistance	287
Fungicide resistance risk components	287
Cross-resistance and multiple resistance	288
Turfgrass fungicide resistance case studies	288
Fungicide resistance management strategies	290
Glossary	295
Index	301

About the Authors

Gary W. Beehag
Gary is semi-retired but is actively involved in the Australian turfgrass industry. Gary has college qualifications in turfgrass management, horticulture and teaching and has held several positions since the 1970s. Gary was employed as Senior Turfgrass Consultant at the Australian Turfgrass Research Institute (ATRI), Sydney (New South Wales (NSW), Australia) and was awarded an NSW Churchill Fellowship investigating turfgrass courses and research in the USA. Gary has written many extension-based articles and is a member of the Australian Sports Turf Managers Association (ASTMA).

Gary has spoken at regional, state, national and international turfgrass conferences throughout Australia and in the USA, New Zealand, Singapore and Scotland. Gary and Jyri (with Dr A. Manners) co-authored *Pest Management of Turfgrass for Sport and Recreation* (CSIRO Publishing, 2016). Gary has a long association with Dr Percy Wong (Australia) and is currently cooperating in the investigation of the biological causes of new-encounter turfgrass diseases in Australia.

Nathan R. Walker
Dr Walker is a Professor, Turfgrass Integrated Pest Management/Turfgrass Pathology at Oklahoma State University (Stillwater, Oklahoma, USA). Dr Walker obtained his Bachelor's degree in Environmental Planning (1993) from Bloomsburg University (Pennsylvania), his Master's degree in Plant Pathology (1996) from Clemson University (South Carolina) and his Doctorate in Plant Science (1999) from the University of Arkansas (Arkansas USA).

Dr Walker joined the Department of Entomology and Plant Pathology at Oklahoma State University in 1999. Key professional responsibilities are Turfgrass IPM/Turfgrass Pathology and his appointment is 78% research, 13% teaching (undergraduate Turfgrass IPM, Graduate IPM, Pesticide Applications and Plant Nematology), and 9% Extension. Among his memberships are The International Turfgrass Society (ITS), European Turfgrass Society (ETS) and the Oklahoma Turfgrass Research Foundation. He is currently the Editor of *International turfgrass*, the newsletter of The International Turfgrass Society.

Dr Walker's research specialization is the identification, biology and integrated management of soil-borne pathogens and diseases of turfgrasses in the United States and elsewhere. Dr Walker has written a book chapter, numerous peer-reviewed papers and non-peered extension articles regarding turfgrass diseases and their ecology and management. He is also the Director of the OSU Turfgrass Disease Diagnostic Laboratory.

Percy T.W. Wong

Dr P.T.W. Wong is an Honorary Associate of the University of Sydney (Camden Campus), NSW, Australia. He is an eminent plant pathologist and mycologist specializing in soil-borne fungal pathogens. Dr Wong obtained his Bachelor of Science in Agriculture (Hons) degree and his Doctorate in Plant Pathology from the University of Sydney. He was employed by the NSW Department of Agriculture (1972–2006) as Principal Research Scientist and worked on the etiology, biology and management (especially biological control) of several diseases of field crops and pastures.

During 1980–1981, he was Visiting Professor at Colorado State University (Fort Collins, Colorado, USA) and the first person to identify the soil-borne pathogen *Gaeumannomyces avenae* from turfgrass in Colorado. Dr Wong has been an Honorary Associate (University of Sydney) for many years and Adjunct Associate Professor (University of Western Sydney). He is a Fellow of the Australasian Plant Pathology Society and a member of the Australasian Mycological Society, the American Phytopathological Society (APS) and the International Turfgrass Society (ITS). Dr Wong's interest in turfgrass diseases began in the 1980s; he has spoken at numerous national and international conferences, co-authored one book and written numerous peer-reviewed papers and non-peered extension articles regarding plant pathogens and diseases in national and international journals.

Jyri Kaapro

Jyri Kaapro is Senior Market Development Specialist for Envu (Australia/New Zealand). Jyri obtained his Bachelor of Natural Resource Management at the University of New England (Armidale, NSW, Australia), a Graduate Diploma in Agriculture at Charles Sturt University (Wagga Wagga, NSW, Australia) and a Master of Agriculture, Turf Management at the University of Sydney (NSW, Australia). Jyri has been involved in the Australian turfgrass industry since the 1990s and was employed at the Australian Turfgrass Research Institute (ATRI), Sydney as Research Manager.

Jyri is a member of the Australian Sports Turf Managers Association (ASTMA) and the International Turfgrass Society (ITS). Jyri has attended and presented at numerous local, state and national seminars and conferences, and has attended international conferences of the ITS. He has written numerous extension-based papers in journals associated with the Australian turfgrass industry. Jyri and Gary (with Dr A. Manners) co-authored *Pest Management of Turfgrass for Sport and Recreation* (CSIRO Publishing, 2016).

Foreword

Much as we would like it to be otherwise, disease seems to be an ever-present threat to the integrity of intensively managed sports turf such as golf greens and bowling greens. The disease pressure eases to some extent with lower levels of management, but turf of this kind, including golf fairways and high-class recreational lawns, is still subject to periodic disease outbreaks that can be disfiguring and destructive.

The scope of this new multi-author book is clearly indicated by its title, namely *Biology and Integrated Management of Turfgrass Diseases*. A comprehensive approach is taken, grouped around the themes of: (i) raising awareness of the physical and biological complexity of turfgrass systems; (ii) understanding the epidemiological importance of the interactions between the host plant, the pathogen and the total environment; (iii) providing knowledge to assist in identifying early signs and later symptoms of the diseases; and (iv) giving guidance on developing integrated disease management strategies to minimize frequency, severity and duration of disease.

Turf diseases occur on many different geographic scales. Some major diseases such as brown patch and dollar spot have almost worldwide distributions and can be expected to appear wherever susceptible host species are managed as turf. Others are more restricted and operate on a continental scale, for example in North America, while others again are localized diseases, heavily restricted in terms of grass species, soil type and locational environment. These distribution patterns raise many interesting issues. A disease confined to the USA at present may have the potential to spread to Western Europe. Based on a threat assessment, importation of host species material from the USA may be prohibited, the material may be admitted subject to quarantine restrictions, or if the threat level is considered very low, it may be allowed free entry in those countries. Until proved otherwise, very localized diseases, especially if capable of attacking more than one turf species, must be regarded as having the potential to spread, a tendency that may well be enhanced by climate change.

The fairway patch disease originating in eastern Sydney (New South Wales, Australia) is a particularly interesting case. The disease appeared suddenly in the year 2000 on a couch (bermudagrass) fairway at a golf club in Sydney's sand belt. Soon it was found at other sand-belt golf clubs nearby, on couch and in one instance on a kikuyugrass fairway. By 2020 it had been diagnosed from several other states of Australia and the North Island of New Zealand on closely mown couch. It is now regarded as a disease with potential to spread throughout the coastal areas of Australia. Much remains to be learnt about the mechanism by which it has spread so widely and so quickly. One suggestion is that interstate movement of infected turf sod could be responsible, but this has yet to be conclusively confirmed. Clearly other countries with extensive areas of closely mown couch turf need to be considering steps to exclude this apparently readily communicable and grossly disfiguring disease. Fairway patch disease is incited by the ectotrophic root-infecting (ERI) fungus *Phialocephala bamuru*. In common with many other soil-borne diseases, management has proved difficult and there are, at present, no effective chemical or cultural management tools available.

Several new patch diseases of turf incited by previously unknown fungal species have been described in Australia in the last 20 years. This has naturally caused interested people to wonder why so many new disease species have come to hand and whether this trend will continue. Some argue that the modern diagnostic methods, especially molecular biology (DNA) techniques, are simply providing finer resolution of aggregate species into forms that have always been there. While this is undoubtedly true in some cases, in others the incitants are either undescribed (new) species or virulent strains of species previously regarded as benign or of low pathogenicity, thereby bringing new 'players' into the game.

New ERI diseases are typically not controlled by existing widely used fungicides. This is probably because the incitant fungi have undergone natural selection over many years in environments regularly exposed to these chemicals. This could be a case of a classic Darwinian response. Some of the less susceptible species may have developed tolerance and with greatly reduced competition from fungi readily controlled by the chemical management programme, the less pathogenic species have been provided with an opportunity to colonize the roots. Once there, selection pressure among them will lead to an advantage for slightly more virulent forms and so on until ultimately a strongly virulent type emerges. Whatever the true explanation, it is a fact that many fungal species previously undescribed or not known to cause disease have now been shown to be serious pathogens of warm-season grasses.

The book does not seek, nor did it seek, to provide laboratory diagnostic information such as spore sizes and isolation techniques. However, it provides a solid account of signs and symptoms in the field as visible to the naked eye and as seen with the aid of a 10× magnification hand lens or a 30× magnification zoom stereo microscope. Good notes on the nature of the infestation in the field supplemented by a few well-chosen photographs are of considerable assistance to the laboratory in making a diagnosis.

Apart from the widespread common diseases, interpreting the disease samples can be very challenging for the pathologist, particularly when complex associations between fungal species and between fungi and nematodes are involved, or the species turn out to be undescribed. As a result, a full diagnosis may take weeks or even months when DNA support is needed, meaning that the provision of timely information for preparing urgently required integrated management plans is often not possible. In this situation it may be feasible to get an interim or indicative diagnosis that will allow the implementation of a provisional plan. However, there is still a need to persist with the diagnostics until a definite result is found, so that in the future, on the basis of the more readily observed signs and symptoms, a rapid provisional diagnosis can be made and treatment applied as appropriate.

With the increasing numbers of newly emerging and multi-organism diseases, not to mention the largely ignored area of virus diseases, the diagnostic process is becoming very demanding and there is an urgent need for additional turf pathologists, soil biologists and nematologists. There is also a need for highly trained turf agronomists to serve as a link between the laboratory specialists and the turf managers, because it is no longer realistic to expect a busy facility manager to single-handedly prepare a wide-ranging disease management plan on the basis of a brief report from a diagnostic laboratory.

Globally, the managed turf industry in all its branches is one of the largest and most valuable components of the amenity horticulture sector. I am sure that this new book, with its decidedly international and systematic approach, will make an important contribution to better management of turf diseases throughout the industry.

Dr Peter Martin, E.D., PhD, MScAgr, FLS
Formerly Director of Amenity Plant Science and
Leader of the Graduate Turf Management Program,
University of Sydney Plant Breeding Institute,
Cobbitty, New South Wales, Australia
27 July 2023

Preface

Biology and Integrated Management of Turfgrass Diseases assembles comprehensive and up-to-date information about microbial pathogens and their infectious diseases on turfgrass wherever managed. Infectious describes diseases incited by a living or biotic entity as opposed to non-infectious conditions simply considered aspects of the wider turfgrass system. For the purpose of this book the term 'turfgrass' is taken to cover a select group of perennial grasses and forbs capable of forming a perennial sward under close and continual defoliation.

Specifically, *Biology and Integrated Management of Turfgrass Diseases* covers the identification, ecology, epidemiology and integrated management of infectious diseases caused by bacterial, fungal, viral and nematode pathogens on mown grasses and forbs worldwide. All turfgrass systems, particularly highly valued swards, require an integrated approach for mitigating against diseases utilizing disease-resistant varieties and proven cultural, biological and chemical options.

Numerous species of grasses and a limited number of forbs (decumbent growing plants) are managed as mown swards for recreational purposes throughout the world. The vast majority of mown swards comprise grasses (family *Poaceae*) commonly called either cool-season or warm-season species. Forbs are members in several dicotyledonous families.

The framework of this book is set against widespread observations that highly managed grasses and forbs subject to mowing, irrigation and nutrition regimes, as opposed to the same plant species growing in natural grasslands, are prone to frequent and severe outbreaks of infectious diseases. Such observations raise two intriguing questions. First, why are highly managed turfgrass systems, as opposed to wild plants in natural ecosystems, subject to the reoccurrence of the same disease? Second, why do certain diseases arise for the first time on certain turfgrasses in new locations?

Biology and Integrated Management of Turfgrass Diseases has been specifically written for the needs of professional turfgrass managers, consultants, technical representatives and college students and others. The book is not encyclopaedic.

The driving force behind the researching and writing of this book was to assemble currently available and credible information about the economically important turfgrass diseases worldwide together with up-to-date information about the emergence and persistence of numerous turfgrass pathogens and their new-encounter diseases arising in many countries. Certain turfgrass diseases are no doubt more widespread than currently documented.

One assumption that requires acknowledgement is that a plethora of microorganisms, mostly benign or beneficial, some parasitic even pathogenic, are a normal part of any healthy turfgrass system. It is when pathogenic microbes become populous and highly competitive that it becomes obvious the ecological balance we wish to see operating as normal, becomes out of balance. Hence why *Biology and Integrated Management of Turfgrass Diseases* approaches infectious turfgrass diseases from an ecological and integrated management perspective.

Symptom expression of infectious turfgrass diseases varies widely and the so-called classic symptom for many diseases not always results. The traditional approach when diagnosing diseased turfgrass has been to assign the biological cause to a single pathogen. However, for certain diseases this assertion may be based on a false premise because it is increasingly obvious that symptomatic turfgrass plants, when examined microscopically, more often than not yield two or more different microorganism species. Hence, in cases of root-infecting pathogens, certain disease maladies need to be reconsidered a complex or syndrome of multiple pathogens – one primary, others secondary, perhaps even acting synergistically.

Biology and Integrated Management of Turfgrass Diseases focuses on the environmental and management-induced factors influencing pathogen growth and development and expression of their respective diseases, providing the basis for formulating integrated disease management plans based on implementation of disease-resistant cultivars and biosecurity and hygiene practices in combination with proven cultural, biological and chemical technologies.

Acknowledgements

This book has been many years in the making. Publication of the book was achieved with the generous cooperation, guidance and assistance of numerous persons worldwide. First and foremost, the authors sincerely thank the book production team at CABI (Rebecca Stubbs, Lucy Pritchard, Lauren Davies, Emma McCann, Shankari Wilford and Rachel Bowen.) for their support and guidance to achieve publication of the book. The authors sincerely thank Dr Peter Martin (University of Sydney, Australia) for enthusiastic guidance and writing of the Foreword.

We are indebted to the numerous scientists, plant pathologists, nematologists and turfgrass agronomists who enthusiastically provided guidance, constructive comments and editing of both relevant sections and entire chapters of the book.

We acknowledge the cooperation and guidance from Dr Michael Fidanza, Dr Paul Giordano, Dr James Hempfling, Dr Richard Latin and Dr Richard Smiley (United States), Dr Tom Hsiang (Canada), Dr Kate Entwistle, Dr Colin Fleming and Dr Ruth Mann (United Kingdom), Alex Glasgow, Brendan Hannan, David Howard, Dr Andrew Mitchell and David Ormsby (New Zealand), Dr Micah Woods (Thailand) and Dr Jordan Bailey, Dr Sophia Callaghan, Craig Campbell, Andrew Daly, Gary Dempsey, Michelle Dickinson, Dr Andrew Geering, Ken Johnston, Albie Leggett, Paul Looby, Dr Don Loch, Peter McMaugh AM, Bruce McPhee, John Neylan, David Nickson, Peter Ruscoe, Dr Allan Smith, Dr Brian Stynes, David Westall and David Worrad (Australia).

We are most grateful for the generous cooperation provided by the numerous bowling green-keepers, golf course superintendents and sportsground and tennis court managers for discussing their personal experiences and observations about turfgrass diseases and allowing photographs of symptoms to be taken at their respective venues.

Finally, we sincerely acknowledge the following persons and organizations who generously provided colour photographs under copyright for inclusion in this book:

- Chris Banks, The LBJ Library, Austin, Texas, USA.
- Andrew Daly, Elizabeth MacArthur Agricultural Institute, Menangle, New South Wales (NSW), Australia.
- Gary Dempsey, Tafe Teacher, Sydney, NSW, Australia.
- Ben Evans, Golf Course Superintendent, Sydney, NSW, Australia.
- Dr Colin Fleming, Maxstim Ltd, Belfast, UK.
- John Forrest, Turf Agronomist, Perth, Western Australia (WA), Australia.
- Dr Andrew D.W. Geering, University of Queensland, Queensland (QLD), Australia.
- Dr Paul Giordano, Harrell's LLC, Canton, Michigan, USA.

- Lawrie Greenup, Sydney, NSW, Australia.
- Brendan Hannan, New Zealand Sports Turf Institute, Palmerston North, New Zealand.
- Mark Hooker, Golf Course Superintendent, Auckland, New Zealand.
- Terry Howe, Golf Course Superintendent, NSW, Australia.
- Dr Tom Hsiang, University of Guelph, Guelph, Ontario, Canada.
- Ken Johnston, Turfgrass Agronomist, Perth, WA, Australia.
- Dr Andrew Mitchell, New Zealand Sports Turf Institute, Palmerston North, New Zealand.
- Dr Roger Shivas, University of Southern Queensland, Toowoomba, QLD, Australia.
- Dr Marcelle Stirling, Brisbane, QLD, Australia.
- Dr Maria Tomaso-Peterson, Mississippi State University, Starkville, Mississippi, USA.
- Nadeen Zreikat, Technical Advisor, Sydney, NSW, Australia.

1

Turfgrass Ecosystems

Abstract

This introductory chapter emphasizes that turfgrass ecosystems are dynamic and are impacted continually by the interaction of numerous environmental factors and management-induced inputs. Monoculture or near-monoculture, as opposed to polyculture, turfgrass swards are generally more prone to disfiguring outbreaks of infectious disease requiring management intervention.

Turfgrass ecosystems are biologically complex and dynamic, comprising an assemblage of mown plants and a suite of living organisms and non-living components all interacting in a common environment. Plant diseases in one form or another remain a natural component in turfgrass systems and remain challenging to effectively manage worldwide. The fungal diseases fairy ring and red thread were among the earliest diseases recognized on mown grass.

The introduction of hand-pushed, cylinder mowers and metal-barrel rollers during the mid-1800s, initially on large estates in Western Europe then elsewhere, made possible the presentation of highly managed turfgrass swards. The Budding mower and later designs allowed low-growing grasses (e.g. bentgrasses and fine fescues) and forbs (e.g. chamomile and clovers) to be maintained at a specific height. Complementary practices (e.g. watering and sand–manure mixes) improved soil moisture, transforming turfgrass swards; thus facilitating the emergence of many turfgrass diseases. Later technologies (e.g. cultivars tolerant of close mowing and synthetic fertilizers) induced a drastic shift in the type and severity of turfgrass diseases (Kerns and Tredway, 2013).

Severe disfiguration on the Kentucky bluegrass lawn of the White House (Washington, DC) in the 1980s (Smiley, 1987) undoubtedly remains the highest-public-profile example of turfgrass disease (Fig. 1.1). Advances in turfgrass pathology, disease-resistant cultivars, aerifying equipment and consumable products (e.g. fungicides and plant growth regulators) have significantly benefited the management of turfgrass diseases worldwide.

Turfgrass Plants

Numerous species of true grasses (family *Poaceae*) and a select number of forbs (decumbent flowering plants) are managed as mown swards in both community and private use throughout the world. Grasses and forbs are classified taxonomically, based on a universally accepted Latin

Fig. 1.1. Disease patches on the Kentucky bluegrass lawn at the White House, Washington, DC. (Courtesy of the LBJ Library, Austin, Texas.)

binomial system of genus and species, and can also be grouped according to the physiology and climatic adaptation. The taxonomy of all plants always remains subject to reclassification. The broader grouping of grasses, based on their physiology, geographic origins and climatic adaptation (i.e. cool-season or warm-season), provides a meaningful understanding for their management. Each grass and forb species has ecological and pathological strengths and weaknesses which partly explain their inherent ability to tolerate (or otherwise) infectious diseases under specific environments.

Primary grasses

Primary grasses managed as mown turfgrass covers the most economically important species managed throughout the world. Approximately two dozen grass species can be categorized this way. The relative importance of each grass species and cultivar varies widely among states and between countries. Grasses in this group generally have long been utilized as mown swards and purposely selected because of numerous attributes and remain the main focus of mainstream breeding programmes.

Primary turfgrasses are members of one of three *Poaceae* subfamilies: *Pooideae*, *Chloridoideae* or *Panicoideae* (Table 1.1). The long history of utilization of the most common species (e.g. bentgrasses, bluegrasses, fescues and bermudagrasses) has resulted in extensive documentation of their overall ecological and pathological strengths and weaknesses as mown turfgrass (Aldous and Chivers, 2002; Bonos and Huff, 2013; Hanna *et al.*, 2013; Loch *et al.*, 2013; Braun *et al.*, 2020).

Secondary grasses

Secondary grasses utilized as mown turfgrass covers numerous species cultivated in a few continents or countries. The precise number of grass species remains unclear because many remain largely unknown outside the country in which they are managed. In many cases, their taxonomy remains unclear and certain species may not have universally accepted common names. None the less, many species demonstrate desirable attributes but have largely escaped the attention of turfgrass breeding programmes.

Most grasses managed as secondary turfgrasses are members of one of four *Poaceae* subfamilies: *Bambusoideae*, *Chloridoideae*, *Panicoideae* and *Pooideae* (Table 1.2). The ecological and pathological strengths and weaknesses of many secondary species managed as secondary turfgrasses remain unclear and largely undocumented (e.g. Duistermaat, 2005; Aldous and Loch, 2013; Loch *et al.*, 2013).

Several species of bluegrasses (*Poa* spp.) and fescues (*Festuca* spp.) and lesser-known members of *Digitaria* spp. and *Paspalum* spp. and other grass species (Fig. 1.2) having desirable attributes are utilized in certain countries.

Experimental grasses

'Novel' grass species from natural grasslands in certain countries continue to be evaluated and some have been commercialized, having demonstrated experimentally at least one attribute for niche markets. Grass species in the genera *Bouteloua*, *Distichlis*, *Hilaria*, *Koeleria* and others are among those that continue to be evaluated (Watkins *et al.*, 2013; Gomez da Silva *et al.*, 2018; Stewart, 2022). The agronomic and pathological strengths and weaknesses of most species remain to be evaluated. Certain species of transgenic turfgrasses have also been developed (Webber, 2000) with limited success.

Table 1.1. Primary turfgrasses by subfamily, genus/species and common name.

Subfamily	Genus/species	Common name
Chloridoideae	*Bouteloua dactyloides*	American buffalograss
	Cynodon spp.	Bermudagrass
	Zoysia spp.	Zoysiagrass
Panicoideae	*Axonopus* spp.	Carpetgrass
	Digitaria didactyla	Queensland blue couch
	Eremochloa ophiuroides	Centipede grass
	Paspalum notatum	Bahiagrass
	Paspalum vaginatum	Seashore paspalum
	Cenchrus clandestinus	Kikuyugrass
	Stenotaphrum secundatum	St. Augustinegrass
Pooideae	*Agrostis* spp.	Bentgrass
	Cynosurus cristatus	Crested dog's tail
	Dactylis glomerata	Orchard grass
	Festuca rubra	Chewing's/Fine fescue
	Festuca spp.	Fescue
	Lolium perenne	Perennial ryegrass
	Poa annua	Annual bluegrass
	Poa compressa	Canada bluegrass
	Poa pratensis	Kentucky bluegrass

Table 1.2. Secondary turfgrasses by subfamily, genus/species and common name.

Subfamily	Genus/species	Common name
Bambusoideae	*Microlaena stipoides*	Weeping grass
Chloridoideae	*Bouteloua gracilis*	Blue grama
	Dactyloctenium australe	Durban grass
	Sporobolus virginicus	Dropseed/Sand couch
Panicoideae	*Bothriochloa macra*	Red leg grass
	Bothriochloa pertusa	Pitted beard grass
	Chrysochloa orientalis	Gingerspike
	Digitaria decumbens	Pangola grass
	Digitaria diversinervis	Richmond grass
	Panicum repens	Torpedo grass
	Paspalum conjugatum	Sour grass
	Paspalum nicorae	Brunswick grass
	Polytrias indica	Java lawngrass
	Stenotaphrum dimidiatum	Pembagrass
Pooideae	*Puccinella distans*	Alkali grass
	Poa compressa	Canada bluegrass
	Festuca longifolia	Hard fescue
	Festuca pratensis	Meadow fescue
	Festuca ovina	Sheep fescue
	Agrostis alba	Redtop bentgrass
	Agrostis canina	Velvet bentgrass
	Phleum pratense	Timothy grass

Fig. 1.2. Durban grass (sweet smother grass) in Australia. (Courtesy of Gary W. Beehag.)

Mown forbs

Forbs are dicotyledonous, flowering plants that have a decumbent growth habit and are a common component of grassland and woodland systems worldwide. A select number of species of forbs from one of several plant families are managed as mown swards in certain countries (Table 1.3).

Table 1.3. Principal mown forbs by genus/species and common name.

Genus/species	Common name	References
Hydrocotyle tripartita	Australian pennywort	Baltensperger and Gaussoin (1985); Harrington (1990); Aldous (1991); Gilbert (1991); Munro (1997); Christians (2004); Ormsby (2005); NZSTI (2008); Bigelow et al. (2020)
Chamaemelum nobile	Chamomile	
Leptinella dioica	Cotula	
Leptinella maniototo	Cotula	
Dichondra repens	Dichondra	
Arachis prostrata	Creeping groundnut	
Trifolium fragiferum	Strawberry clover	
Trifolium repens var. Pirouette	Microclover	
Plantago triandra	Starweed	
Veronica filiformis	Creeping speedwell	
Phyla nodiflora	Lippia	
Bouteloua dactyloides	American buffalograss	

Most mown forb species with few exceptions are managed as domestic or estate lawns.

Certain forb species are relatively well known (e.g. chamomile and dichondra) while others are managed as a playing surface (e.g. cotula bowling greens) in New Zealand. Chamomile has long been cultivated in Europe and remains a component of the lawn at Buckingham Palace in England (Gilbert, 1991). Mown forbs with few exceptions (chamomile, clover and cotula) have escaped plant breeding programmes.

Physiological Classification, Geographic Origin and Climatic Adaptation

Grasses fall into one of two physiological groupings in accordance with how they capture light energy, assimilate atmospheric carbon dioxide and manufacture carbohydrates during photosynthesis. The geographic distribution and climatic adaptation of managed grasses is a reflection of the environmental responses between the two groups. Each physiological process can be manipulated, to lesser or greater degree, by one or more cultural management practices which, in turn, may impact the incidence of turfgrass diseases. A synopsis only of the physiological classification, geographic origin and climatic adaptation is considered appropriate for the purpose of this book.

The Calvin–Benson photosynthetic or C3 pathway and the Hatch–Slack cycle or C4 pathway and its three variants are the two photosynthetic pathways of grasses (Sinclair, 2002). As a group, C3 grasses are genetically adapted to cool–cold regions and possess low efficiencies of water and nitrogen use. In contrast, C4 grasses are genetically adapted to warm–hot and drier regions and possess relatively high efficiencies of water and nitrogen use (Bell, 2011). Suffice to say, awareness of the physiological classification and differential response of grasses to temperature, light and rainfall has practical significance when grasses are managed in locations marginal to or well outside their regions of natural distribution.

The geographic centres of origin of most turfgrass species can be broadly traced to the cooler regions of Europe, the warm and humid regions of Asia or the warmer regions of southern Africa (Fig. 1.3).

Grasses in the genera *Agrostis* (bentgrass), *Festuca* (fescue), *Lolium* (ryegrass) and *Poa* (bluegrasses) and many others have their greatest concentration north of 30°N or south of 30°S latitude (Moser and Hoveland, 1996) and in the high mountainous regions of the tropics (Beard et al., 2014). Grasses originating from subalpine and temperate regions are often called temperate or more commonly 'cool-season' turfgrasses. C3 grasses grow actively during spring and autumn during cooler periods. Cool-season grasses primarily produce a form of carbohydrate

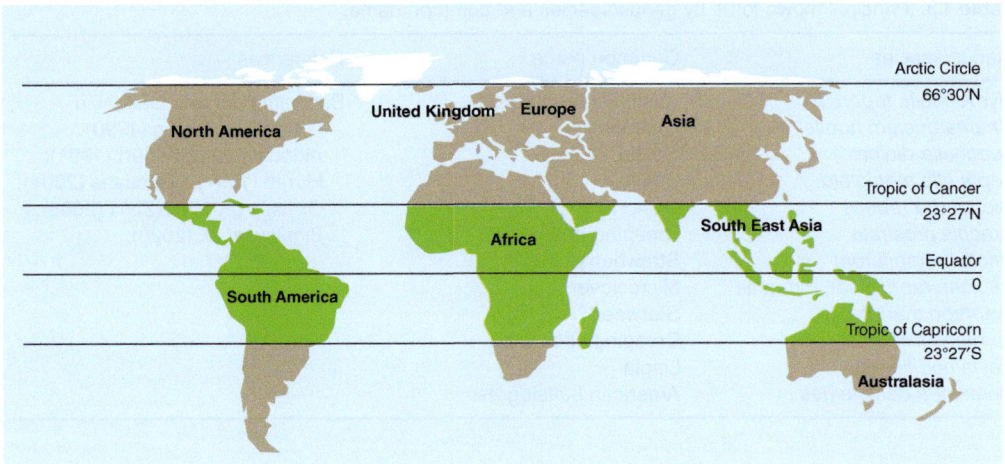

Fig. 1.3. Global map showing temperate and tropical regions.

(fructosan) which is water soluble, aiding tolerance to freezing temperatures (McCarty, 2010).

Grasses in the genera *Axonopus* (carpetgrass), *Digitaria* (fingergrasses), *Eremochloa* (centipede grass), *Cynodon* (couchgrass or bermudagrass), *Paspalum* (paspalum), *Stenotaphrum* (St. Augustinegrass) and *Zoysia* (zoysiagrass) and many others are usually found between 30°N and 30°S latitudes even up to 55°N (Moser *et al.*, 2004). Grasses originating from tropical and subtropical regions are commonly called 'warm-season' turfgrasses. C4 grasses grow actively from late spring through to late summer. Grasses with C4 photosynthesis are able to maintain high photosynthetic rates at high temperatures (McCarty, 2010).

All turfgrass species have definite tolerance ranges to climatic and environmental factors based on their places of origin. Provided the critical tolerance limits of a turfgrass species are not exceeded for extended periods, the respective turfgrass species will persist provided there are no limiting factors. When non-adaptable turfgrass species are cultivated well outside their normal range of distribution, their persistence is governed by the level of cultural management intervention.

Turfgrass Microbiome

A turfgrass microbiome comprises a community of microorganisms living in a common environment. Certain aspects of the microbiome and impacts of certain cultural practices have been investigated on golf courses (e.g. Stingl *et al.*, 2021) and institutional lawns (e.g. Crouch *et al.*, 2017) but many aspects of their biological processes remain unclear.

Co-evolution in natural plant systems has resulted in each group of microorganisms performing a functional role or 'ecological niche', some of which occur in turfgrass ecosystems. The functional roles of microorganisms may be transitory anywhere between the extremes of saprophytic to antagonistic, depending on circumstances not fully understood.

Endophytes, mycorrhizae and nitrogen-fixing bacteria that form on plant roots are examples of symbiotic relationships (Powell and Klironomos, 2007). Accelerated microbial biodegradation of pesticides also involves microbial species (Felsot and Shelton, 1993). The phenomenon of bilateral extinction of fairy ring formation, observed very early in Great Britain when two fairy rings intersect, is an example of microbial antagonism (Smith *et al.*, 1989).

Turfgrass Pathosystem

Plant pathosystems are defined by the types of microbial parasitism and provide some understanding of specific turfgrass host–parasite relationships. Turfgrass systems, unlike annual crops, provide a continual food source for microorganisms of live hosts and their decomposing tissues. Many turfgrass pathogens can be described as perennial co-inhabitants of the turfgrass pathosystem.

Populations of fungal pathogens interact with host plants in such a way that they exert mutual selection pressures on each other (Zadoka, 1987). Many turfgrass pathogens have coexisted for many years without one eliminating the other (Schumann and Wilkinson, 1992) and are favoured during conditions conducive for disease development (Endo, 1972).

Diseases in Monoculture versus Polyculture Turfgrass Systems

Plant monocultures and near-monocultures are considered rare in nature. Turfgrass systems are established and maintained either as polyculture or monoculture systems (Box 1.1) depending on the specific management objectives.

Turfgrass monocultures are structurally and functionally very different from polycultures in natural plant systems. Monocultures are generally viewed as being largely ecologically unstable and resulting in increased disease problems over time. Much time and effort are expended in managing turfgrass monocultures (Fig. 1.4) while attempting to moderate natural ecological forces. Infectious diseases of most crops worsen with continued monocultures since pathogen populations tend to increase (Endo, 1972). Plant ecologists have long attempted to quantify the widespread assertion that plant diseases are seemingly less frequent and less severe in populations of 'wild' or natural plant species than in 'managed' or unnatural plant populations.

Box 1.1. Difference between monoculture and polyculture systems.

Monoculture (mono-specific) and polyculture describe the botanical composition of a turfgrass sward. A monoculture is one species of grass or forb or a cultivar. A polyculture is two or more species or cultivars. An undesirable plant species is termed a weed or could be an unwanted cultivar.

The distinction between monoculture and polyculture systems is based on the premise that multiple copies of the same plant species (or cultivar) in a confined space lack genetic diversity (i.e. defence mechanisms) and thus are often highly susceptible to disease.

Plant diseases are considered by many a normal biological component regulating plant populations in natural ecosystems. Some authors have stated that in undisturbed, natural ecosystems, relationships between pathogens and their hosts tend to attain a level of stability (Raikes et al., 1996) and diversity of natural ecosystems ensures severe disease epidemics are rare (Garrett and Mundt, 1999). Studies into long-term mono-cropping have shown a decline of soil-borne pathogens due to interactions between the cause-and-effect cycles between the host and beneficial microbes. That take-all decline in crop systems results from an increased population and activity of select groups of antagonistic soil-borne (but generally not known) microbes is well documented (Weller et al., 2002).

Immature turfgrass monocultures or near-monocultures are often regarded as lacking microbial biodiversity. A limited number of studies strongly suggest soil microbial biomass increases in mature turfgrass in parallel with increasing soil organic matter (Shi et al., 2007). An increasing weight of evidence suggests that crops grown in chemically fertilized systems, as opposed to being grown in organically rich and biologically active soils, are more susceptible to diseases (van Bruggen, 1995; Finckh et al., 2015). Thus, it could be concluded that the ability of crop plants to resist or tolerate pathogens is associated partly to the physical, chemical and biological properties of the soil.

The concept of organic turfgrass management has been discussed particularly on golf courses (Carlson, 2006; Marshall et al., 2015; Howard and Ormsby, 2017). The concept being to reduce, even eliminate, synthetic fertilizer and pesticide applications (Rossi and Grant, 2009).

Factors influencing monoculture diseases

Numerous factors have been proposed to influence the propensity of monocultured crops to develop disease (Box 1.2). The same factors could be said of turfgrass monocultures.

Fig. 1.4. A highly managed bermudagrass (couchgrass) monoculture lawn. (Courtesy of Gary W. Beehag.)

Box 1.2. Factors influencing disease in monoculture systems.

- Host longevity
- Multi-line cultivars and cultivar mixtures
- Closely spaced plants and small-sized host plants
- Retarded plant growth by increased frequency of monocultured cropping
- Impact of organic matter
- Synthetic-nitrogen fertilizers and chemical pesticides

Shift in the microbial equilibrium following application of soil fumigants has long been reported in certain monocultured crops. Biocidal effects on beneficial and detrimental microbial species may range from outright death of some, to minimal to slight inhibition of others allowing their survival and thus multiplication of the species. The detrimental or so-called 'non-target' effects of pesticides on turfgrass microbiology have been widely documented (e.g. Tredway *et al.*, 2023). Irrigation regimes, as opposed to renovation practices, may have a greater impact on microbial activity (Mu and Carroll, 2013).

The term 'disease trading' or exchange describes the net effect of shifts in the microbial equilibrium (Gupta, 2004). In other words, under environmental changes whereby the primary causal agent becomes more or less suppressed and another becomes dominant, even producing different symptoms. The effects of disease trading, although limited, have been observed on turfgrass. A pioneering Australian study reported the stimulation of an unknown soil-borne fungus in turfgrass following the application of the early fungicide benomyl (Smith *et al.*, 1970). Results of a more recent study suggested increased growth or stimulation of the fungus that causes dollar spot when fungicide concentrations were very low (Pradham *et al.*, 2019).

Turfgrass surface-renovation practices (i.e. physical rejuvenation of the rootzone without turfgrass replacement) are commonplace using surface-penetrating equipment (Fig. 1.5). A combination of topdressing (e.g. sand and/or organic media) and chemical amendments (e.g. gypsum and lime) partly moderates the rootzone microbiome. Certain turfgrass management practices, if incorrectly implemented or timed, may inadvertently cause a shift in the microbial equilibrium; thus causing accentuation of certain pathogens.

New-Encounter Turfgrass Pathogens and Diseases

Newly described microbial pathogens continue to be documented in turfgrass systems in many countries (Table 1.4). The net result is the formation of new pathogen–host combinations leading to newly emergent pathogens or novel disease.

Pathogens of cultivated plants have evolved from natural plant systems and new pathogens appearing *de novo* are extremely rare (Zadoka, 1987). Documentation of new-encounter pathogens in turfgrass systems raises many questions about the circumstances and possible environmental factors that have contributed to their emergence (Fig. 1.6). Even more intriguing is the initial emergence followed in later years with the apparent decline in prevalence of certain pathogens and their diseases.

Several factors potentially contribute to the emergence of new-encounter turfgrass diseases (Box 1.3). Unlawful, accidental or even legislated

Fig. 1.5. A tractor-mounted, mechanical drilling machine on a bermudagrass bowling green in Australia. (Courtesy of Gary W. Beehag.)

Table 1.4. Nominated new-emergent turfgrass pathogens by country.

Pathogen name	Country	References
Acidovorax avenae	Japan	Furuya *et al.* (2009)
Candidacolonium cynodontis	USA	Bronzato-Badial *et al.* (2020)
Labyrinthula terrestris	UK	Entwistle *et al.* (2006)
Ophiosphaerella agrostis	USA	Camara *et al.* (2000)
Phialocephala bamuru	Australia	Wong *et al.* (2015)

Fig. 1.6. A newly emergent patch disease caused by *Magnaporthiopsis dharug* on a fine fescue lawn in Australia. (Courtesy of Percy T. Wong.)

Box 1.3. Possible factors of new-encounter diseases.

- Transport of infected turfgrass plants into new regions
- Transport of rootzone into new regions
- Improper, stress-inducing management regimes
- Application of certain sources of irrigation water
- Changes in microbial composition in new environments

introduction of infected turfgrasses from their region of origin offers a potential source of new plant–pathogen combinations (Kerns and Tredway, 2013; Settle and Fidanza, 2014).

Many years may pass before infected turfgrass plants become recognizably diseased. Hence the need for national and international quarantine and biosecurity protocols and legislation covering the transport of turfgrass between countries. The specific circumstances behind most new-encounter turfgrass diseases will undoubtedly remain unknown.

Economic Significance of Turfgrass Diseases

Seasonal frequency and severity among turfgrass diseases ranges anywhere between the extremes of cosmetic to severe disfiguration. The economic cost to effectively manage frequently occurring disfiguring diseases can be considerable, particularly on high-value turfgrass sites (Fig. 1.7), depending on epidemiological conditions.

The economic importance of turfgrass diseases can be categorized (Table 1.5), which partly determines in making decisions about the need and level of their management.

On one hand, a widely known assortment of fungal pathogens and their diseases of one sort or another occur primarily during either cool to cold and wet or warm to hot and dry conditions, depending on host–pathogen combination and underlying environmental conditions. Numerous fungal diseases of cool-season and warm-season grasses are widely distributed and well known in many countries throughout the northern and southern hemispheres and are capable of causing significant and prolonged damage in the absence of effective management practices and strategies.

The most widespread, frequent and major fungal diseases (e.g. brown patch, dollar spot, microdochium patch and others) are economically important on turfgrass and have largely been the focus of turfgrass pathology research. Hence, why knowledge of the epidemiology and management of these fungal diseases is relatively well understood and has been widely documented.

Fig. 1.7. Harvesting high-quality turfgrass on a commercial sod farm. (Courtesy of Gary W. Beehag.)

Table 1.5. Categorization of the economic importance of turfgrass diseases.

Categorizing factor	Definition/Characterization
Geographic distribution	**Widespread** – disease known to have a wide geographic distribution throughout a state or country **Restricted** – disease whose known distribution is limited to a specific region within a state or country
Frequency of occurrence	**Frequent** – the disease causes high infections or occurs in high populations every year within the same geographic region **Infrequent** – the disease does not cause high infections or occur every year or only in low populations within the same geographic region
Severity of damage	**Major** – the disease causes unacceptable damage to the host plant which may result over a short time frame or a relatively long period unless appropriate management intervention is implemented while damage is relatively minor **Minor** – the disease causes only superficial damage or partial injury to the host plant and which injury may recover over time of its own accord with minimal or no management intervention

On the other hand, a lesser-known suite of fungi also occur on cool-season and warm-season grasses in both hemispheres during either cool to cold or warm to hot conditions, again depending on host–pathogen combination and environmental conditions. This second group of restricted, infrequent and minor turfgrass fungal diseases (e.g. Septoria leaf spot) is lesser known and generally causes only cosmetic or minimal damage, except during highly favourable conditions, and normally does not require management actions. Most of these fungal diseases are not economically important and consequently much about their identity, epidemiology and management remains imprecise.

Disease management versus control

General observations and experience by turfgrass managers raise the question about adoption of the terms management versus control of turfgrass diseases (Box 1.4). Generally, plant pathologists have the view that it is probably impossible in the long term to separate cultivated plants from their pathogens; hence the reason for adopting the term disease management as opposed to control in this book.

Layout of This Book

Biology and Integrated Management of Turfgrass Diseases has been structured with three overarching objectives. The book's first objective is for readers is to understand that turfgrass systems are ecologically complex and dynamic communities. The book's second objective is for turfgrass managers to understand how the environment and host plants interact. Chapter 2 outlines how the environment and host plants interact with parasites and pathogens. The book's third objective is to empower turfgrass managers with the necessary knowledge to identify the early warning signs and symptoms of turfgrass diseases; thus being able to make informed decisions about manipulating the turfgrass environment using proven and predictable turfgrass practices and to minimize the frequency, severity and duration of turfgrass diseases.

The underlying factor determining a successful integrated approach to turfgrass disease management is to fully understand when existing environmental and management-induced conditions in a turfgrass system fall short of providing optimal performance of the nominated species or cultivar. Chapter 3 details the ecological groups of microbes and the environmental conditions affecting their activities. Chapter 4 focuses on disease monitoring and forecasting together with sampling procedures and how microbial agents and their diseases can be confidently diagnosed in order to make informed decisions about their management. Chapter 5 details knowledge of the impact of cultural and biological practices on turfgrass diseases which provides the basis of how to formulate and develop a personalized integrated disease management plan.

For disease identification purposes, details of each fungal disease have been alphabetically arranged according to the season of the year in which symptoms first appear. Chapter 6 covers cool-weather fungal diseases of grasses; Chapter 7 covers warm-weather fungal diseases of grasses; and Chapter 8 covers all fungal diseases of forbs. Chapter 9 focuses on turfgrass bacterial and viral diseases. Chapter 10 covers plant-parasitic nematodes of turfgrasses, both of true grasses and forbs. Photographic images showing symptoms of turfgrass diseases are included to further assist field identification.

Disease epidemics in highly valued turfgrass often require application of chemical pesticides. Fungicides (biofungicides) and nematicides (bionematicides) are specifically registered for turfgrass application in most countries. Chapter 11 details the chemical properties and biological characteristics, application and the environmental behaviour and fates of fungicides, as well as how to minimize or avoid non-target effects.

Box 1.4. Turfgrass disease management versus control.

Throughout this book the authors have purposely adopted the term management rather than control with respect to infectious turfgrass diseases. Populations of turfgrass pathogens (e.g. fungi and nematodes) may be managed using appropriate practices but never eliminated (i.e. controlled) from the host in the long term. Turfgrass pathogens do re-occur on their hosts, which indicates that their population – and thus the disease – has only been managed in the short term.

Finally, the inclusion of scientific terms has been minimized in an attempt to balance science and practice. Where used, the relevant technical terms have been included to facilitate greater understanding of the principal terms frequently used by plant pathologists and nematologists. The meaning of all technical terms is included in the glossary placed at the end of this book.

References

Aldous, D.E. (1991) *Lawn Care and Lawn Alternatives*. Lothian Publishing Company Pty Ltd, Melbourne, Australia.

Aldous, D.E. and Chivers, I.H. (2002) *Sports Turf & Amenity Grasses; A Manual for Use and Identification*. Landlinks, Melbourne, Australia.

Aldous, D.E. and Loch, D.S. (2013) A review of the biology, adaptation and management of *Dactyloctenium australe* Steud. (Durbangrass) as a turfgrass with particular reference to Australia. *International Turfgrass Society Research Journal* 12, 135–142.

Baltensperger, A.A. and Gaussoin, R.E. (1985) 'Fresa' strawberry clover, *Trifolium fragiferum* L. in reduced maintenance polystands and as monostand ground cover. In: Lemaire, F. (ed.) *Proceedings of the Fifth International Turfgrass Research Conference, Avignon, France, 1–5 July 1985*. INRA, Paris, pp. 311–316.

Beard, J.B., Beard, H.J. and Beard, J.C. (2014) *Turfgrass History and Literature*. Michigan State University Press, East Lansing, Michigan.

Bell, G.E. (2011) *Turfgrass Physiology and Ecology: Advanced Management Principles*. CAB International, Wallingford, UK.

Bigelow, C.A., Macke, G.A., Johnson, K. and Richmond, D.S. (2022) Cool-season lawn performance as influenced by 'Microclover' inclusion and supplemental nitrogen. *International Turfgrass Society Research Journal* 14, 121–132. DOI: 10.1002/its2.19

Bonos, S.A. and Huff, D.R. (2013) Cool-season grasses; biology and breeding. In: Stier, J.C., Horgan, B.P. and Bonos, S.A. (eds) *Turfgrass: Biology, Use, and Management*. Agronomy Monograph No. 56. American Society of Agronomy, Inc., Crop Science Society of America, Inc. and Soil Science Society of America, Inc., Madison, Wisconsin, pp. 591–600.

Braun, R.C., Patton, A.J., Watkins, E., Koch, P.L., Anderson, N.P., et al. (2020) Fine fescues: a review of the species, their improvement, production, establishment, and management. *Crop Science* 60, 1142–1187. DOI: 10.10002/csc2.20122

Bronzato-Badial, A., King, J. and Tomaso-Peterson, M. (2020) Monitoring ectotrophic root-infecting fungi associated with bermudagrass putting greens using quantitative multiplex assays. *Plant Heath Progress* 21, 144–151.

Camara, M.P.S., O'Neill, N.R., van Berkum, P., Dernoeden, P.H. and Palm, M.E. (2000) *Ophiosphaerella agrostis* sp. nov. and its relationship to other species of *Ophiosphaerella*. *Mycologia* 92, 317–325. DOI: 10.2307/3761568

Carlson, J.W. (2006) An organic approach to golf course management. *USGA Greens Section Record* 44, 13–16.

Christians, N. (2004) *Fundamentals of Turfgrass Management*, 2nd edn. Wiley, Hoboken, New Jersey.

Crouch, J.A., Carter, Z., Ismaiel, A. and Roberts, J.A. (2017) The US national mall microbiome: a census of rhizosphere bacteria inhabiting landscape turf. *Crop Science* 57, S-341–S-348. DOI: 10.2135/cropsci2016.10.0849

Duistermaat, H. (2005) *Field Guide to the Grasses of Singapore (Excluding the Bamboos)*. Supplement of The Gardens' Bulletin Singapore, Vol. 57. National Parks Board, Singapore.

Endo, R.M. (1972) The turfgrass community as an environment for the development of facultative fungal parasites. In: Younger, V. (ed.) *The Biology and Utilization of Grasses*. Elsevier, Amsterdam, pp. 171–202.

Entwistle, C.A., Olsen, M.W. and Bigelow, D.M. (2006) First report of a *Labyrinthula* spp. causing rapid blight of *Agrostis capillaris* and *Poa annua* on amenity turfgrass in the UK. *New Disease Reports* 11, 30.

Felsot, A.S. and Shelton, D.R. (1993) Enhanced biodegradation of soil pesticides. Interactions between physicochemical processes and microbial ecology. In: Linn, D.M., Carski, F.H., Brusseau, M.L. and Chang, T.-H. (eds) *Sorption and Degradation of Pesticides and Organic Chemicals in Soils*. SSA Special Publication No. 32. Soil Society of America, Inc. and American Society of Agronomy, Inc., Madison, Wisconsin, pp. 227–251.

Finckh, M.R., van Bruggen, A.H.C. and Tamm, L. (2015) (eds) *Plant Disease Their Management in Organic Agriculture*. The American Phytopathological Society, St. Paul, Minnesota.

Furuya, N., Ito, T. and Tsuchiya, K. (2009) Occurrence of bacterial brown stripe of creeping bentgrass on golf course green in Kyushu. *Journal of the Faculty of Agriculture Kyushu University* 54, 13–17.

Garrett, K.A. and Mundt, C.C. (1999) Epidemiology of mixed host populations. *Phytopathology* 89, 984–990.

Gilbert, O.L. (1991) *The Ecology and Urban Habitats*. Chapman and Hall, London.

Gomez da Silva, S.C.A., Santos, A.G., Silva, S.S.L., Loges, V., de Souza, F.H.D. and Castro, A.C. (2018) Characterisation and selection of Brazilian native grasses for use as turfgrass. *Acta Horticulturae* 1215, 255–258. DOI: 10.17660/Actahortic.2018.1215.45

Gupta, G.P. (2004) *Text Book of Plant Diseases*. Discovery Publishing House, New Delhi.

Hanna, W., Raymer, P. and Schwartz, B. (2013) Warm-season grasses: biology and breeding. In: Stier, J.C., Horgan, B.P. and Bonos, S.A. (eds) *Turfgrass: Biology, Use, and Management*. Agronomy Monograph No. 56. American Society of Agronomy, Inc., Crop Science Society of America, Inc. and Soil Science Society of America, Inc., Madison, Wisconsin, pp. 543–590.

Harrington, K. (1990) Understanding hydrocotyle. In: *Proceedings of the Fourth New Zealand Sports Turf Convention, 13–16 August 1990*. Massey University, Palmerston North, New Zealand, pp. 62–63.

Howard, D. and Ormsby, D. (2017) Organic course management – is it viable? *New Zealand Turfgrass Management Journal* 34, 20–23.

Kerns, J.P. and Tredway, L.P. (2013) Advances in turfgrass pathology since 1990. In: Stier, J.C., Horgan, B.P. and Bonos, S.A. (eds) *Turfgrass: Biology, Use, and Management*. Agronomy Monograph No. 56. American Society of Agronomy, Inc., Crop Science Society of America, Inc. and Soil Science Society of America, Inc., Madison, Wisconsin, pp. 733–776.

Loch, D.S., McMaugh, P. and Scattini, W. (2013) A review of *Digitaria didactyla* Willd., a low-input warm-season turfgrass in Australia; biology, adaptation and management. *International Turfgrass Society Research Journal* 12, 1–14.

McCarty, L.B. (2010) *Best Golf Course Management Practices: Construction, Watering, Fertilizing, Cultural Practices and Pest Management Strategies to Maintain Golf Course Turf with Minimal Environmental Impact*, 3rd edn. Prentice-Hall, Upper Saddle River, New Jersey.

Marshall, S., Orr, D., Bradley, L. and Moorman, C. (2015) A review of organic lawn care practices and policies in North America and the implications of plant lawn diversity and insect pest management. *HortTechnology* 25, 437–446.

Moser, L.E. and Hoveland, C.S. (1996) Cool-season grass overview. In: Moser, L.E., Buxton, D.R. and Casler, M.D. (eds) *Cool-Season Forage Grasses*. Agronomy Monograph No. 34. American Society of Agronomy, Inc., Crop Science Society of America, Inc. and Soil Science Society of America, Inc., Madison, Wisconsin, pp. 1–14.

Moser, L.E., Burson, B.L. and Sollenberger, L.E. (2004) Warm-season (C_4) grass overview. In: Moser, L.E., Burson, B.L. and Sollenberger, L.E. (eds) *Warm-Season (C4) Grasses*. Agronomy Monograph No. 45. American Society of Agronomy, Inc., Crop Science Society of America, Inc. and Soil Science Society of America, Inc., Madison, Wisconsin, pp. 1–14.

Mu, Y. and Carroll, M.J. (2013) Thatch and soil microbial activity in recently cultivated turfgrass. *International Turfgrass Society Research Journal* 12, 567–574.

Munro, P. (1997) The dichondra lawn (Mercury Bay weed): low maintenance or not? *New Zealand Turf Management Journal* 11, 5–7.

NZSTI (2008) *Establishment and Management of Natural Bowling Greens in New Zealand*. New Zealand Sports Turf Research Institute, Palmerston North, New Zealand.

Ormsby, D. (2005) Starweed. *New Zealand Turf Management Journal* 20, 29–31.

Powell, J. and Klironomos, J. (2007) The ecology of plant–microbial mutualisms. In: Paul, E.A. (ed.) *Soil Microbiology, Ecology and Biochemistry*, 3rd edn. Elsevier, New York, pp. 257–282.

Pradham, S., Miller, L., Marcillo, V.F., Koch, A.R., Grachet, N.G., *et al.* (2019) Hormetic effects of thiophanate-methyl in multiple isolates of *Sclerotinia homeocarpa*. *Plant Disease* 103, 89–94.

Raikes, C., Lepp, N.W. and Canaway, P.M. (1996) The effect of dual species mixtures and monocultures on disease severity on winter sports turf. *Journal of the Sports Turf Research Institute* 72, 67–71.

Rossi, F.S. and Grant, J.A. (2009) Long term evaluation of reduced chemical pesticide management of golf course putting turf. *International Turfgrass Society Research Journal* 9, 77–90.

Schumann, G.L. and Wilkinson, H.T. (1992) Research methods and approaches to the study of diseases in turfgrass. In: Stier, J.C., Horgan, B.P. and Bonos, S.A. (eds) *Turfgrass: Biology, Use and Management*.

Agronomy Monograph No. 56. American Society of Agronomy, Inc., Crop Science Society of America, Inc. and Soil Science Society of America, Inc. Madison, Wisconsin, pp. 653–688.

Settle, D. and Fidanza, M. (2014) Troublesome and emerging turf disease of golf course greens maintained on constructed rootzones (abstract). *Applied Turfgrass Science* 10, 1. DOI: 10.2134/ATS-2013-0026BC

Shi, W., Bowman, D. and Rufty, T. (2007) Soil microbial community composition and function in turfgrass systems. *Bioremediation, Biodiversity and Bioavailability* 1, 72–77.

Sinclair, R. (2002) Ecophysiology of grasses. In: Mallett, K. and Orchard, A. (eds) *Flora of Australia – Vol. 43 Poaceae 1 – Introduction and Atlas*. CSIRO Publishing, Melbourne, Australia, pp. 133–156.

Smiley, R.W. (1987) The etiologic dilemma concerning patch diseases of bluegrass turfs. *Plant Disease* 71, 774–781.

Smith, A.M., Stynes, B.A. and Moore, K.J. (1970) Benomyl stimulates growth of a basidiomycete on turf. *Plant Disease Reporter* 54, 774–775.

Smith, J.D., Jackson, N. and Woolhouse, A.R. (1989) *Fungal Diseases of Amenity Turf Grasses*, 3rd edn. E. & F.N. Spon, London.

Stingl, U., Choi, C.J., Dhillon, B. and Schiavon, M. (2021) The lack of knowledge on the microbiome of golf courses impedes the development of successful microbial products. *Agronomy* 12, 71. DOI: 10.3390/agronomy12010071

Stewart, A. (2022) Domesticating native grasses in New Zealand. *International Turfgrass Society Research Journal* 14, 1067–1069. DOI: 10.1002/its2.130

Tredway, L.P., Tomaso-Peterson, M., Kerns, J.P. and Clarke, B.B. (2023) *Compendium of Turfgrass Diseases*, 4th edn. The American Phytopathological Society. St. Paul, Minnesota.

van Bruggen, A.H.C. (1995) Plant disease severity in high-input compared to reduced-input and organic farming systems. *Plant Disease* 79, 976–984.

Watkins, E., Brilman, L. and Kopec, D. (2013) Development of native grasses for turf. In: Stier, J.C., Horgan, B.P. and Bonos, S.A. (eds) *Turfgrass: Biology, Use, and Management*. Monograph No. 56. American Society of Agronomy, Inc., Crop Science Society of America, Inc. and Soil Science Society of America, Inc., Madison, Wisconsin, pp. 661–682.

Webber, A.P. (2000) Transgenic turfgrasses. *USGA Green Section Record* 38, 31–33.

Weller, D.M., Raaijmakers, J.M., McSpadden Gardener, B.B. and Thomashow, L.S. (2002) Microbial populations responsible for specific soil suppresiveness to plant pathogens. *Annual Review of Phytopathology* 40, 309–348. DOI: 10.1146/annurev.phyto.40.030402.110010

Wong, P.T.W., Dong, C., Martin, P.M. and Sharp, P.J. (2015) Fairway patch – a serious emerging disease of couch (syn. bermudagrass) [*Cynodon dactylon*] and kikuyu (*Pennisetum clandestinum*) turf in Australia caused by *Phialocephala bamuru* Wong, P.T.W. & C. Dong. sp. nov. *Australasian Plant Pathology* 44, 545–555. DOI: 10.1007/s13313-015-0369-0

Zadoka, J.C. (1987) The function of plant pathogenic fungi in natural communities. In: Andel, J.V., Bakker, J.P. and Snaydon, R.W. (eds) *Disturbance in Grasslands: Causes, Effects and Processes*. Dr W. Junk Publishers, Dordrecht, The Netherlands, pp. 201–212.

2

Environment, Host and Pathogen: Disease Epidemiology

Abstract
This chapter emphasizes the complex interactions between environments, host and pathogen and the study of how disease epidemics occur is called disease epidemiology. Disease epidemiology involves complex weather, biochemical and physiological interactions of the host plant and pathogen, far from fully understood. Many management-induced factors impact on plant pathogens and subsequent diseases.

Turfgrass diseases can only occur during a series of simultaneously overlapping host, pathogen and environmental conditions (Fig. 2.1). Once pathogens have infected host tissues, complex interactions are triggered between the host and pathogen. Knowledge of all these influences has practical significance in the understanding of how, when and why turfgrasses experience disease and, most importantly, for effective disease management.

Knowledge of the epidemiological factors associated with common turfgrass diseases is relatively well understood. Under environmental conditions that are not conducive for disease, in a host having elevated susceptibility and in the presence of a virulent pathogen, diseases do not occur or occur at a reduced level.

Favourable Environment

A favourable environment is generally acknowledged as the primary factor governing the incidence, severity and duration of plant diseases. A multitude of environmental and management-induced factors continually and simultaneously interact with each other and may have a positive, neutral or negative effect on disease.

The overall impact of the environment on turfgrasses and their respective diseases to tolerate (or otherwise) diseases is clearly demonstrated between the extremes of subalpine inland (Fig. 2.2) and subtropical coastal (Fig. 2.3) regions. Certain diseases tend occur more commonly in one or the other environment.

Turfgrass systems comprise three distinct physical regions (Fig. 2.4), each of which is impacted by a range of natural and management-induced inputs.

Phyllosphere

The phyllosphere or the aboveground portion of turfgrass forms a habitat for populations of inflorescence- and foliage-inhabiting microorganisms under conditions favourable for their existence. The physical architecture of the turfgrass phyllosphere is determined largely by the mowing regime by altering its depth (verdure) and density. Mowing results in cut leaf ends and abrasion that are possible sites of infection. Key environmental factors of the turfgrass phyllosphere are canopy temperature, moisture, air movement, light intensity and quality.

Thatch-mat region

Thatch-mat accumulation is an organic-based region between the phyllosphere and the rhizosphere occurring in natural grasslands (Gibson, 2009) and managed turfgrass (Gaussoin et al., 2013). Most topdressed swards (Fig. 2.5) typically have alternating layers of decomposing organic matter and topdressed sand, as opposed to non-topdressed swards (Fig. 2.6), and clearly indicate that accumulation of thatch-mat is a management-driven process (Beehag and McMaugh, 2018).

The thatch-mat region provides a highly favourable habitat for stem- and root-inhabiting microorganisms. Temperature, moisture, aeration and fertility input are the key environmental factors influencing the populations of microbes in the thatch-mat. The architecture of the thatch-mat region can be successfully moderated by the combination of physical practices (Gaussoin et al., 2013).

Rhizosphere

The rhizosphere, or the region associated with the rootzone adjacent to the external surface of plant roots within the rootzone, provides a highly favourable habitat for populations of root-inhabiting microorganisms. Turfgrass rootzones vary widely in their physical make-up depending on the turfgrass site. Rootzones range from near-pure sands to sand–organic mixtures containing additives (i.e. plant- or animal-derived) or inorganic (e.g. pearlite, zeolite or plastic fibre matrix) products (Bigelow and Soldat, 2013).

Grassed cricket wicket rootzones are unique (Fig. 2.7). These rootzones comprise a high-cracking

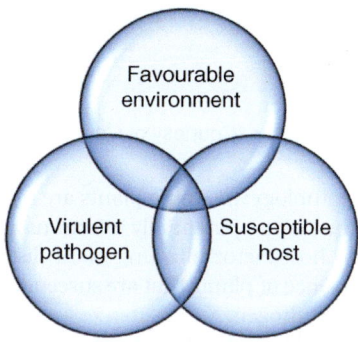

Fig. 2.1. Factors of disease epidemiology.

Fig. 2.2. A subalpine environment with turfgrass partly under snow. (Courtesy of Gary W. Beehag.)

Fig. 2.3. A subtropical environment with turfgrass adjacent a coastal region. (Courtesy of Gary W. Beehag.)

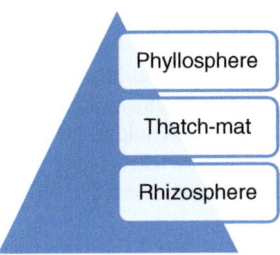

Fig. 2.4. Physical regions of a turfgrass system.

clay or vertisol having a clay fraction content ranging between 30 and 50% or more (Tainton and Klug, 2002; Beehag and Gmehling, 2010). Cricket wicket profiles are highly prone to water–oxygen imbalances during prolonged wet periods.

Key rhizosphere variables are soil temperature, nutrient status, acidity/alkalinity, organic matter content and oxygen–moisture balance. In addition, plant roots naturally exude an extensive array of complex biochemical compounds (allelochemicals), which influences the growth and development of soil-borne microorganisms.

Susceptible Host

Turfgrasses, whether grasses or forbs, have a level of immunity or tolerance to most invading microbes, giving rise to one of the central principles in plant pathology; meaning plants are resistant or immune to most potentially infectious agents. The main host factor affecting plant disease is the occurrence of plants that are susceptible to a particular pathogen.

Susceptible and resistant are relative terms existing as a continuum of dynamic responses ranging from immunity, highly resistant to highly susceptible (Box 2.1).

Primary, secondary and reservoir hosts

Turfgrass plants can be categorized on how they react to the presence of pathogens (Table 2.1). Certain turfgrass cultivars may remain outwardly unaffected (i.e. resistant). Other cultivars may succumb (i.e. susceptible) to the pathogen, resulting in disease.

The abovementioned definitions must be seen as dynamic. A turfgrass cultivar may possess resistance against one disease but be susceptible to another in the same or a different location due to changing environmental conditions.

Inherited plant factors

Host plant susceptibility to pathogens is governed largely by inherited characteristics of both the pathogen and the plant, controlled by specific genes. Disease susceptibility may be greater

Fig. 2.5. Thatch-mat region beneath a mature creeping bentgrass putting green. (Courtesy of Gary W. Beehag.)

Fig. 2.6. Organic matter accumulation beneath a 30-plus-year-old St. Augustinegrass (buffalo grass) lawn. (Courtesy of Gary W. Beehag.)

in immature and lesser in mature plants. This partly explains why seedlings of certain species suffer damage from disease and mature plants of the same species often are generally less affected. Opposite disease scenarios may also apply.

Management-induced factors

The degree of susceptibility or alternatively resistance to disease can be influenced by the intensity of cultural management of a turfgrass sward (Table 2.2).

Turfgrasses managed under low cultural intensity (i.e. infrequent mowing) have relatively taller-growing plants all with larger/longer-sized leaves, as opposed to plants managed under high cultural intensity (i.e. very frequent mowing at very low height) which have very small/short-sized leaves intimately close together and often in close contact with leaves from other plants.

The point here is: the reduction of leaf area (i.e. frequent removal of photosynthesizing tissue) can result in closely spaced and 'relatively immature' leaves and may partly explain the

Fig. 2.7. A cracking clay soil profile of an Australian cricket wicket. (Courtesy of Gary W. Beehag.)

Box 2.1. Distinction between immune, susceptible and resistant hosts.

Immunity – lack of disease symptoms.
Highly resistant – low expression of disease symptoms.
Highly susceptible – high expression of disease symptoms.

Table 2.1. Categorization of primary, secondary and reservoir hosts.

Category	Definition
Primary	Pathogen reaches maturity and reproduces to infect the host
Secondary	Host harbours the pathogen for a short period during which some developmental stage of the pathogen is reached
Reservoir	Host harbours a pathogen indefinitely without ill effects and is non-symptomatic

observation that some turfgrass diseases may be greater as the intensity of cultural management increases. The degree of genetic susceptibility or resistance between species and among cultivars also must be considered.

Turfgrass stress as a disease factor

Plant stress is a complex topic far from understood in turfgrass systems. Highly managed bowling and golf greens and cricket wickets are often subject to the highest degree of environmental and management-induced stress factors. Clipping removal is an important factor affecting plant growth rates and the relative biomass quantities of roots, stems and leaves (Younger, 1969).

Plant stress leading to a diseased condition is not always easily observed. Use of normalized difference vegetation index (NDVI) is commonplace in cropping systems to differentiate between healthy and unhealthy plants for whatever reason. Prolonged exposure to one or more environmental factors may induce a stress condition. If the stress factor is removed, the stress effect can be reversed and the affected plants typically recover, i.e. an elastic response. On the other hand, if the affected plant does not recover despite the stress being removed, the condition is called plastic (Levitt, 1980). For certain conditions (e.g. temperature extremes) plants have evolved tolerance mechanisms when subject to the stress (Box 2.2). Genetic tolerance, resistance and avoidance can also be used in relation to turfgrass diseases.

Hardiness, as applied to the ranking of turfgrasses to temperature stress, is a further term

Table 2.2. Categories of turfgrass cultural intensity.

Primary use	Intensity of turfgrass use			Reference
	High	Medium	Low	
Active sports – high wear	Bowling greens, cricket wickets, croquet lawns, elite sportsgrounds, golf greens, golf tees, tennis courts	Golf fairways, horse racecourses, football fields, polo fields		Adapted from Beehag et al. (2016)
Passive recreation – medium wear		Commercial lawns, domestic lawns, forest reserves, institutional lawns, municipal parklands, nature reserves, suburban road verges		
Functional use – no wear			Airfield verges, golf roughs, motorway verges, rural road verges	

Box 2.2. Resistance, avoidance and tolerance.

Resistance – ability to stop infection and disease development.
Avoidance – prevention by multiple methods to prevent disease.
Tolerance – prevention or reduction in infection and disease.

---

that implies tolerance or acclimatization rather than stress avoidance. From a turfgrass disease management perspective, the ongoing challenge for turfgrass managers is to first identify individual turfgrass stress factors (e.g. heat and drought) and second, if possible, minimize those effects (e.g. wilting) to prevent disease occurrence. A simple concept but complex in implementation particularly for highly managed, monocultured swards.

Host defence mechanisms: how turfgrasses defend against pathogens

Turfgrasses, like all plants, lack immune systems and antibodies. Rather, through evolution plants can possess structural and complex biochemical or physiological mechanisms to prevent potential pathogens from causing disease. Detailed knowledge of plant defence mechanisms among grasses is limited; hence the understanding of defence mechanisms in turfgrasses is largely extrapolated from what is known to occur in crop plants.

The structural, biochemical and physiological functions and responses underlying the defence mechanism between resistant plant and pathogen are complex and far from completely understood. Hence, a detailed discussion about how turfgrasses prevent pathogen infection is well beyond the scope of this book. However, breeding for plant defence mechanism has practical significance and is important in the development and commercialization of disease-resistant turfgrasses.

Defensive plant mechanisms are generally categorized as constitutive or induced based on their existence prior to or following interaction with a pathogen (Table 2.3). When challenged, plant cells must first recognize and then respond to invading pathogens. Resistant plants then activate a series of defensive responses to minimize penetration, infection and colonization by a pathogen.

Structural mechanisms

Certain anatomical structures are preformed plant defence mechanisms against invading pathogens. Among turfgrasses, the existence of high silica content (e.g. zoysiagrasses), cuticular waxes (e.g. ryegrasses) or leaf salt glands (e.g. paspalum and St. Augustinegrass) may assist in minimizing leaf moisture accumulation, thereby increasing leaf water repellence and thus reducing the potential for infection by a pathogen.

Phytochemicals

Plants can biosynthesize an extensive array of phytochemicals in various tissues. The main phytochemicals associated with plant defences against invading pathogens are hormones, phytoanticipins, phytoalexins and elicitors (Tiwari and Rana, 2015; Taiz *et al.*, 2018). Certain phytochemical compounds and molecules may have multiple defence roles.

Hormones are chemical compounds that influence biochemical and physiological processes. The term hormone or plant growth substance encompasses natural plant hormones and their synthetic derivatives. The relative amount, degree of synergy and internal mobility of each hormone in plant tissues vary widely. Certain phytohormones have been shown to have inhibitory or antimicrobial action against certain pathogens. The awareness of natural plant hormones has practical significance in the development and commercialization of synthetic analogues in 'biostimulants'.

Phytoanticipins and phytoalexins are plant metabolites synthesized in extremely low concentrations that have inhibitory or antimicrobial activity. The distinction between phytoanticipins and phytoalexins has largely been based on the occurrence of their formation. In other words, whether they exist prior to infection (i.e. phytoanticipins) or post-infection (i.e. phytoalexins).

Biochemical elicitors simulate an infection of pathogens by acting as signalling agents for phytoanticipins or phytoalexins. Elicitors may be natural biochemical compounds or synthetic analogues able to activate defence responses, improve plant growth and rate of photosynthesis. Numerous natural and synthetic compounds are known to act as elicitors or plant disease resistance activators (Table 2.4). Several synthetic compounds that induce systemic acquired resistance in plants (e.g. acibenzolar-S-methyl, benzothiadiazide, butanediol, ethephon, jasmonic acid and salicylic acid) have been evaluated against fungal diseases on creeping bentgrass and perennial ryegrass (Cortes-Barco *et al.*, 2010; Rahman *et al.*, 2014) with varying degrees of success.

Plant disease resistance

Resistance in plants can be induced by infection or external treatment that results in resistance

Table 2.3. Constitutive and induced defensive mechanisms of plants.

Mechanism	Example
Constitutive	Structural (leaf hairs, cuticle waxes, cell walls)
	Phytochemical (phytoanticipins, plant activators and elicitors)
Induced	Phytochemical (phytoalexins, phytohormones, antimicrobial proteins)
	Physiological (induced systemic signalling and systemic acquired resistance)

Table 2.4. Natural and synthetic elicitors.

Origin	Chemical compounds	References
Natural	Abscisic acid (ABA), gibberellic acid (GA), jasmonic acid (JA), salicylic acid (SA), harpins	Thakur and Sohal (2013); Bektas and Eulgem (2015)
Synthetic	Acibenzolar-S-methyl (ASM), β-aminobutyric acid (BABA), benzothiadiazide (BTH), butanediol (BD), imprimatins, indole-3-acetic acid (IAA), isotianil, methyl jasmonate (MeJa), phosphorous acid, probenazole (PBZ), prohexadione-Ca, sulfonamides, tiadinil	

mechanisms against pathogens. Induction of disease resistance among plants occurs primarily by one of three mechanisms: hypersensitive reaction (HR), induced systemic resistance (ISR) and systemic acquired resistance (SAR) (Box 2.3). Expression of induced systemic resistance and systemic acquired resistance is similar and can result in reduced disease severity, pathogen vigour and tissue colonization (Agrios, 2005).

The hypersensitive reaction is associated with production of reactive oxygen species and has been investigated in turfgrass following fungal infection (Flores et al., 2017). Systemic acquired resistance is a response that results in resistance developing distant from the infected plant parts, usually leaves, throughout the entire plant. Induced systemic resistance is often the result of colonization of roots by certain non-pathogenic microbes, primarily plant-growth-promoting rhizobacteria.

Phytopathogenic Microbes

Phytopathogens or plant pathogens remain the most problematic of all plant pests associated with turfgrasses worldwide. A plant pathogen is an organism that lives on or in some other organism (host) and obtains its food from this organism (host), causing a disease. Several factors associated with plant pathogens are known to affect disease development in plants (Agrios, 2005). These factors can also be applied to turfgrasses (Box 2.4).

Virulence describes the genetic potential of a particular pathogen to cause disease. More inoculum (e.g. spores) produced over shorter time periods generally increases the chances of infection. The frequency of pathogens to reproduce or multiply varies widely, depending on the species.

Host–pathogen relationship

A host–pathogen relationship is the result of a pathogen obtaining nutrition from a host plant that may result in a diseased condition. The precise biological mechanisms of host plant–pathogen relationships at the molecular and cellular level in turfgrass systems are far from fully understood. Host–pathogen relationships are genetically driven events resulting in a series of interacting genetic processes and biochemical compounds or 'signalling pathways' between the plant and pathogen (Schumann and D'Arcy, 2010). Host–pathogen relationships explain partly why some turfgrass species (cultivars) are susceptible while others are tolerant of certain diseases.

Disease pressure

Disease pressure is a poorly defined term widely used to describe the degree of disease, i.e. high, medium to low. High disease pressure infers a combination of a highly susceptible turfgrass host, highly populated and virulent pathogen, and highly favourable environmental conditions. In other words, high disease pressure is expressed as a relatively high amount of disease (i.e. infection centres) in a given area (Fig. 2.8). In contrast, low disease pressure encompasses the opposite combination and a relatively low number of symptoms over the same area.

Pathogen attack mechanisms: how pathogens infect turfgrass hosts

Plant pathogens have evolved numerous mechanisms enabling attachment, penetration, infection and colonization of host tissues (Box 2.5). The adopted attack mechanism depends on the host–pathogen relationship.

Box 2.3. Distinction between HR, ISR and SAR mechanisms.

HR – induced by pathogens, resulting in rapid death of infected plant cells.
ISR – induced mainly by non-pathogenic microbes and plant-growth-promoting rhizobacteria.
SAR – induced by invading non-pathogens and pathogens and accumulation of specific molecules.

Box 2.4. Pathogen factors influencing disease development.

- Level of virulence
- Quantity of inoculum
- Reproduction method of pathogen
- Environmental requirements
- Ecology and mode of spread of pathogen
- Host dormancy

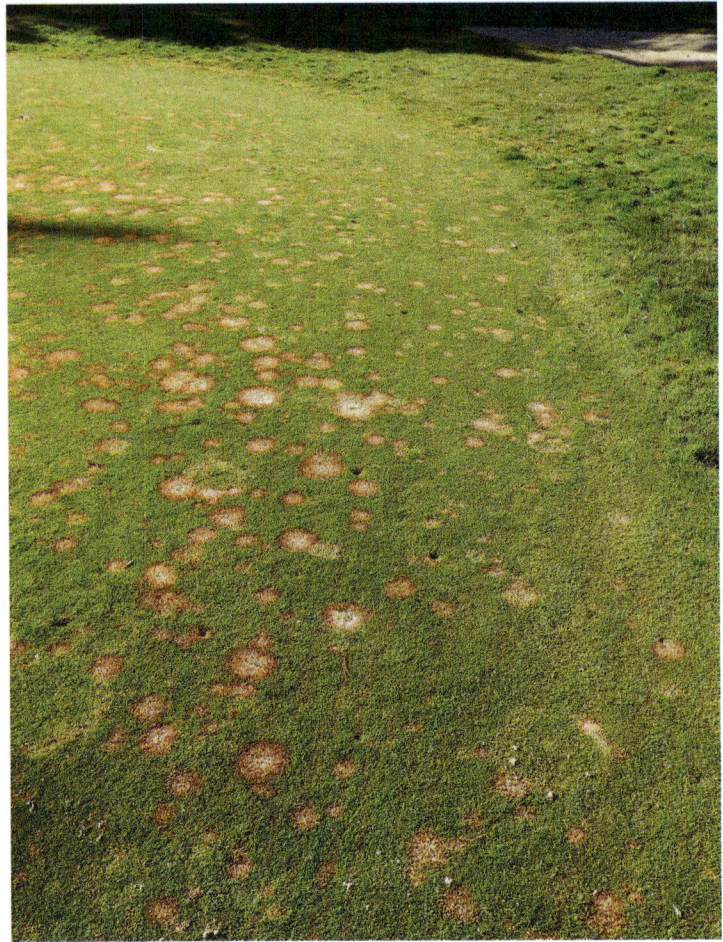

Fig. 2.8. High disease pressure of microdochium patch on a golf green in southern Australia. (Courtesy of Gary W. Beehag.)

Box 2.5. Pathogen attack mechanisms.

- Mechanical force
- Natural openings
- Wounds
- Biochemical compounds
  - Enzymes and proteins
  - Toxins
  - Growth regulators

Mechanical

Mechanical attachment and infection of plant surfaces is commonly used by pathogenic fungi. Physical penetration by fungal pathogens is normally associated with secretion of various enzymes to penetrate through leaf waxes and cell walls. Specialized structures when produced by fungal pathogens (e.g. hyphopodia) can aid in identification of the pathogen.

Biochemical compounds

Biochemical compounds synthesized by infecting pathogens can have numerous effects on the host. The net effect of pathogen infection is often biochemically driven by production and secretion of various biochemical substances and physiological reactions between the pathogen and the host. Biochemical compounds, their concentrations and the effects of each compound vary with the host–pathogen interaction.

Pathogens also synthesize a vast array of chemical and growth-regulating compounds that can cause abnormal growth of the host. In addition, certain plant pathogens synthesize substances that act as pathogenicity factors that can suppress activation of plant defences by disrupting activation of plant biochemical pathways.

Certain toxins contributing to disease may be host-specific, others non-specific. Members of the fungal genera *Alternaria*, *Ascochyta*, *Drechslera*, *Fusarium* and *Pyricularia* produce mycotoxins (Ismaiel and Papenbrock, 2015). The toxic effects of unspecified mycotoxins secreted by *Rhizoctonia solani* on creeping bentgrass have implications in the development of disease-resistant bentgrass cultivars (Tomaso-Peterson and Krans, 1993).

How invading pathogens disrupt their host

Key plant physiological processes (Box 2.6) are all disrupted by pathogens, to varying degrees, depending on epidemiological conditions (Agrios, 2005). The extent and duration of disruption to the host's processes vary according to all interacting environmental factors prior to, during and after pathogen infection. Disruption to one function can affect all other functions to lesser or greater degrees. The net result of the disruption to key physiological processes by plant pathogens varies from temporary chlorosis to permanent damage to turfgrass plants, depending on the infecting pathogen and environmental conditions.

Photosynthesis

Photosynthesis (an oxygen-releasing process) occurs in the green parts of the plant using carbon dioxide taken in through the stomates, the actual site of photosynthesis being the microscopic green chloroplasts in the cells. Any disruption to photosynthesis interferes with the ability of turfgrass plants to produce, transport and store carbohydrates.

Microbial pathogens disrupt photosynthesis resulting in the lack of 'normal' green colour or leaf yellowing (chlorosis) or leaf death (necrosis). Disruption by pathogens results in destruction of leaf tissue, reduction in the photosynthetic-producing surface of leaves, increased water loss through the damaged leaf cuticle, alteration of carbon dioxide absorption, production of toxins that inhibit production of plant growth hormones and enzymes, and destruction of chloroplasts with an associated reduction of chlorophyll.

Respiration

Respiration (an oxygen-consuming process) transforms glucose produced during photosynthesis, releasing carbon dioxide and water via the stomates to the atmosphere. Respiration usually increases in response to an invading pathogen. The rapid rise in respiration in resistant varieties is due to the requirement for energy to rapidly produce and mobilize defence mechanisms. Any disruption to respiration therefore interferes with the ability of turfgrass plants to conduct a wide range of cell functions.

Transpiration

Transpiration (a water-loss process) occurs directly through the cuticle, stomates and cut leaf ends. Stomates open in daylight in response to light and carbon dioxide and close during darkness. Transpiration by turfgrass is a cooling process and particularly significant for C3 turfgrasses; high transpiration rates are associated with certain turfgrass diseases. Transpiration usually increases in the presence of leaf-infecting pathogens as a result of alteration to cuticle or stomate function.

Box 2.6. Key turfgrass plant functions.

Photosynthesis – combines carbon dioxide from the atmosphere and water from the growing medium in the presence of light to produce carbohydrates.
Respiration – transformation of carbohydrates for utilization by the plant in the presence of oxygen and releasing carbon dioxide and water vapour.
Transpiration – unavoidable loss of water vapour through the stomates.
Solute absorption and transport – uptake and translocation of water and soluble nutrients via the xylem tissues.

Solute absorption and translocation

Turfgrasses absorb water and solutes (i.e. inorganic nutrients and chemicals) primarily in their root system and translocate them upwards (acropetally), generally in their xylem (water-conducting) vessels, into the leaves. Concurrently, plants translocate organic nutrients and compounds from their leaves downwards (basipetally) in their phloem (food-conducting) vessels. Many microbial pathogens disrupt the absorption and translocation or water and nutrients, resulting in disruption of the root system, blockage of the movement of water and nutrients in the xylem or phloem, or the redirection of host nutrients.

Changes to root and root-hair morphology and function by pathogens include alteration of cell permeability leading to uncontrolled loss of compounds and proliferation and enlargement of abnormal cells (root knots and galls). Other pathogens (e.g. bacteria) may fill the vessels with the body of the pathogen or substances they secrete. Hence, any disruption to solute movement will therefore reduce the ability of turfgrass plants to absorb and translocate water and organic and inorganic compounds throughout parts or the entire plant.

Plant growth and reproduction

In turfgrass systems, photosynthesis exceeds respiration under non-stressed conditions. Destruction and disruption to key biochemical and physiological functioning regions inevitably results in reduced growth, vigour and size of infected plants. In addition, plant pathogens have a direct adverse effect on plant reproduction or the propagation of their host. Pathogens may disturb the hormone balance by releasing plant hormones or by inducing increased or decreased synthesis (production) or degradation of plant hormones.

Probable Impacts of Climate Change on Turfgrass Diseases

Climate change has been defined as 'a movement in the climate system because of internal changes within the climate system or in the interaction of its components, or because of changes in external forcing either natural factors or anthropogenic activities' (International Panel on Climate Change, 1996).

> **Box 2.7.** Environmental changes on crop production.
>
> - Increasing atmospheric carbon dioxide concentration
> - Global warming associated with 'greenhouse gases'
> - Increasing concentrations of ozone
> - Anthropogenic soil salinization

Specific predictions regarding climate change and turfgrass diseases are difficult to state with accuracy. The combinations of environmental and management-induced variables affecting low-, medium- and high-intensity turfgrass make long-term predictions challenging. Climate change impacts on certain turfgrass diseases may prove to increase, decrease or have no effect on disease incidence or severity.

Ultimately, the degree of the impact of climate change will partly depend on the specific turfgrass and pathogen species involved. Climate change also has implications on crop production (Box 2.7) and pathogens that may be less aggressive in natural plant communities could devastate crop monocultures grown in close proximity (Chakraborty et al., 2011).

Probable impacts on diseases of cereal and pasture grasses are under constant review (e.g. Chakraborty and Newton, 2011; Elad and Pertot, 2014). However, few papers have been published covering probable impacts of climate change on turfgrasses and their associated diseases (e.g. Hatfield, 2017; Neylan, 2018).

Changes in temperature and rainfall patterns are predicted to cause poleward movement of wild and cultivated plant species; thus resulting in changing host–pathogen relationships. Some pathogens and diseases may prove to be more problematic and others to a lesser degree. More aggressive strains of pathogens with broader host ranges (e.g. *Pythium*, *Rhizoctonia* and *Sclerotium*) and the possibility of new pathogens from either agricultural crops or natural plant communities entering turfgrass systems remain concerns.

References

Agrios, G.N. (2005) *Plant Pathology*, 5th edn. Elsevier, Amsterdam.

Beehag, G.W. and Gmehling, E. (2010) Wicket soils put to the test. *Australian Turfgrass Management Journal* 12(6), 14–20.

Beehag, G.W. and McMaugh, P.E. (2018) An investigation of thatch-mat accumulation beneath new bentgrass (*Agrostis* spp.) putting greens. In: *Proceedings of the Australasian Turfgrass Conference & Trade Exhibition 2018, Wellington, New Zealand, 24–28 June 2018* [USB stick]. Australian Golf Course Superintendents Association, Melbourne, Australia and New Zealand Golf Course Superintendents Association, Auckland, New Zealand.

Beehag, G.W., Kaapro, J. and Manners, A. (2016) *Pest Management of Turfgrass for Sport and Recreation*. CSIRO Publishing, Melbourne, Australia.

Bektas, Y. and Eulgem, T. (2015) Synthetic plant defence elicitors. *Frontiers in Plant Science* 5, 804. DOI: 10.3389/fpls.2014.00804

Bigelow, C.A. and Soldat, D.J. (2013) Turfgrass root zones; management, construction methods, amendments and characterization and use. In: Stier, J.C., Horgan, B.P. and Bonos, S.A. (eds) *Turfgrass: Biology, Use and Management*. Agronomy Monograph No. 56. American Society of Agronomy, Inc., Crop Science Society of America, Inc. and Soil Science Society of America, Inc., Madison, Wisconsin, pp. 383–424.

Chakraborty, S. and Newton, A.C. (2011) Climate change, plant disease and food security: an overview. Special Issue: Climate Change and Plant Disease. *Plant Pathology* 60(1), 2–14.

Chakraborty, K., Tiedemann, A.V. and Teng, P.S. (2011) Climate change, plant diseases and food security: an overview. *Plant Pathology* 60, 2–14.

Cortes-Barco, A.M., Hsiang, T. and Goodwin, P.H. (2010) Induced systemic resistance against three foliar diseases of *Agrostis stolonifera* by (2R,3R)-butanediol or an isoparafin mixture. *Annals of Applied Biology* 157, 179–189.

Elad, Y. and Pertot, I. (2014) Climate change impacts on plant pathogens and plant diseases. *Journal of Crop Improvements* 28, 99–139. DOI: 10.1080/15427528.2014.865412

Flores, F.J., Marek, S.M., Anderson, J.A., Mitchell, T.K., Moreno-Zambrano, M. and Walker, N.R. (2017) Reactive oxygen species and alternative hosts of spring dead spot-causing fungi. *International Turfgrass Society Research Journal* 13, 213–224.

Gaussoin, R.E., Berndt, W.L., Dockrell, C.A. and Drijber, R.A. (2013) Characterization, development and management of organic matter in turfgrass systems. In: Stier, J.C., Horgan, B.P. and Bonos, S.A. (eds) *Turfgrass: Biology, Use and Management*. Agronomy Monograph No. 56. American Society of Agronomy, Inc., Crop Science Society of America, Inc. and Soil Science Society of America, Inc., Madison, Wisconsin, pp. 425–456.

Gibson, D.J. (2009) *Grasses and Grassland Ecology*. Oxford University Press, Oxford, UK.

Hatfield, J. (2017) Turfgrass and climate change. *Agronomy Journal* 109, 1708–1718.

International Panel on Climate Change (1996) *Climate Change 1996. Impacts, Adaptations and Mitigation on Climate Change*. Cambridge University Press, Cambridge, UK.

Ismaiel, A.A. and Papenbrock, J. (2015) Mycotoxins: producing fungi and mechanism of phytotoxicity. *Agriculture* 5, 492–537. DOI: 10.3390/agriculture5030492

Levitt, J. (1980) *Responses of Plants to Environmental Stress*. Academic Press, New York.

Neylan, J. (2018) Turf management in a changing climate. In: *Proceedings of the Australasian Turfgrass Conference & Trade Exhibition 2018, Wellington, New Zealand, 24–28 June 2018* [USB stick]. Australian Golf Course Superintendents Association, Melbourne, Australia and New Zealand Golf Course Superintendents Association, Auckland, New Zealand.

Rahman, A., Kuldau, G.A. and Uddin, W. (2014) Induction of salicylic acid-mediated defence response in perennial ryegrass against infection by *Magnaporthe oryzae*. *Phytopathology* 104, 614–623.

Schumann, G.L. and D'Arcy, C.J. (2010) *Essential Plant Pathology*, 2nd edn. The American Phytopathological Society, St. Paul, Minnesota.

Tainton, N. and Klug, J. (2002) *The Cricket Pitch and Its Outfield*. University of Natal Press, Pietermaritzburg, South Africa.

Taiz, L., Zeiger, E., Moller, I.M. and Murphy, A. (2018) *Plant Physiology and Biochemistry*, 6th edn. Sinauer Associates, Inc., Sunderland, Massachusetts.

Thakur, M. and Sohal, B.S. (2013) Role of elicitors in inducing resistance in plants against pathogen infection: a review. *ISRN Biochemistry* 2013, 762412. DOI: 10.1155/2013/762412

Tiwari, R. and Rana, C.S. (2015) Plant secondary metabolites: a review. *International Journal of Engineering Research and General Science* 3, 661–670.

Tomaso-Peterson, M. and Krans, J. (1993) Recovery of *Rhizoctonia solani* resistant creeping bentgrass germplasm using the host–pathogen interaction system. In: *1993 USGA Progress Report Executive Summary*. United States Golf Association, Liberty Corner, New Jersey, pp. 40–41.

Younger, V.B. (1969) Physiology of growth and development. In: Hanson, A.A. and Juska, F.V. (eds) *Turfgrass Science*. Agronomy Monograph No. 14. American Society of Agronomy, Inc., Madison, Wisconsin, pp. 187–216.

# 3

Ecological Groups and the Environment of Turfgrass Microorganisms

Abstract

The turfgrass microbiome comprises a multitude of different microflora and macrofauna groups, each performing one or more functional roles. Most microbes associated with turfgrass systems are not pathogenic. The complex parasitic and pathogenic relationships between different groups partly explain how they gain nutrition and why certain microorganism species offer potential in the biological management of turfgrass diseases.

Microflora and Microfauna Groups

Microorganism diversity in turfgrass systems is driven largely by interactions between the plant and the environment. Taxonomic groups of microorganisms occurring in turfgrass systems can be grouped as either microflora or microfauna (Box 3.1). Microfauna are organisms classified as either eukaryotes (i.e. having cells possessing a membrane-bound nucleus and other structures) or prokaryotes (i.e. cells lacking membrane-bound structures).

Prokaryotes (e.g. actinomycetes and bacteria) are primitive microbes being either aerobic or anaerobic, with certain types playing important roles in nitrification. Actinomycetes are aligned to primitive bacteria and possess certain features of fungi. Actinomycetes play important roles in organic matter decomposition, recycling of organic nutrients and biosynthesis of antibiotics.

Bacteria are microscopic, single-celled microbes that reproduce rapidly by fission or simple cell division and perform essential roles in numerous biochemical processes associated with plant roots. Bacterial populations are highly diverse in mature turfgrass systems (Elliott *et al.*, 2008; Crouch *et al.*, 2017). Cyanobacteria are photosynthesizing microbes similar to algae and often called blue-green algae (Baldwin and Whitton, 1992).

Fungi are eukaryotes, mostly multicellular organisms, providing beneficial roles in organic matter decomposition, nutrient recycling and acquisition of nutrients, while some are even predatory on other microorganisms. Fungi lack the ability to synthesize carbohydrates but derive their nutrition from either dead or living organisms. Most fungi are filamentous, possessing hyphae giving rise to mycelium. Fungi reproduce sexually (perfect stage) or asexually (imperfect stage) or a combination of both. Fungal-like organisms, the oomycetes (e.g. *Pythium* and *Phytophthora*), are taxonomically distantly related to true fungi. Fungi cause the majority of described turfgrass diseases.

Box 3.1. Microflora and microfauna groups.

Microflora – actinomycetes, archaea, bacteria, fungi and fungal-like organisms.
Microfauna – nematodes and rotifers.

© G.W. Beehag *et al.* 2024. *Biology and Integrated Management of Turfgrass Diseases* (G.W. Beehag *et al.*)
DOI: 10.1079/9781789246230.0003

A suite of primitive fungus-like eukaryotes (e.g. *Olpidium brassicae*, *Rhizophydium graminis*, *Polymyxa graminis* and others) infect but rarely damage grass roots (Tredway *et al.*, 2023). *P. graminis*, causal agent of Polymyxa root rot, has been documented on symptomatic mown grass (Couch, 1995; Forrest, 2014).

Other microorganism groups (e.g. mollicutes, protozoans and rickettsias) associated with wild grasses have not been widely investigated on managed turfgrass. The identity of many of these respective microbes and their ecology and epidemiology remain uncertain and are not covered in this book.

Rotifers are microscopic, omnivorous animals mainly associated with aquatic systems. Nematodes are ubiquitous, transparent microbes directly or indirectly impacting turfgrass health by engaging a range of functional roles in plant systems depending on species. Free-living nematodes gain their nutrition by consuming or feeding on fungi (i.e. fungivorous) or bacteria (i.e. bacterivorous). Entomopathogenic nematodes obtain nutrition by feeding on soil-borne insects.

Viruses and virus-like microbes are submicroscopic particles of various sizes and shapes. All plant-associated viruses multiply in host's cells and in some situations cause ill-health but seldom cause death of turfgrasses. Viruses cannot be cultured on artificial media and can only be observed using high-resolution electron microscopy; hence are extremely difficult to diagnose.

Microbial Lifestyles: Parasitic, Pathogenic and Saprophytic

Microbial populations associated with turfgrass systems live and gain their nutrition by one of three ways (Box 3.2). These relationships can vary greatly and may be transient depending on many environmental factors.

Certain fungal species, under conducive environmental conditions, may temporarily transfer from a saprophytic to a pathogenic phase. Certain fungal species (e.g. fairy ring fungi) which normally grow saprophytically may moderate specific environmental conditions which alter normal plant growth.

Strictly speaking, the terms 'pathogen' and 'parasite' should not be interchanged. The terminology of biotroph, hemibiotroph and necrotroph

---

Box 3.2. Parasitic, pathogenic and saprophytic lifestyles.

Parasite – an organism that lives on or in some other organism (host) and obtains its food from this organism (host).
Pathogen – an organism that lives on or in some other organism (host) and obtains its food from this organism (host), causing disease.
Saprophyte – an organism that lives on decomposing plant tissues or some form of organic matter.

(Glazebrook, 2005) is commonly used to define a plant–pathogen interaction (Table 3.1).

Biotrophs (previously called obligate parasites) are highly specialized parasites and their distribution is determined by their host plants. Rust fungi, smut fungi and primitive root microbes are common biotrophs associated with turfgrasses. Hemibiotrophs (previously known as facultative parasites or facultative saprophytes) are common in turfgrass systems. Hemibiotrophs are the dollar spot, anthracnose and snow mould fungi. Necrotrophs are less specialized parasites and can have a wide host range. *Pythium* fungi and certain *Rhizoctonia* species are generally considered to be necrotrophs (Tredway *et al.*, 2023).

Symbiotic Relationships: Endophyte, Mycorrhizae and Rhizobacteria

Certain groups of microbes form one of several symbiotic relationships with host plants whereby one or both partners benefit (Box 3.3). Endophyte and mycorrhizae relationships are common in grassland systems, forming temporary or permanent mutualistic lifestyles with host plants.

Endophytic bacteria and fungi are transmitted maternally from infected seeds to grow systemically throughout the host. Endophytes are obligate biotrophs and the symbiosis remains largely non-pathogenic and symptomless, provided the relationship remains balanced. Endophyte–grass relationships have been documented for bacteria (e.g. Kandel *et al.*, 2017) and fungi (e.g. Cheplick and Faeth, 2009). Fungal endophytes are the most well known in grassland systems and numerous genera are associated with cool-season and warm-season grasses worldwide (Table 3.2). Bacterial and fungal endophytes biosynthesize a wide range of secondary metabolites and confer a degree of environmental stress tolerance on their host plants, depending on environmental conditions. *Epichloe* and *Neotyphodium* are the most commonly studied fungal endophytes among cool-season turfgrasses.

The mycorrhizal relationship is mostly mutually beneficial to the fungus and colonized roots. The fungi facilitate increased absorption of water and nutrients otherwise unavailable to the plant; thus improve tolerance to environmental stresses (Brundrett, 1991). Mycorrhizal associations are common, particularly in nutrient-deficient sands. Members of the fungal genera *Acaulospora, Entrophospora, Glomus, Gigaspora* and *Scutelospora* are common arbuscular mycorrhizal fungi in turfgrass systems (Murakoshi *et al.*, 1996; Koske *et al.*, 1997).

The current weight of available information strongly suggests cool-season grasses, having greater numbers of smaller-diameter roots with greater branching and root-hair development, are less mycorrhizal-dependent, as opposed to warm-season grasses having larger-diameter roots with lessened branching, which are more dependent on mycorrhiza fungi for nutrient absorption (Brundrett, 1991).

Plant-growth-promoting rhizobacteria (PGPR) are beneficial, free-living bacteria enhancing plant growth. *Agrobacterium, Bacillus, Pseudomonas, Rhizobia, Streptomyces* and *Trichoderma* are among common rhizobacteria genera (Backer *et al.*, 2018).

The Environment of Turfgrass Microbes

Temperature and moisture, in combination, are the primary forces driving the growth of microorganisms and disease development in turfgrass

Table 3.1. Biotroph, hemibiotroph and necrotroph lifestyles and microbial examples.

Lifestyle	Characterization	Microbial examples
Biotroph	• Synchronous growth with their hosts • Typically do not cause extensive damage	*Blumeria, Puccinia, Sclerophthora, Ustilago*
Hemibiotroph	• Initially gain their nutrition as biotrophs then transfer to a necrotrophic phase • Certain species can survive on decaying tissues	*Clarireedia, Colletotrichum, Microdochium, Pyricularia*
Necrotroph	• Parasites that gain their nutrition and survive on dead hosts or decaying tissues	*Pythium, Rhizoctonia*

Box 3.3. Endophyte, mycorrhizae and rhizobacteria relationships.

Endophyte – bacteria or fungi living inside a plant.
Mycorrhizae – relationship between fungi and host plant.
Rhizobacteria – beneficial, free-living bacteria.

Table 3.2. Grasses and their fungal endophytes.

Grass group	Grass genera	Fungal endophyte	Reference
Cool-season	*Danthonia* *Agrostis, Dactylis, Poa* *Festuca, Lolium, Poa*	*Atkinsonella* *Epichloe* *Neotyphodium*	Modified from Cheplick and Bacon (2009)
Warm-season	*Axonopus* *Panicum, Paspalum, Sporobolus* *Panicum, Paspalum* *Bothriochloa, Cynodon*	*Fusarium* *Balansia* *Myriogenospora* *Nigricornis*	

systems. Other factors of rootzone physiochemical properties, host nutrition, microbial competition and antagonism also come into play with specific microbes in the presence of applied chemicals (Box 3.4). Ultimately, all environmental forces act in combination often with unpredictable outcomes on microbial growth and development.

Temperature

Plant pathogens can be characterized based on the temperature range favouring their growth (Table 3.3). Pathogens that naturally occur nearer equatorial regions, as opposed to other plant pathogens in temperate regions having evolved to tolerate more variable temperatures, tolerate a relatively narrow temperature range. Most turfgrass fungal pathogens are mesophilic species that grow at moderate temperatures. Temperature impacts on viruses are relatively unpredictable.

Fungal pathogens adapted to grow at relatively low temperatures (psychrophilic), as opposed to other fungal species best adapted to survive at relatively high temperatures (thermophilic), are capable of causing significant disease to warm-season turfgrasses managed in temperate or transition zones limited by low temperatures.

Under laboratory-controlled conditions, fungal species and strains slowly commence growth at a minimum temperature, increase rapidly throughout an optimum range then slowly decline at a maximum temperature called cardinal temperature points (Smith et al., 1989). Outside these temperature ranges, microbial growth becomes restricted or ceases, depending on the geographic origin of the fungal isolate. Nutrient and moisture level of agar cultures may impact the growth of the fungal isolate under test at any given temperature.

However, the optimum temperature range for growth of a specific fungal species, as measured in the laboratory, often differs considerably from the temperature range of infection and disease progression caused by the same fungus. This important point is illustrated in relation to two common turfgrass fungal pathogens (Table 3.4). The optimum temperature range for infection of most fungal pathogens often falls within the cardinal range for the same fungus.

Onset of turfgrass diseases is often related to the differential response between the host and the pathogen at that temperature. Generally, disease development occurs most rapidly at temperatures favouring the growth and development of the pathogen but in a range which hinders the growth and vigour of the particular turfgrass species or cultivar. Diurnal temperatures between daytime and night-time as well as 'cold hardening' also come into play (Agrios, 2005).

From a fungal disease management perspective, the temperature at which infection occurs and the virulence of the pathogen are more useful for disease prediction. The virulence of the infecting pathogen to cause damage to the foliage or roots may vary at different temperatures. Progression to higher temperatures results in fungal pathogens becoming more active and infecting their hosts provided other conditions (e.g. moisture) remain favourable for infection.

Box 3.4. Key environmental factors influencing microbial growth.

- Temperature
- Moisture
- Solar radiation
- Oxygen
- Nutrition
- Chemicals

Table 3.3. Microbial temperature categorization.

Temperature categorization	Temperature range (°C)	Reference
Psychrotolerant	< 5	Deakin (2006)
Psychrophiles	16–20	
Mesophiles	10–40	
Thermophiles	20–50	
Hyperthermophiles	> 60	

Moisture

All forms of environmental moisture to lesser or greater degrees impact the growth and development of turfgrass pathogens and their diseases. Moisture exists internally inside the host (i.e. water-conducting tissues or xylem), externally on plant parts (i.e. foliage, stem and roots) and in their decomposing tissues.

External free moisture originates from the host (e.g. transpiration water and dew formation) and from the atmosphere (e.g. condensation, rainfall and frost) (Fig. 3.1). Surface and subsurface irrigation water further contributes to free environmental moisture for microbial growth.

Observations over extended periods strongly indicate that the frequency of occurrence of many foliage-borne pathogens in a particular region is associated with seasonal rainfall events, dew formation or with frequently irrigated turfgrass. Other diseases are associated more so during drier periods or on non-irrigated turfgrass. Certain fungi may be described as water-stress-tolerant, others as stress-intolerant; with the requiring of greater moisture allowing their categorization on the basis of moisture requirements (Table 3.5).

Canopy moisture

Canopy moisture directly impacts foliage-borne pathogens. Most foliage-borne fungi (e.g. *Clarireedia* and *Colletotrichum*) require extended periods

Table 3.4. Temperature ranges for growth and infection of nominated fungal pathogens.

Pathogen/disease	Cardinal range (°C)	Infection range (°C)	References
Athelia rolfsii (southern blight)	8; 25–35; 40	30–35	Smith *et al.* (1989);
Pyrenophora poae (melting-out)	3; 18; 30–35	18–24	Couch (1995); Tredway
Limonomyces roseipellis (pink patch)	4; 20–23; 31	16–21	*et al.* (2023)
Monographella nivalis (microdochium patch)	–6; 21–25; 30	0–18	

Fig. 3.1. Early-morning frost around a golf green complex. (Courtesy of Gary W. Beehag.)

of high humidity to initiate spore germination. Intermittent wetting and drying cycles may promote sporulation of certain foliage pathogens (e.g. *Bipolaris* and *Pyrenophora*), particularly in the upper thatch-mat layer. Frozen water as frost and snow aids the growth of certain cold-tolerant fungi (*Typhula* and *Monographella*). Small-sized water droplets as fog or condensation contribute to turfgrass canopy humidity, which is a significant factor in foliage-borne fungal diseases. Accumulation of dew, guttation fluid and leaf exudates singly or in combination contributes to leaf wetness.

Leaf wetness

The importance of leaf wetness duration in combination with high relative humidity and elevated canopy temperature forms the basis of warning systems and predictive models for several foliar diseases. Leaf wetness refers to the presence of water droplets (e.g. fog, rain or irrigation). Minimum periods of leaf wetness are required for spore germination of many foliar pathogens.

Leaf wetness duration generally decreases with increasing temperatures. Duration of leaf wetness and temperature during the infection process are the two most critical microclimatological factors determining disease incidence and severity of foliage-borne pathogens.

Guttation fluid, strictly speaking, refers to the exudation of droplets of liquid water through structures (hydathodes) located around the margins and tips of leaves (Fig. 3.2). Guttation fluid has been associated with the growth of several fungal pathogens (e.g. *Bipolaris*, *Pyrenophora* and *Rhizoctonia*).

Generally, for a given foliage-borne fungal species, a reduced leaf wetness period is necessary for infection under optimum temperatures, as opposed to greater wetting periods required for infection outside the optimum temperature range for infection.

Rhizosphere moisture

Prolonged rainfall events may cause temporary or prolonged, saturated conditions of the turfgrass

Table 3.5. Categorization of moisture requirements of nominated fungi.

Moisture requirement	Fungal pathogen
Higher	*Bipolaris, Clarireedia, Colletotrichum, Curvularia, Pythium, Rhizoctonia, Ustilago*
Lower	*Gaeumannomyces, Laetisaria, Ophiosphaerella*

Fig. 3.2. Predominance of guttation fluid on grass leaf margins and tips. (Courtesy of Ben Evans.)

surface, thatch-mat and the rootzone. Rootzone moisture impacts soil-borne pathogens (e.g. *Gaeumannomyces*, *Magnaporthiopsis* and *Ophiosphaerella*). Excessive thatch-mat moisture favours the development of *Pythium* and *Rhizoctonia* (Tredway *et al.*, 2023). Conversely, prolonged periods of limited moisture favour other disease maladies (e.g. fairy ring formation).

Rootzone oxygen

Separation of the detrimental effects of reduced oxygen from those of elevated moisture levels on soil-borne pathogens is virtually impossible to interpret. Most fungal species are aerobic, requiring oxygen for almost all stages of their life cycle. However, many fungi are obligate aerobes (e.g. *Gaeumannomyces avenae*) and their growth is reduced with lowered oxygen levels. Anaerobic conditions of a lack of soil oxygen are associated with the phenomenon known as 'black layer' in putting greens which is indicative of the presence of sulfur-reducing bacteria (Berndt, 2016).

Optimum oxygen levels and tolerance to carbon dioxide vary between fungal species. Studies have indicated that *Bipolaris sorokiniana* and *Clarireedia* spp. may be susceptible to elevated carbon dioxide, *Rhizoctonia solani* has a higher tolerance to carbon dioxide and *Gaeumannomyces graminis* is sensitive to low oxygen, while *Fusarium culmorum* is tolerant of both excess carbon dioxide and deficient oxygen (Endo, 1972). Low oxygen (high carbon dioxide) associated with high rootzone moisture levels favours certain facultative pathogens (e.g. *Pythium* spp. and *Phytophthora* spp.). This categorization partly explains why *Pythium* and *Phytophthora* are associated with freshwater bodies and saturated rootzones.

Organic matter accumulation

On one hand, absence or minimal organic matter content in 'immature' rootzones may encourage increased growth and development of certain soil-borne pathogens having an obligate parasitic lifestyle. On the other hand, thatch-mat accumulation beneath 'mature' swards provides opportunities for growth and development of certain facultative parasites (Endo, 1972).

Rootzone pH

Most fungal species have a broad pH optimum (pH 5.0–7.0) with some species being acid-tolerant (e.g. *Pythium*). Assessment of the effect of pH and nutrients on soil-borne pathogens is indirect because pH of the rootzone is often related to availability or balance of plant essential and non-essential elements (Smith *et al.*, 1989).

Host nutrition

Generally, the impact of nutrition on plant resistance is relatively small in highly susceptible or highly resistant cultivars but can be substantial in moderately susceptible or partially resistant cultivars. Nutrient-deficient turfgrass plants have often reduced tolerance to disease which in many instances can be corrected by judicious application of deficient nutrients.

Hydrophobicity

Hydrophobicity (water repellence) is significant in turfgrass systems (Fig. 3.3) because of the association with excessive thatch-mat accumulation and certain basidiomycete fungi. Hydrophobicity is a common phenomenon on ageing bowling and golf greens caused by a coating of complex organic compounds that envelopes sand grains (York and Canaway, 2000; Gaussoin *et al.*, 2013).

Disease-suppressive soils

Natural disease-suppressive soils occur worldwide and are associated with numerous soil-borne pathogens (Weller *et al.*, 2002). Suppressive soils are when the pathogen does not establish or persist, establishes but causes little or no damage, or establishes and causes disease for a while but thereafter the disease is less important, although the pathogen may persist in the soil. In contrast, conducive (non-suppressive) soils are soils in which disease readily occurs.

Fig. 3.3. Severe hydrophobicity symptoms on a creeping bentgrass golf green. (Courtesy of Gary W. Beehag.)

Solar radiation

Reduced light because of shading effects from buildings or trees is a worldwide and common limitation of turfgrass systems (Fig. 3.4). Prolonged shading is commonly associated with certain fungal diseases (e.g. microdochium patch and powdery mildew) while others (e.g. summer patch and necrotic ring spot) are more common in full sun, depending on other environmental factors.

Numerous fungal species (e.g. *Colletotrichum*, *Puccinia* and *Ustilago*) have evolved to detect and respond to the intensity, duration and quality of light. Generally, reduced light appears to have greater impact on fungal sporulation (Deacon, 2006). The impacts of light intensity, duration and quality on many turfgrass pathogens remain to be investigated.

Wind

Air movement in the form of wind immediately above the turfgrass canopy influences foliage-borne plant pathogens by expediting leaf surface drying but at the same time aiding fungal spore dispersal. Wind-driven rain assists in the release and dispersal of bacterial and fungal pathogens before being deposited on wet foliage.

Turfgrass Disease Cycles

Occurrence of turfgrass disease is a programmed sequence of events under highly favourable conditions between a virulent pathogen and susceptible host, called the disease cycle (Fig. 3.5). Individual stages and time periods of each event depend on host–pathogen interaction and prevailing environmental conditions.

Inoculation describes the initial contact between pathogen inoculum and the host surface. Inoculum may be sclerotia, spores or mycelium (i.e. fungi) or entire individuals (i.e. bacteria, viruses). Inoculation may occur on any part of the host plant and more than one pathogen may be present at any one time on the same infected turfgrass plant. Entry by fungal pathogens may involve natural openings (e.g. stomates) and physical wounds (e.g. cut leaf ends).

Ecological Groups and the Environment of Turfgrass Microorganisms 37

Fig. 3.4. A highly shaded recreational lawn. (Courtesy of Gary W. Beehag.)

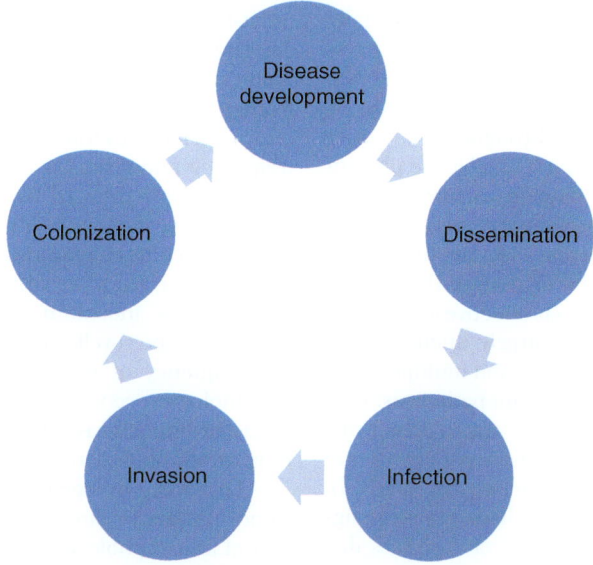

Fig. 3.5. Simplified events during a turfgrass disease cycle.

The mechanisms by which plant pathogens penetrate and then enter external plant surfaces have practical consequences on frequently mown turfgrass. Ordinarily, direct penetration of intact plant surfaces is generally considered the most likely means of penetration by plant pathogens. This assumption is probably correct given the fact that co-evolution between plants and their pathogens occurred in natural plant ecosystems. Many species of pathogenic fungi (i.e. obligate parasites) use mechanical force enabling direct penetration through the intact cuticle.

Natural plant surface openings (e.g. stomates and hydathodes), even when closed, offer ready access for many plant pathogens. Stomates are open during daylight hours and located on the underside of leaves. Many common grass species have stomates on both leaf surfaces. Of the rest, some species have them only on the upper surface and others have them only on the lower surface.

None the less, certain fungal pathogens (e.g. powdery mildew) may grow over open stomates without entering. Infection describes a successful contact and procurement of nutrients from host tissues by the pathogen. Incubation period is the time period between inoculation and appearance of symptoms, which varies widely depending on host–pathogen relationship. Disease symptoms may not appear immediately depending on host–pathogen relationship.

Pathogen invasion describes the different ways and extents to which pathogens internally and externally invade their hosts. Certain bacterial and fungal species may remain largely on the external infected surface, others partially or wholly (i.e. systemically) throughout the entire plant. Some pathogens may inhabit the phloem or xylem vessels, depending on microbial species. Colonization describes pathogen growth, development and reproduction. Filamentous fungi may produce spores and mycelium; thus aiding an increase in mass and reproductive ability. Bacteria and viruses remain largely unchanged in size and shape. Reproduction and multiplication rate varies widely between plant pathogens.

Dissemination describes the mechanisms and extent to which turfgrass pathogens are transported away from their host plant (Box 3.5). All means of dissemination have practical significance in the management of turfgrass diseases. Establishment of new turfgrass sites using infected vegetative material (e.g. stolons or sod)

Box 3.5. Pathogen dissemination means.

- Wind
- Moving water
- Insect and mite vectors
- Infected plant tissues
- Infected plant growth media
- Anthropogenic transport

---

provides an immediate opportunity for dissemination of root-inhabiting pathogens (e.g. fungi).

Overwintering/oversummering of pathogens

Fungal pathogens, during unfavourable conditions and seasons, may either overwinter or oversummer, depending on species, until resumption of conditions favourable for growth and development. There is a wide range of means by which fungi can successfully survive unfavourable environmental conditions.

Disease Epidemics, Complexes and Succession

The foregoing discussion highlights that turfgrass pathogens react to environmental and management-induced inputs and rarely act alone in causing diseases, which are cyclic by nature. Many diseases commence relatively slowly, infecting a limited number of turfgrass plants in one or perhaps several locations. Only when the frequency and intensity of a turfgrass disease increase over time to the point of causing unacceptable damage over greater areas, does it become of concern requiring intervention.

The frequency at which plant pathogens produce new inoculum for infection may be monocyclic, polycyclic or polyetic, depending on the frequency of generations (Box 3.6). The terminology is generally based on a single crop cycle and distinction is not always absolute but can be generally applied to most turfgrass pathogens.

Epidemics or epiphytics occur under the same interactive circumstances initiating disease of a susceptible host, virulent pathogen and favourable environment (Fig. 3.6). Under highly conducive conditions, 'classic' textbook disease symptoms may occur. Disease occurrence can be further qualified by use of the terms disease incidence, severity and duration.

Many turfgrass diseases are a complex of more than one pathogen as attested often by the isolation of multiple fungal species or strains from the same region or the same infected plant. A combination of different fungal pathogens raises the question of distinguishing the primary and secondary pathogens (Fig. 3.7). Certain pathogen–pathogen combinations may be additive or synergistic, complicating the overall expression of disease symptoms.

Box 3.6. Monocyclic and polycyclic disease.

Monocyclic – fewer than or one disease cycle per year.
Polycyclic – more than one disease cycle per year.
Polyetic – several years to undergo all life cycle stages.

Fungal succession in turfgrass diseases occurs over time whereby different species and strains of parasites and pathogens compete against each other for available food resources from the host plants under changing environmental conditions. Different plant pathogens may be isolated from the same diseased turfgrass plant or adjacent infected plants (Fig. 3.8) depending on environmental conditions and time of sampling between inoculation and first symptom expression.

Isolation of species and strains of *Bipolaris*, *Curvularia* or *Pyrenophora* from infected foliage together with species of *Gaeumannomyces*, *Magnaporthiopsis* or *Ophiosphaerella* from roots of the same infected plants gives rise to the possibility of a foliar–root fungal disease complex.

From a turfgrass disease management perspective, existence of disease complexes and pathogen successions are problematical in the selection of disease-resistant cultivars, implementation of effective cultural practices and the selection of appropriate fungicides and biofungicides.

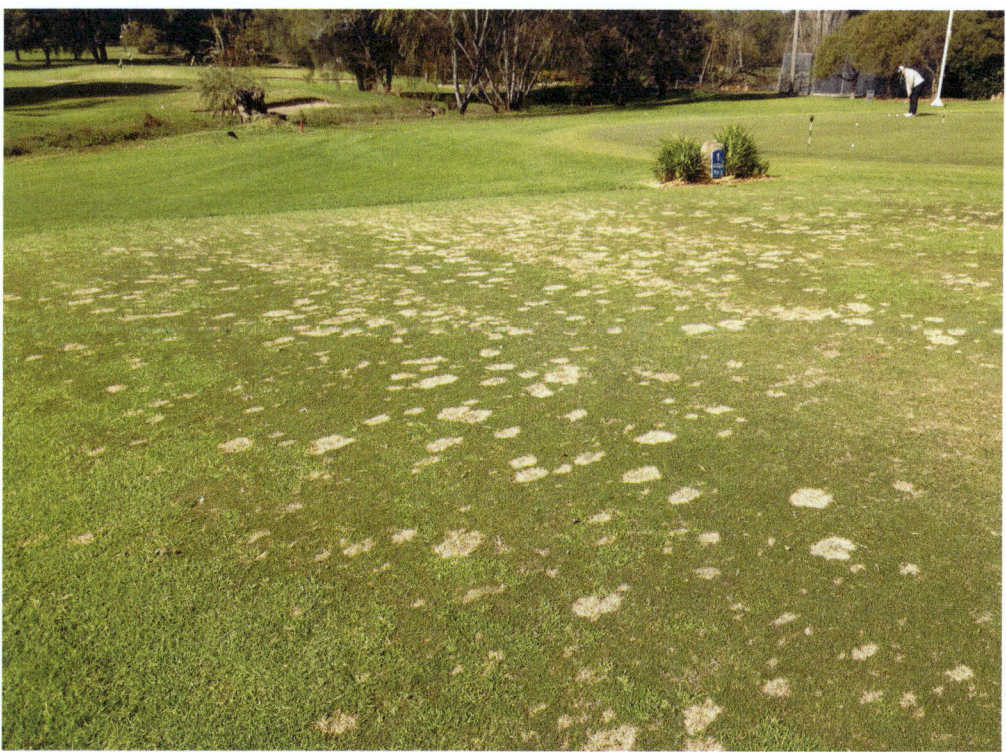

Fig. 3.6. A microdochium patch epidemic on a kikuyugrass tee in Australia. (Courtesy of Nadeem Zreikat.)

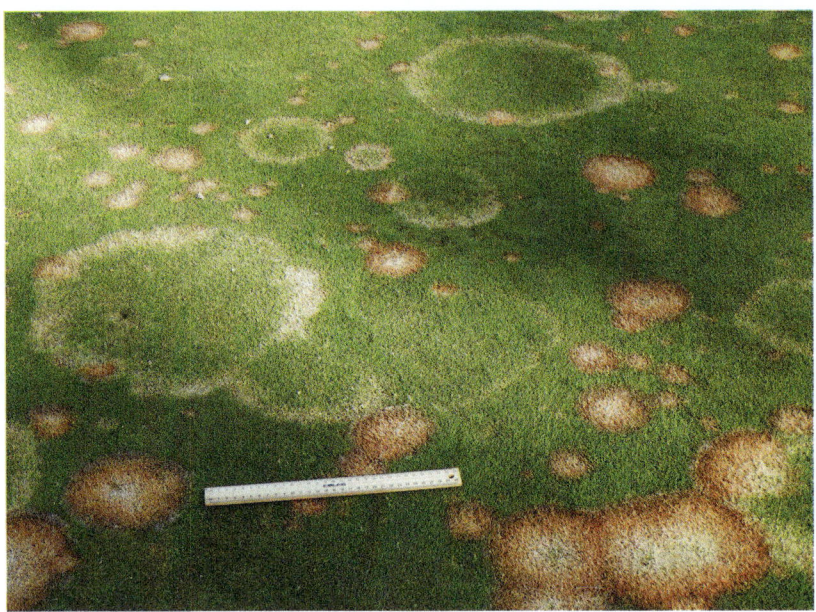

Fig. 3.7. A disease complex of microdochium patch and a *Rhizoctonia* sp. on a mature annual bluegrass–creeping bentgrass golf green in southern Australia. (Courtesy of Gary W. Beehag.)

Fig. 3.8. White-coloured symptoms of a foliage-borne pathogen around a patch perimeter on couchgrass previously infected by root-infecting fungi. (Courtesy of Gary W. Beehag.)

References

Agrios, G.N. (2005) *Plant Pathology*, 5th edn. Elsevier, Amsterdam, the Netherlands.

Backer, R., Rokem, J.R., Ilangumaran, G., Lamont, J., Praslickova, D., *et al.* (2018) Plant growth-promoting rhizobacteria: context, mechanisms of action, and roadmap to commercialisation of biostimulants for sustainable agriculture. *Frontiers in Plant Science* 9, 1473. DOI: 10.3389/fpls.2018.01473

Baldwin, N.A. and Whitton, B.A. (1992) Cyanobacteria and eukaryotic algae in sports turf and amenity grasslands: a review. *Journal of Applied Phycology* 4, 39–47.

Berndt, W.L. (2016) Redox potential and black layer. *Golf Course Management* 84, 86–91.

Brundrett, M. (1991) Mycorrhizas in natural ecosystems. In: Begon, M., Fitter, A.H. and Macfadyen, A. (eds) *Advances in Ecological Research, Vol. 21*. Academic Press, London, pp. 171–313.

Cheplick, G.P. and Faeth, S.H. (2009) *Ecology and Evolution of the Grass Endophyte Symbiosis*. Oxford University Press, Oxford, UK.

Couch, H.B. (1995) *Diseases of Turfgrasses*, 3rd edn. Krieger Publishing Company. Malabar, Florida.

Crouch, J.A., Carter, Z., Ismaiel, A. and Roberts, J.A. (2017) The US National Mall Microbiome: a census of rhizosphere bacteria inhabiting landscape turf. *Crop Science* 57, S-341–S-348. DOI: 10.2135/cropsci 2016.10.0849

Deacon, J.W. (2006) *Fungal Biology*, 4th edn. Blackwell Publishing, Malden, Maryland.

Elliott, M.L., McInroy, J.A., Xiong, K., Kim, J.H., Skipper, H.D. and Guertal, E.A. (2008) Taxonomic diversity of rhizosphere bacteria in golf course putting greens at representative sites in the southeastern United States. *HortScience* 43, 514–518.

Endo, R.M. (1972) The turfgrass community as an environment for the development of facultative fungal parasites. In: Younger, V. (ed.) *The Biology and Utilization of Grasses*. Elsevier, Amsterdam, the Netherlands, pp. 171–202.

Forrest, J. (2014) 'New' disease hits bowls club. *TurfCraft International* 154, 38–42.

Gaussoin, R.E., Berndt, W.L., Dockrell, C.A. and Drijber, R.A. (2013) Characterization, development and management of organic matter in turfgrass systems. In: Stier, J.C., Horgan, B.P. and Bonos, S.A. (eds) *Turfgrass: Biology, Use, and Management*. Agronomy Monograph No. 56. American Society of Agronomy, Inc., Crop Science Society of America, Inc. and Soil Science Society of America, Inc., Madison, Wisconsin, pp. 425–456.

Glazebrook, J. (2005) Contrasting mechanism against defence of biotrophic and neotrophic pathogens. *Annual Review of Phytopathology* 43, 205–227. DOI: 10.1146/annurev.phyto.43.040204.135923

Kandel, S.L., Jourbert, P.M. and Jackson, N. (2017) Bacterial endophyte colonization and distribution within plants. *Microorganisms* 5, 77. DOI: 10.3390/microorganisms5040077

Koske, R.E., Gemma, J.N. and Jackson, N. (1997) A preliminary survey of mycorrhizal fungi in putting greens. *Journal of Turfgrass Science* 73, 2–8.

Murakoshi, T., Tojo, M., Ueda, T. and Ichitani, T. (1996) Two *Glomus* species from rhizosphere soil of a bent grass nursery in Japan. *Mycoscience* 37, 233–235.

Smith, J.D., Jackson, N. and Woolhouse, A.R. (1989) *Fungal Diseases of Amenity Turf Grasses*, 3rd edn. E. & F.N. Spon, London.

Tredway, L.P., Tomaso-Peterson, M., Kerns, J.P. and Clarke, B.B. (2023) *Compendium of Turfgrass Diseases*, 4th edn. The American Phytopathological Society. St. Paul, Minnesota.

Weller, D.M., Raaijmakers, J.M., McSpadden Gardener, B.B. and Thomashow, L.S. (2002) Microbial populations responsible for specific suppressiveness to plant pathogens. *Annual Review of Phytopathology* 40, 309–348. DOI: 10.1146/annurev.phyto.40.030402.110010

York, C.A. and Canaway, P.M. (2000) Water repellent soils as they occur in UK golf greens. *Journal of Hydrology* 231–232, 126–133.

4

Monitoring, Forecasting, Symptomology, Sampling and Diagnosis

Abstract

This chapter highlights that continual monitoring of regional weather conditions and recognition of early warning signs and symptoms of turfgrass diseases are key to their successful management before they become problematic. A meaningful laboratory diagnosis of a disease outbreak requires correct sampling and submission procedures.

Environmental conditions conducive for pathogen infection often precede and continue through disease outbreaks or epidemics, whether foliage- or soil-borne. Intuitive monitoring and forecasting of weather patterns and recognition of obvious disease symptoms are fundamental skills required by turfgrass managers. Once disease outbreaks occur, knowing how to correctly collect, preserve and submit unhealthy turfgrass samples to a pathology laboratory, if and when required, is paramount. Clients have high expectations of a meaningful diagnosis, perhaps recommendations and a management plan in resurrecting turfgrass health.

Computer-based disease models have aided the forecasting of common turfgrass diseases. Increasing utilization of on-site weather stations, hand-held moisture- and temperature-probing devices, smartphone technology and aerial drones with hyperspectral imaging cameras (Fig. 4.1) has enabled the better prediction of turfgrass health and disease outbreaks before they become problematic and potentially unmanageable.

Surveillance and Biosecurity

Surveillance and biosecurity protocols for notifiable diseases and pests are commonplace in agricultural and horticultural crops but not always common for turfgrass systems. The importance of biosecurity and surveillance protocols in the global turfgrass industry is critical because of the high risk of introducing exotic turfgrass pathogens and pests into new regions where they are unknown. Biosecurity and surveillance of turfgrass diseases is particularly critical for soil-borne fungal pathogens and plant-parasitic nematodes. Consideration also needs to be given to sanitizing using water-based products on contracted, loaned or used machinery.

Monitoring of Turfgrass Quality and Key Weather Factors

Management of overall turfgrass health and disease incidence by skilled visual observation of

Fig. 4.1. Aerial drone with front-attached camera. (Courtesy of the NZSTI.)

aboveground and belowground sward characteristics has become increasing aided by data-driven decisions. Observations of changes in foliage colour and rootzone moisture are commonplace procedures when accessing turfgrass health.

Monitoring and compilation of key weather (e.g. humidity, rainfall and temperature) and aboveground (e.g. foliage colour and growth) and belowground (e.g. rootzone moisture, salinity and temperature) factors can be quantified by an increasing range of techniques and technologies. Monitoring of changing weather patterns aids in meaningful warning systems or forecasting of epidemics for some common problematic diseases. Access to local meteorological data via internet sources in conjunction with on-site weather stations (Fig. 4.2) and satellite technology is available for golf courses, sportsgrounds and other turfgrass sites.

Permanently installed, belowground units recording key rootzone data (e.g. moisture, temperature and salinity) located strategically beneath nominated sites are less common. Portable, hand-held devices are routinely used on golf greens and other high-value turfgrass sites to monitor key rootzone data (e.g. moisture using time-domain reflectometry or TDR) for proactive and problem-solving purposes.

Regional weather and aboveground and belowground data must be recorded, allowing for easy access and analyses. Ultimately, record keeping assists in making informed decisions about improving fungicide and biofungicide timing as a means of reducing unnecessary applications and minimizing fungicide resistance risk, particularly for polycyclic diseases.

Disease Prediction and Forecasting

Most turfgrass diseases generally occur around the same time each year. Theoretically, prediction and forecasting of certain disease epidemics is possible based on the interpretation of key weather data and predictive models that have been developed for key fungal diseases (Table 4.1). A disease predictive model is a mathematical calculation based on weather conditions and the assumption that the pathogen is present. Most disease predictive models have yet to incorporate host plant's genetic disease resistance.

Predictive disease models are developed over the course of several seasons based on historical weather data and disease incidence from specific regions. Adoption of a predictive disease

Fig. 4.2. Weather station on a golf course. (Courtesy of Gary W. Beehag.)

Table 4.1. Published disease models of nominated turfgrass diseases.

Fungal disease	Key criteria	References
Anthracnose Brown patch Dead spot Dollar spot Gray leaf spot Pythium blight Rhizoctonia blight	Temperature, leaf wetness, humidity, light, rainfall	Shane (1994); Fidanza (2007); Uddin (2003); Kaminski *et al.* (2007); Smith *et al.* (2018)

model outside the region for which it was developed brings a level of uncertainty requiring field testing and validation.

Weather-based disease models are computer-generated utilizing two or more variables and developed for research-based purposes but a few are utilized by turfgrass managers. Disease models are not highly accurate in the case of diseases caused by a pathogen having multiple strains and one which has a wide range of environmental conditions (e.g. temperature and moisture) governing pathogen activity. Independent evaluations of many fungal disease models generally have shown that inconsistency and variability in predicting disease events (e.g. Palmieri et al., 2006) has been due to weather variables.

Turfgrass disease models are weather-based systems and are unable to determine the continued efficacy of a previously applied fungicide for the disease in question. Disease models cannot predict if the fungal inoculum is capable of causing infection under those measured conditions. Scouting for disease patterns or symptoms and changes in turfgrass quality is necessary. Variations in topography, vegetation and turfgrass cultivars, singly and in combination, impact the first appearance of disease symptoms and knowledge is gained through accumulated experience.

Disease Symptomology

Symptomology of plant diseases describes the association between symptoms of a specific disease and physical signs of the pathogen(s). Recognition of the association of disease symptoms with pathogen signs is procedural-based, requiring keen observation and interpretative skills acquired over time. Certain turfgrass diseases tend to have similar signs and symptoms while other biotic (e.g. plant-feeding insects or mites) or abiotic disorders (e.g. frost damage) may mimic an infectious disease.

Symptom recognition involves inspection of the entire affected area in overall view followed by close examination of individual affected plants. Recognition of early disease symptoms and possible signs on the part of turfgrass managers significantly aids later identification of the pathogen(s) in the laboratory. In some cases, when symptoms become apparent, the disease may be the final result, not initial cause of the problem (Fig. 4.3).

An infectious disease symptom is a complex host reaction exhibited by the infected plant from biochemical and physiological interactions resulting in damage or death of plant organs and interference to normal plant metabolism. In some cases, overall symptoms of certain foliage-borne turfgrass diseases may be barely visible without causing damage. In other cases, disease symptoms may be dramatic and disfiguring, caused by highly virulent pathogen(s) during conducive conditions (Fig. 4.4). Occurrence of turfgrass disease is often dynamic varying in intensity, duration and frequency.

Symptom appearance ranges in colour, location on plant, distribution, size and shape over time depending on host–pathogen combination, mowing height, and environmental and host plant's stress conditions. Colour of infected foliage can vary from shades of yellow, orange, pink to red, purple and black. Certain disease symptoms may be localized to one plant part whether on the foliage, crown or stems. Other symptoms may be more generalized in multiple plant parts, depending on the host–pathogen combination. Certain host reactions may be the result of an interaction of a biotic (i.e. living) agent and an abiotic (i.e. non-living) cause, complicating symptom expression.

On one hand, certain symptoms may be expressed as single or multiple, roughly circular, small-sized spots (Fig. 4.5) as exemplified on closely mown bowling and golf greens and croquet lawns. Over time, certain spots caused by foliage-borne fungal diseases may coalesce forming slightly larger areas of infected plants.

On the other hand, symptoms of soil-borne diseases often appear as multiple, roughly circular and larger-diameter, entirely blighted spots or patches. A striking feature of most patch-type symptoms is the outer perimeter delineating healthy and unhealthy plants (Fig. 4.6). Patch distribution of the same disease is random and its size and growth rate may vary between seasons.

Other patch symptoms may be expressed as multiple 'banana'-shaped arcs or 'doughnut'-shaped patches generally with unaffected green centres. Different shapes and sizes of patch symptoms may simultaneously occur in the one turfgrass site. Over time certain patches

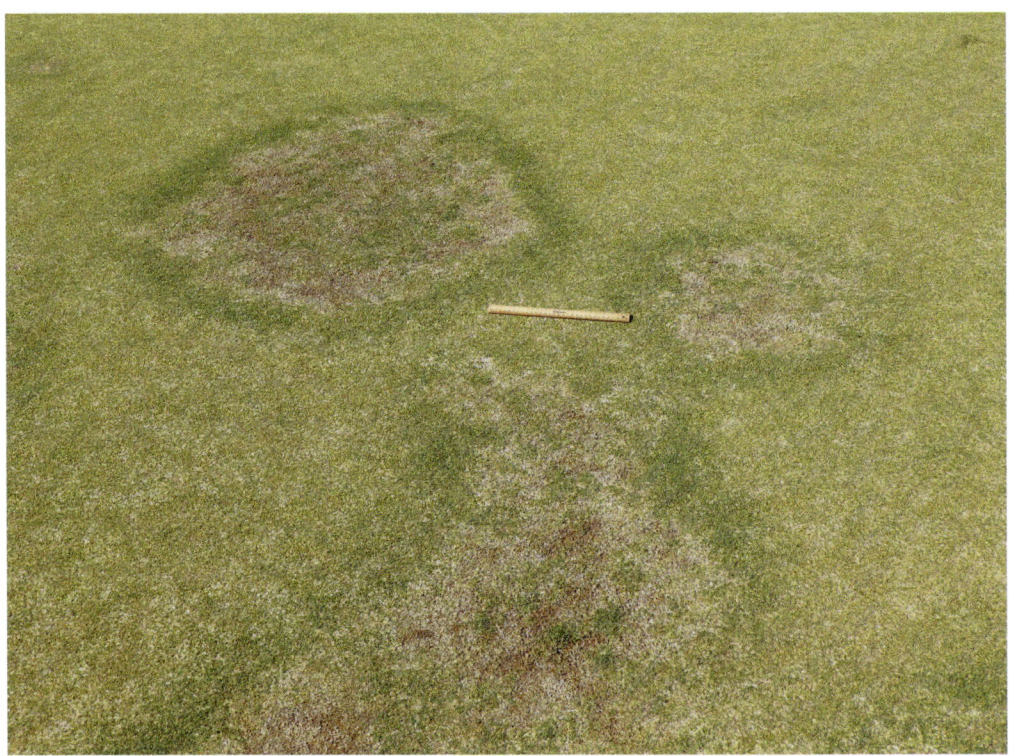

Fig. 4.3. Leaf chlorosis and necrosis caused by foliage-infecting fungi subsequent to root damage by ground pearl (*Eumargarodes laingi*) on a 'Tifdwarf' couchgrass bowling green. (Courtesy of Gary W. Beehag.)

Fig. 4.4. Disfiguring symptoms on a semi-dormant couchgrass bowling green caused by unconfirmed foliar fungal pathogen(s). (Courtesy of Gary W. Beehag.)

Fig. 4.5. Dollar spot symptoms on a creeping bentgrass golf green. (Courtesy of Jyri Kaapro.)

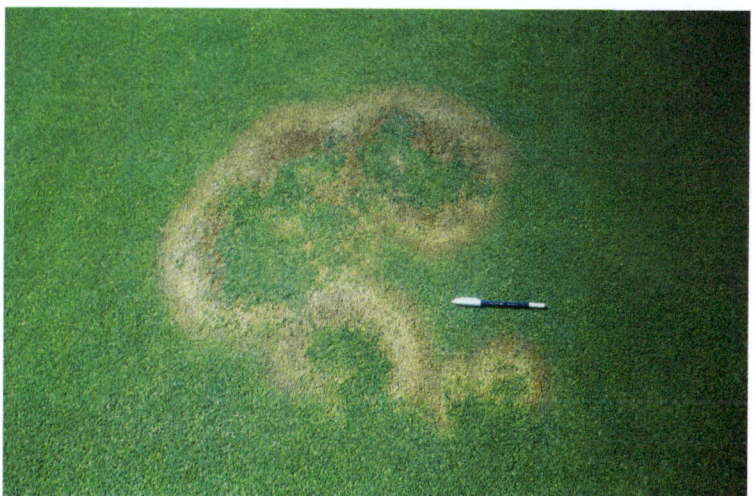

Fig. 4.6. Patch symptoms of southern blight (Rolf's disease) on a creeping bentgrass golf green. (Courtesy of Gary W. Beehag.)

may coalesce, forming much larger areas of disfigured turfgrass. Explanation of the distribution and size of turfgrass patch diseases remains unknown. The specific biological reasons explaining the development of patch distribution and size have been investigated for Rhizoctonia bare patch of cereals (Macnish, 1996). The importance of understanding plant stresses (e.g. low moisture, high salinity, etc.) as disease factors cannot be ignored, requiring additional examination and tests.

Recognition of early, advancing and late-stage symptoms

Most turfgrass diseases undergo a progression of symptom expression over time from an early stage through an advancing phase finally to the late stage. Commencement of early-stage symptoms signifies when a pathogen, normally the primary causal agent, is highly active causing host infection. The appearance and growth of new leaves generally indicates a lack of symptom expression resultant of management intervention or altered environmental factors. Confident recognition of early-stage symptoms (Fig. 4.7) can be problematic and challenging among many soil-borne turfgrass diseases.

Overall disease symptomatology can be similar for certain host–fungal pathogen combinations and different for others. Patch symptoms are generally caused by root- and crown-infecting fungi while blighting and other leaf symptoms are caused by foliar-infecting fungi, bacteria, foliar-feeding nematodes and viral pathogens. Recognition of certain disease symptoms can eliminate other diseases. Early-stage symptoms of certain foliar-borne fungal diseases (e.g. copper spot and dollar spot) and soil-borne fungal diseases (e.g. summer patch and take-all patch) are similar in appearance and subject to misdiagnosis, depending on their seasonal occurrence.

Symptom progression from the early stage through an advancing phase indicates the prevailing environmental conditions remain conducive for the pathogen(s). Coalescing of patch symptoms also indicates highly conducive conditions. Symptom development and damage can occur rapidly for certain fungal diseases (e.g. anthracnose and Pythium blight) and slower among others (e.g. spring dead spot and summer patch) depending on environmental conditions.

Progression from the early to late-stage symptoms among many patch-type diseases can result in the formation of a completely blighted patch or doughnut-shaped rings (Fig. 4.8) depending on the fungal pathogen(s). Late-stage symptoms result in severe disruption to plant metabolic and physiological processes and destruction of plant tissues or total plant death.

Textbook-pictured, 'classic' late-stage symptoms of many turfgrass diseases are not always expressed and their appearance is subject to the host–pathogen combination and underlying environmental conditions between regions. Expression of late-stage symptoms can be partly or fully disguised by multiple pathogens, either primary or secondary, on any plant part (e.g. root-feeding nematodes).

Symptom expression can be further complicated in cases where unhealthy turfgrass is initially infected by a primary, root-infecting pathogen causing patch-type symptoms to subsequently be infected by a secondary, foliar-infecting pathogen.

Fig. 4.7. An early-stage symptom of a root decline disease on 'Tifdwarf' couchgrass. (Courtesy of Gary W. Beehag.)

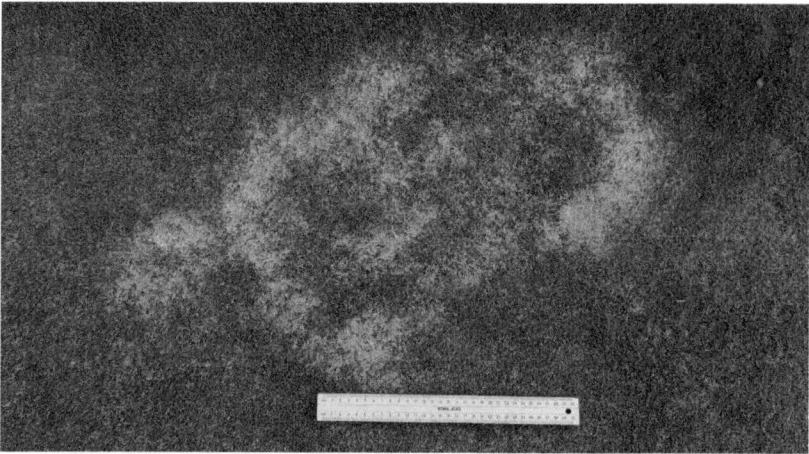

Fig. 4.8. Late-stage symptom of a root decline disease on a 'Tifdwarf' couchgrass bowling green. (Courtesy of Gary W. Beehag.)

It must always be assumed symptoms of any malady could be the result of more than one or a combination of biotic and abiotic causes, complicating a diagnosis.

Detection of disease signs

Disease signs are detectable structures of the causal agent(s) which may be fungal mycelium (Fig. 4.9), spores or sclerotia, possibly present in lesions (Fig. 4.10) aiding a tentative diagnosis. Fungal mycelium is best observed during early morning in the presence of dew. Signs of bacterial diseases may appear less obvious as slight chlorosis or leaf streaking. Observation for the possible presence of any obvious disease signs can be initially conducted in the field by turfgrass managers. Signs (when present) characteristic of certain fungal diseases (e.g. dollar spot) may be readily observed by the naked eye.

Observing the presence of other fungal signs on infected leaf or stem tissues (e.g. acervuli and sclerotia) requires some form of magnification using either a simple 10× hand lens (Fig. 4.11), a laptop USB-attached camera microscope or similar devices. Several areas within the one turfgrass site should be examined for the presence of fungal signs on all plant parts. Foliar structures (when present) of foliar pathogens can be readily observed by carefully using microscopic devices. It is important to observe the condition and colour of leaf tips and the lower leaves which may indicate other problems (e.g. nutrient deficiency or senescence). Examination of the crown, stem and root tissues of suspect plants for fungal structures requires careful pre-washing to remove soil and organic material.

Detection of individual disease signs during the very early stages of infection by certain pathogens can be problematic. Portable immunoassay kits, based on colour intensity in the presence of the pathogen, are available for the detection of a few pathogens only. Each immunoassay kit is subject to false positives and is pathogen-specific and should be used only on foliage and not crown or stem tissues. Where the test proves negative for one pathogen the result provides no guidance for other biological causes. These kits will be replaced by those that utilize much more specific DNA sequence analysis in the future.

Detection and identification of bacteria and viruses (when present) requires high-magnification equipment (e.g. electron microscopes) by highly trained persons normally employed by university, government or research organizations.

Turfgrass managers should also take the opportunity to detect for the presence of other signs of biotic origin (e.g. chewed or bunched leaves) and the physical presence of small-sized plant-feeding insects (e.g. billbugs and weevils) and mites while using magnification devices, as outlined in turfgrass entomological texts (e.g. Beehag *et al.*, 2016; Vittum, 2022).

Fig. 4.9. Fungal mycelium on foliage in early morning. (Courtesy of Jyri Kaapro.)

Fig. 4.10. Multiple fungal lesions on infected leaves of St. Augustinegrass (buffalo grass). (Courtesy of Ben Evans.)

Turfgrass Disease Diagnosis

A meaningful turfgrass disease diagnosis begins with collating key information (Box 4.1). This information allows for the problem to be narrowed down to several possibilities which can be further investigated. Conventional methods of turfgrass disease diagnosis relied on microscopy techniques to identify the signs and symptoms followed by microbial culturing. Advances in molecular technology have seen the introduction of enzyme-linked immunosorbent assay (ELISA), polymerase chain reaction (PCR), multiplex PCR, DNA sequencing and isothermal techniques of loop-mediated amplification (LAMP), and others, for specific turfgrass diseases (Stackhouse et al., 2020). A meaningful diagnosis of turfgrass diseases and their pathogens will require a combination of microscopy techniques and molecular technologies.

Correct identification of the turfgrass species and cultivar (if known) and plant age (i.e. seedlings or mature) are key to disease diagnosis. Recognition of infected species particularly in mixed swards is critical because certain fungal pathogens are

---

Box 4.1. Documentation of key information.

- Identification of infected turfgrass species (cultivar)
- Documentation of weather and environmental conditions
- Recognition of 'healthy' turfgrass morphology
- Documentation of field signs and symptoms

Fig. 4.11. Hand-held magnifying lens with illumination. (Courtesy of Gary W. Beehag.)

Fig. 4.12. Turfgrass sampling devices. (Courtesy of Gary W. Beehag.)

associated with specific turfgrass species. The botanical and common names of the infected turfgrass should be recorded to avoid confusion. Knowing the 'normal' or healthy appearance of mown turfgrasses allows to recognize plants in the sward which display 'abnormal' or unhealthy growth. This is where experience in turfgrass identification comes into play.

Recent weather conditions (e.g. temperature, rainfall and humidity) and alteration of cultural practices (e.g. extreme close mowing, extended irrigation or lime and fertilizer application) which may have acted as predisposing or stress-induced factors must be recorded. Certain fungal diseases (e.g. dollar spot and copper spot) are commonly associated with both dry and nitrogen-deficient plants, some (e.g. brown patch) are more severe under humid, extended periods of leaf wetness and excess-nitrogen conditions, while others (e.g. take-all patch on cool-season grasses) become problematic under alkaline conditions. Recording of the timing and appearance of field symptoms (i.e. spots or patches) supported by clear photographs aids in the ruling in of certain diseases and dismissal of others.

Sample collection, preservation, packaging and submission

Submission of carefully prepared samples to a plant pathology laboratory may be required when a field diagnosis remains unclear or if the cause is unknown. Time is critical particularly when managing polycyclic, foliage-borne diseases requiring a tentative diagnosis and some form of informed, affirmative action while awaiting a final diagnosis.

Correct sampling, preservation, packaging and submission of unhealthy turfgrass is essential to obtain a meaningful laboratory diagnosis. It is imperative to collect samples during the early stage when symptoms are first observed and may coincide with the time when the primary pathogen is highly active, if a disease agent is present. Sampling of late-stage symptoms or when damaged plants appear to be recovering may only reveal secondary or saprophytic microbes. Sampling should be conducted before any intended pesticide application which may prevent the laboratory obtaining meaningful results.

Numerous designs of sampling devices are available (Fig. 4.12). When sampling, factors to consider are: (i) specific locations; (ii) number of samples; and (iii) type and size of samples. Sampling of necrotic or dead turfgrass must be avoided. Dead and dying plants are often invaded by saprophytes or secondary pathogens and generally may overwhelm the primary pathogen.

Sampling of symptomatic plants displaying general chlorosis, leaf blighting or streaking can be taken anywhere at the edges of infected regions. Symptoms displaying patches or rings require sampling to account for the entire disease profile (i.e. foliage, crown, stems and roots). Typically, active disease occurs on locations (i) at the outside of symptoms in what appear to be unaffected areas, (ii) at the outer edge of the existing symptomatic region or (iii) in the inside of the symptomatic area. Sampling patch-type symptoms needs to be conducted in all these areas (Fig. 4.13) allowing for a comparative assessment of potentially pathogenic microbes

Fig. 4.13. Sampling an advancing phase of a patch disease symptom. (Courtesy of Gary W. Beehag.)

between the three areas. Replicate samples will increase chances of a meaningful diagnosis.

Collected samples must be carefully wrapped in clean, moistened paper. Samples with wet leaves should be carefully dried using clean paper. Packaging samples need to be secure to ensure they are intact on arrival at the laboratory. Samples must be kept cool (e.g. room temperature or refrigerated) but not frozen to avoid becoming desiccated. Samples must be correctly labelled in accordance with sampling locations.

Submission of samples by the quickest available means should be done preferably early in the week, if possible, to avoid weekend storage in the parcel delivery facilities or laboratory. Prior consultation with the pathology laboratory or service provider is advisable for any specific sampling or submission requirements.

Laboratory diagnostic methodologies

The detection and identification of disease-causing microbes utilizes various laboratory equipment and a sequence of steps (Table 4.2). Most procedures apply to fungi, being the most common turfgrass disease-causing agent. Detection and identification of bacteria, viruses and other non-fungal microbes requires more sophisticated methodologies and specialist expertise. Clients will generally have high expectations on the part of the laboratory diagnostician or plant pathologist to identify the causal agent(s), an interpretation for the reasons underlying the disease occurrence and, most likely, a management action plan to resolve the problem.

The range of laboratory diagnostic methodologies continues to expand from 'conventional' microscopy-based techniques for observing any signs of the pathogen to attempts to isolate the pathogen in culture towards more 'advanced' genetic-based DNA sequencing and other techniques (Stowell and Gelernter 2001; Stackhouse et al., 2020). Adoption of laboratory procedures depends largely on the reason for the diagnosis, whether problem-solving (e.g. unexplained turfgrass loss), purely routine (e.g. seasonal diagnosis) or research purposes. The range of equipment in the diagnostic laboratory and the technical expertise and experience in turfgrass diseases of employees play a major part.

Most turfgrass samples submitted to a laboratory for diagnosis are for primarily 'problem-solving' reasons. In many cases, sampling may have been conducted during an advancing phase or late stage of symptom development. At the time of sample arrival, turfgrass health may have already declined, to greater or lesser degrees, and it is imperative a diagnosis be made as soon as possible. A tentative diagnosis and informed action plan are required relatively quickly (i.e. within 12–24 h of arrival) for high-quality turfgrass (e.g. bowling and golf greens and elite sportsgrounds) in decline, because of their high value.

Table 4.2. Laboratory-based procedures of fungal disease diagnosis.

Laboratory procedure	Technique and purpose
Microscopic examination of unhealthy plant parts	Plant parts (i.e. leaves, stems and/or roots) are carefully washed in water and observed under the microscope for the presence of fungal structures (i.e. spores and mycelium) and fruiting and survival bodies (i.e. acervuli and sclerotia)
Dissection and staining of plant parts	Plant parts are dissected using scalpels into small-sized pieces then placed on glass slides to observe fungal spores (e.g. *Pythium*) and microbial filaments (e.g. cyanobacteria)
Culturing of fungal isolates	Additional plant parts are dissected using sterile scalpels into small-sized pieces and placed in Petri dishes on agar medium to stimulate fungal mycelium for identification
Culturing of pure fungal isolates	Following sterile dissection of mycelium growth, isolates are placed in Petri dishes on agar medium to stimulate mycelium growth of the pure fungal cultures for a final identification or for storage or proof by Koch's postulates
Advanced technologies	Adoption of serological, molecular and genetic-based techniques for differentiation between genera and among species, particularly for isolates which cannot be identified by conventional methods or are unknown

Most cases of a sudden and unexplained loss of turfgrass of biotic origin involve foliar-infecting fungi and generally an identification of fungi to the genus level based on morphological characteristics (e.g. spores and mycelium) is usually sufficient. Time is of the essence and, in many diagnoses, identification by the laboratory down to species level inevitably will take longer, may not be warranted and could be considered more academic than of practical significance. On the other hand, identification of fungal pathogens for survey or research purposes requires identification to species level.

A conventional laboratory diagnosis typically involves microscopy, staining and plate culturing. Skilful use of dissecting or compound microscopes (Fig. 4.14) allows for a close-up examination and assessment of overall plant health (e.g. foliage colour, stem and root morphology) and presence of any fungal inoculum (e.g. small-sized spores, white or coloured mycelium) and fruiting bodies (e.g. sclerotia or perithecia) otherwise not readily visible by the naked eye.

Microscopic examination of chemically stained, dissected foliage and stem sections is used to detect the presence of *Pythium* spp. oospores. Mounting of dissected leaves in water droplets placed on thin glass slides and viewing microscopically may reveal mass streaming of bacterial spores. Recognition of fungal structures aiding a tentative identification of the fungus to genus level is based on experience on the part of the diagnostician in consultation with fungal taxonomic keys (e.g. Williams-Woodford, 2001).

Failure to detect and identify any obvious fungal structures brings into play skilful techniques of fungal isolating and culturing. These techniques are adopted primarily for the isolation and identification of stem- and root-infecting fungi. Carefully washed and prepared stem and/or root sections are placed in Petri dishes containing agar culture medium in a dedicated and sterilized work area. Surface sterilization is typically done using a 10% sodium hypochlorite solution or bleach prior to placement on to the agar medium and helps to avoid bacterial and saprophytic fungal contamination.

Laboratory incubation times may range from several days to weeks at temperatures of 20–30°C depending on fungal growth rates. Different laboratories may use a specific temperature and more than one type of agar medium, depending on circumstances. A combination of the carbohydrate-rich agar medium, acting as a fungal food source, and a temperature- and moisture-controlled environment encourages

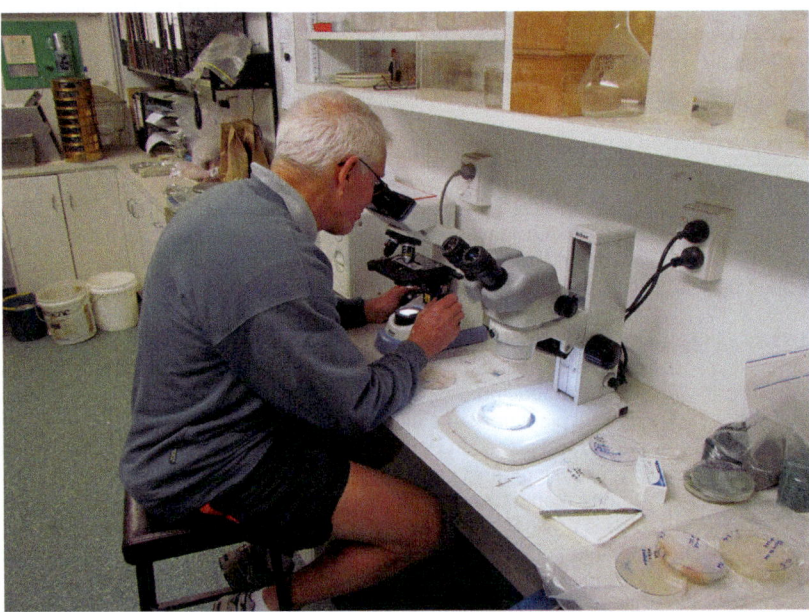

Fig. 4.14. A high-quality laboratory microscope. (Courtesy of the NZSTI.)

fungal sporulation and mycelium growth at increased rates. The Petri dishes containing the agar medium are typically placed under controlled temperature, moisture and light conditions for daily inspection.

A small section of fungal mycelium dissected to obtain a 'pure culture' of the fungal isolate is then transferred to allow further growth (Fig. 4.15). Fungal structures may be clearly seen microscopically using special chemical stains that can aid a more accurate identification using taxonomic keys. However, the presence of one or more known or unknown isolates does not necessarily mean the fungus is pathogenic. Proof of pathogenicity requires further procedures mostly conducted by research-based laboratories.

Laboratory efforts may reveal bacterial, fungal or viral species and strains previously not known to cause disease and, in some situations, possibly result in a 'novel' turfgrass pathogen. This outcome requires diverse efforts and often genetic-based techniques (e.g. DNA sequencing and others) adopted by university and government laboratories.

Koch's postulates

Koch's postulates, a science-based procedure which is universally accepted among plant pathologists, is used to verify the cause and pathogenicity between an isolated microbe and the host plant (Agrios, 2005). Proving Koch's postulates comprises four steps (Box 4.2). The procedure requires expertise and patience to fulfil. Identification to species level of a fungus isolated from infected plant tissues, without satisfying steps 3 and 4, does not prove the isolated fungus is the causal agent of the disease in question.

Box 4.2. Koch's postulates.

1. **Association** – the suspected microbe must be constantly associated with the plant disease.
2. **Isolation** – the suspected microbe must be isolated from the symptomatic host and grown in pure culture under laboratory-controlled conditions.
3. **Inoculation** – when a disease-free but susceptible host plant is inoculated with the microbe grown in pure culture, the same symptoms of the original disease must develop on the inoculated host.
4. **Re-isolation** – the same microbe is re-isolated from the inoculated host plant and when compared to each other, the two microbes must be identical.

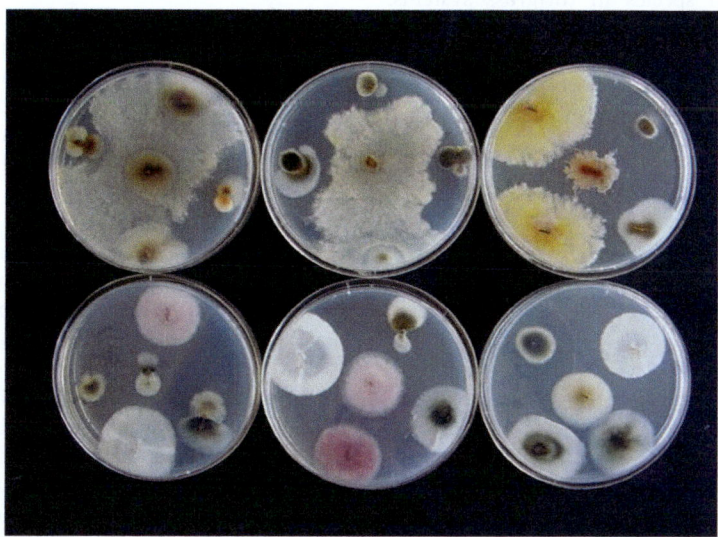

Fig. 4.15. Mycelium growth of different fungal species in agar culture. (Courtesy of Percy T. Wong.)

Understanding and Interpretation of Disease Diagnostic Reports

From a client's perspective, the format and language adopted in disease diagnostic reports need to be understood and be clear and precise. Definitive identification of causal agent(s) and interpretation of the underlying management-induced and environmental factors leading to the stated disease requires caution on the part of the laboratory. Several aspects are key from an understanding and interpretive perspective (Box 4.3).

Turfgrass disease common names will vary between countries and among regions. Thus, it is critical that both the current scientific and common names are clearly defined, should any pathogen be identified. Taxonomic nomenclature of plant-associated microbes is subject to constant revision sometimes making it difficult to be aware of their current, accepted scientific name.

Diagnostic reports stating a genus–species combination implies involvement by a qualified and experienced diagnostician or plant pathologist familiar with turfgrass diseases. Naming of fungi only to genus level (e.g. *Curvularia* sp., *Pythium* sp. or *Rhizoctonia* sp.) suggests the species name is either uncertain or unknown.

Listing the names of more than one fungus is not unusual, reflecting co-colonization of multiple fungal species in turfgrass systems. However, documentation of the level of fungal inoculum should be viewed with a degree of suspicion. Many laboratory reports omit stating the form (i.e. spores or mycelium) and location of the fungi present. The inference being that the fungal species having the greatest presence is likely the causal agent. However, this does not always hold true.

Listing of a fungus only to the genus level raises the question of its lifestyle. In other words, the isolated fungus could be pathogenic, parasitic or a saprophyte. Several questions arise: which if any fungal species is the primary causal agent, are there secondary causal agents and is the cause a disease complex of multiple pathogens or stress-related? Proof of pathogenicity of a fungal isolate requires adoption of Koch's postulates. This is where high-quality photographs and detailed information supplied with the samples come into play in aiding a meaningful diagnosis.

For many clients, their expectations from pathogen diagnosis are the inclusion of informative recommendations and increasingly a meaningful management plan to treat the problem. Multiple strategies are possible in the case of 'problem-solving' diagnoses. Short-term remedial actions against foliage-borne fungal diseases may involve certain inputs (e.g. fertilizer and/or pesticide treatment). Statements of any findings must be made in consideration with the overall turfgrass health and environmental conditions for a meaningful interpretation.

Suggestions only or definitive recommendations nominating specific pesticides (e.g. fungicides, biofungicides, etc.) and other synthetic chemicals (e.g. plant-growth regulators, etc.) or biologically based compounds (e.g. biostimulants, etc.) are commonplace in disease diagnostic reports. From a legal perspective, a 'suggestion' infers the cause is unclear or unknown and the stated remedial actions may not work. A 'recommendation' infers the cause is known and the stated remedial actions will work. As clients, turfgrass managers always have the option of rejecting partly (even fully) any suggestions or recommendations contained in a diagnostic report.

Should significant doubt exist on the part of the turfgrass manager, an alternative laboratory provider or turfgrass pathologist should be consulted. It must be remembered that implementation of any remedial action plan or integrated management programme, whether culturally, chemically and/or biologically based, rests entirely with the turfgrass manager.

Box 4.3. Key aspects of turfgrass pathology reports.

- Use of correct scientific and disease names
- Documenting fungi to genus or species level
- Listing multiple fungal microbes and level of inoculum
- Statements about plant health and thatch-mat accumulation
- Statements of findings and diagnosis
- Interpretation, discussion and relevance of findings
- Written suggestion or recommendation programme

Understanding Turfgrass Disease Trial Data

A plethora of science and extension-based information concerning all aspects of turfgrass diseases is continually published. Comparisons of turfgrass disease susceptibility, fungicide and biofungicide efficacy, and impact of microbial antagonists on disease suppression are among important research investigations to understand. Credible science-based research data contains key elements (Box 4.4).

Although obvious, the purpose or objective of research-based experiments must be stated at the beginning of the report. It is important to state the recognized industry standard (e.g. fungicide used), a cultural procedure (e.g. hollow tining) or another input as a treatment. In any experiment, valid comparisons can only be made when replicates of 'treated' and 'non-treated' (control) plots are included. Replicates account for any 'experimental error' which may arise in the case of differences in disease symptoms between plots. More replication of plots or trials generally provides for greater accuracy in the results.

Assessments or measurements should be made through the duration of the experiment, depending on the time frame. The duration of experiments must be clearly stated be it weeks to years. Experiments conducted over several years allow for differences between years because of variable climatic or environmental influences. Methodologies utilized need to be clearly stated but may not be relevant to the turfgrass manager.

Most importantly, caution must be exercised when reading results gained from relatively short-length studies (i.e. several weeks or months) based on artificially controlled environments. While artificially controlled environments are useful to evaluate potential impacts of a single influence (e.g. temperature or moisture) on disease progression, the results often do not replicate real-life turfgrass conditions. Thus, extrapolation and expectation of the results, either positive or negative, gained from artificially environmentally controlled experiments may be unreliable as other very important factors could have been excluded such as environmental stresses.

Evaluation of turfgrass quality is somewhat subjective and characterized in terms of colour, sward density and uniformity. Commonly adopted assessment scales tend to be based on a scale of 1 to 9, where an assignment of 1 is unacceptable and 9 is excellent (www.ntep.org, accessed 10 June 2022). Assessment scales are routinely used for evaluation of turfgrass cultivar performance, biostimulants, fertilizers, plant-growth regulators, disease or insect injury, or other chemical inputs. The study data are generally reported using tables, charts or images, depending on the author.

Box 4.4. Critical elements of research-based data.

- Statement of objective of trial
- Industry standard product or procedure
- Untreated plots or controls
- Number of replicate plots
- Frequency of observations or measurements
- Methods of assessment
- Period over which trial was evaluated
- Number of locations where trial was conducted
- Format of trial results (tables, charts, etc.)
- Statistical analysis of results

References

Agrios, G.N. (2005) *Plant Pathology*, 5th edn. Elsevier, Amsterdam, the Netherlands.
Beehag, G.W., Kaapro, J. and Manners, A. (2016) *Pest Management of Turfgrass for Sport and Recreation*. CSIRO Publishing, Melbourne, Australia.
Fidanza, M.A. (2007) Predicting Rhizoctonia blight with 'risk models'. *Turfgrass Trends* 7, 1–5.
Kaminski, J.E., Dernoeden, P.H. and Fidanza, M.A. (2007) Environmental modelling and exploratory development of a predictive model for dead spot of creeping bentgrass. *Plant Disease* 91, 565–573.
Macnish, G.C. (1996) Patch dynamics and bare patch. In: Sneh, B., Jabaji-Harte, S., Neate, S. and Dijst, G. (eds) *Rhizoctonia Species: Taxonomy, Molecular Biology, Ecology, Pathology and Disease Control*. Springer, Dordrecht, the Netherlands, pp. 217–226.

Palmieri, R., Tredway, L., Niyogi, D. and Lackman, G.M. (2006) Development and application of a system for fungal disease in turfgrass. *Meteorological Applications* 13, 405–416. DOI: 10.1017/S1350482706002428

Shane, W.W. (1994) Use of disease models in turfgrass management decisions. In: Leslie, A.R. (ed.) *Handbook of Integrated Pest Management for Turf and Ornamentals*. CRC Press, Boca Raton, Florida, pp. 397–404.

Smith, D.L., Kerns, J.P., Walker, N.R., Payne, A.F., Horvath, B., *et al.* (2018) Development and validation of a weather-based warning system to advise fungicide applications to control dollar spot on turfgrass. *PLoS ONE* 13, e0194216. DOI: 10.1371/journal.pone.0194216

Stackhouse, T., Martinez-Espinoza, A.D. and Eli, M.D.E. (2020) Turfgrass disease diagnosis: past, present and future. *Plants* 9, 1544. DOI: 10.3390/plants9111544

Stowell, L.J. and Gelernter, W.D. (2001) Diagnosis of turfgrass diseases. *Annual Review of Phytopathology* 39, 135–155. DOI: 10.1146/annurev.phyto.39.1.135

Uddin, W. (2003) Temperature and moisture – forecasters of gray leaf spot. *Golf Course Management* 71, 113–116.

Vittum, P.J. (2022) *Turfgrass Insects of the United States and Canada*. Cornell University Press, Ithaca, New York.

Williams-Woodford, J. (2001) *Simplified Fungi Identification Key*. Special Bulletin 37. The University of Georgia Cooperative Extension Service, College of Agricultural & Environmental Sciences, Athens, Georgia.

5

Genetic, Cultural and Biological Management of Turfgrass Diseases

Abstract
Plant disease development occurs during environmental conditions that favour the virulent pathogen(s) at the expense of the susceptible host(s). A systems approach to implementing proven genetic, cultural and biological technologies and practices in combination attempts to favour the host by manipulation of the turfgrass environment.

Disease resistance among improved turfgrass cultivars, cultural management practices, consumable chemicals and biologically derived or organic amendments, singly and in combination, impact the occurrence of turfgrass pathogens and the diseases they cause (Fig. 5.1). The integration of available genetic, cultural and biological management options is a widely practised strategy to minimize the occurrence and severity of turfgrass diseases. The intended benefits or detrimental impacts of many cultural practices and organic-derived products are relatively well understood and predictable for the most common diseases.

Disease-Resistant Cultivars

Turfgrass breeding and development programmes have realized huge advances among cool-season and warm-season grasses towards abiotic and biotic genetic resistance and stress tolerance. Adoption of disease-resistant cultivars is the first step in durable and low-cost turfgrass disease management (Fig. 5.2). Disease resistance describes the inherent ability of turfgrass to recognize a pathogen and implement inherent biochemical pathways to prevent disease. In reality, most turfgrasses fall within the two extremes of complete resistance and susceptibility.

Commercialization of disease-resistant turfgrasses has focused largely on the more economically important species and their commonly associated fungal diseases (Table 5.1). Adoption of blends or mixtures is commonplace with seeded cool-season grasses as a way to create genetic diversity in the turfgrass sward; thereby reducing the development of widespread disease. Resistance to one disease should not be viewed as compete resistance to all potential diseases. Overall performance and degree of disease resistance vary widely between turfgrass species and among cultivars, governed by existing environmental and management-induced factors.

Turfgrass managers are strongly advised, given the continual release and differential susceptibilities of new turfgrass cultivars to many of the most economically important diseases, to consult the results of independently conducted trials of the National Turfgrass Evaluation

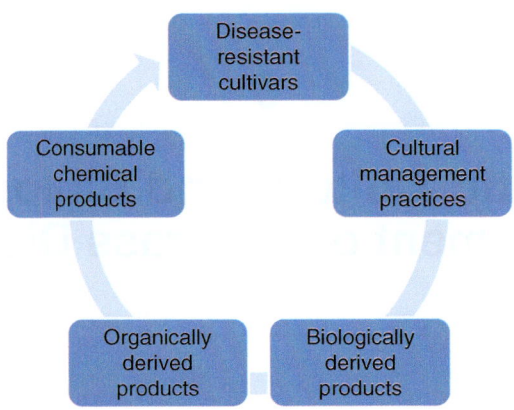

Fig. 5.1. Turfgrass integrated disease management practices.

Fig. 5.2. Plots of fine fescue cultivars in the United States for assessing disease resistance to summer patch. (Courtesy of Nathan R. Walker.)

Program (NTEP) in the USA, the Scandinavian Turfgrass and Environmental Research Foundation (STERF) in Europe or the Australian National Turfgrass Evaluation Program (ANTEP) in Australia, or similar data, when assessing disease susceptibility among new turfgrass cultivars. Caution is required when assessing data from only one year and only consider resistance to diseases that occur in your region (Morris, 2000).

Sanitation

Sanitation practices are aimed at minimizing the risk of introducing plant pathogens via turfgrass (seed or vegetative material), sands and soils (construction and topdressing) and turfgrass machinery and foot traffic. Exclusion of turfgrass diseases in the long term is highly problematic given certain pathogens may be inadvertently introduced via vegetative turfgrass (e.g. root-infecting fungi), irrigation water (e.g. cyanobacteria and *Pythium*) and others seed-borne (e.g. *Curvularia*) or potentially wind-borne.

Primary Cultural Practices

Primary cultural practices are key to managing turfgrass systems worldwide (Box 5.1). Their single and interactive effects have been extensively documented for many common fungal diseases. Mowing, irrigation and fertilization are key primary cultural practices.

Mowing regime

Low and frequent mowing generally favours most turfgrass diseases on high-quality turfgrass systems (Fig. 5.3). Newer creeping bentgrass, fine fescue, bermudagrass and seashore paspalum cultivars can tolerate exceptionally low mowing heights, provided there are no limiting factors, but may remain susceptible to certain diseases. Increased mowing height and reduced mowing frequency, a commonly stated strategy to reduce disease occurrence, can be implemented, but only for short time periods, given the very narrow range of mowing height considered acceptable for certain high-quality turfgrasses.

Mowing regimes have direct (i.e. wound creation and release of leaf exudates) and indirect (i.e. reduced host competitiveness) turfgrass effects. The effects of mowing have been documented for several common diseases (Table 5.2).

Increased occurrence of foliar-borne diseases, based on the premise that foliage wounds caused by mowing provide entry sites for pathogens which may otherwise be incapable of penetration, remains circumstantial (Fig. 5.4). Several studies investigating foliar-borne pathogens have shown conflicting outcomes (Muchovej and Couch, 1987; Khan and Hsiang, 2003). Release of exudates from cut leaf ends as a source of nutrition (Endo, 1972) may partly explain colonization of plants by certain fungal pathogens.

Demonstrating the effects of mowing with and without leaf clipping catchers on disease incidence has yielded variable and inconsistent results (Williams *et al.*, 1996; Settle *et al.*, 2001). Mowing potentially disseminates inoculum of the foliar-infecting diseases (e.g. red thread, dollar spot, grey leaf spot and Pythium blight) (Smith *et al.*, 1989; Tredway *et al.*, 2023). The

---

Box 5.1. Primary turfgrass cultural practices.

- Mowing
- Irrigation
- Fertilization
- Rolling

Table 5.1. Documented examples of disease-resistant turfgrasses and respective diseases.

Turfgrass species	Fungal disease	References
American buffalograss	Leaf blotch, leaf rust	Bonos *et al.* (2006); Bonos and Huff (2013); Hanna *et al.* (2013); Meyer *et al.* (2017); Fulkerson *et al.* (2021)
Annual bluegrass	Anthracnose, dollar spot	
Bermudagrass	Spring dead spot	
Kentucky bluegrass	Leaf spot, necrotic ring spot, powdery mildew, stipe smut, summer patch	
Kikuyugrass	Kikuyu yellows, black leaf spot	
Perennial ryegrass	Grey leaf spot	
Seashore paspalum	Dollar spot	
St. Augustinegrass	Brown patch, grey leaf spot	
Tall fescue	Brown patch, grey leaf spot	
Zoysiagrass	Brown patch	

Fig. 5.3. A multiple-cylinder mower on a golf fairway. (Courtesy of Gary W. Beehag.)

Table 5.2. Documented examples of mowing impacts on specific diseases.

Turfgrass disease	References
Anthracnose	Inguagiato et al. (2008)
Brown patch	Madison (1966)
Dollar spot	Putman and Kaminski (2011)
Spring dead spot	Martin et al. (2001)

potential of mowing during wet conditions and uncollected leaf clippings as sources of inoculum dissemination of other foliar pathogens cannot be ignored.

Irrigation scheduling

Frequent irrigation water favours certain diseases and infrequent scheduling favours others. Computer-controlled irrigation systems are commonplace for managed turfgrass systems worldwide (Fig. 5.5). Irrigation, probably more so than any other cultural practice, has the

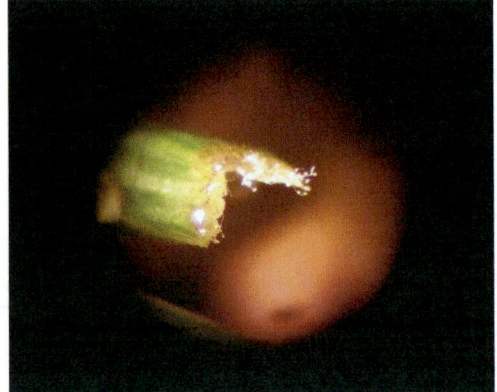

Fig. 5.4. Microscopic image of fungal mycelium protruding from a cut leaf end. (Courtesy of Percy T. Wong.)

potential to be one of the greatest influences on disease occurrence in a turfgrass microclimate.

Impacts of irrigation regimes have been documented for certain fungal diseases (Table 5.3). However, it is problematic to separate the

Fig. 5.5. Automatic sprinkler system on a high-quality sportsground. (Courtesy of Gary W. Beehag.)

impacts of moisture from that of other factors in the rootzone when considering a soil-borne disease. Excessive irrigation creating near-saturated rootzone conditions favours certain diseases (e.g. brown patch, Pythium blight, Typhula blight) while lower moisture levels may favour others (e.g. dollar spot, red thread). Furthermore, water chemistry (i.e. pH level) favours some diseases (e.g. take-all patch) and water quality may favour others (e.g. *Pythium*) depending on water source.

Fertilization regime

Suboptimal or imbalanced levels of certain major, minor and non-essential elements impact turfgrass diseases by moderating the competitiveness of the turfgrass host and may favour a pathogen. Carbon and nitrogen are most likely to be limiting for pathogenic fungi in high-sand/low-organic content rootzones. Nutritional impacts can be relatively small in highly susceptible or highly resistant cultivars but can be substantial in moderately susceptible or partially resistant cultivars (Huber *et al.*, 2012).

Nitrogen, the element required in the greatest amount in frequently mown and irrigated turfgrass systems, represents the most investigated element with respect to its impact on turfgrass fungal diseases. Studies have confirmed certain fungal diseases are more commonly associated under nitrogen-deficient conditions while other fungal diseases are more common under nitrogen-excess conditions (Table 5.4). Nutritional requirements of obligate pathogens in concert with altered turfgrass physiology partly explain the increase of disease for excessively nitrogen-fertilized turfgrass. Nitrogen forms and source (i.e. water soluble, granular slow release or organic-based) have been shown to impact certain fungal diseases (Frank and Guertal, 2013).

The impacts of other major elements, aside from nitrogen, on turfgrass disease have not been precisely determined and remain debated. Limited studies have shown imbalances of certain major

Table 5.3. Documented cases of irrigation regime impacts on nominated turfgrass diseases.

Disease	References
Brown patch	Fidanza and Dernoeden (1996)
Dollar spot, Pythium blight	Couch (1980)
Necrotic ring spot	Melvin and Vargas (1994)
Summer patch	Kackley et al. (1990)

elements to be associated with specific fungal diseases (Table 5.5). Reduction of certain diseases (e.g. microdochium patch, spring dead spot, summer patch and take-all patch) by sulfur may be a direct result as it has fungicidal properties or indirectly as a plant nutrient.

Limited studies have shown experimentally certain minor and non-essential elements to be associated with a select number of fungal diseases (Table 5.6). For example, the precise

Table 5.4. Documented examples of nitrogen regime impacts on nominated turfgrass diseases.

Nitrogen level	Turfgrass disease	References
Higher	Brown patch	Turner and Hummel (1992); McCarty (2010); Frank and Guertal (2013); Espevig et al. (2018)
	Microdochium patch	
	Pythium blight	
	Spring dead spot	
Lower	Anthracnose	
	Dead spot	
	Dollar spot	
	Red thread	
	Summer patch	
	Take-all patch	

Table 5.5. Documented cases of major element imbalance on nominated turfgrass diseases.

Plant element	Fungal disease	References
Phosphorus (P)	Microdochium patch	Kowalewski et al. (2018)
	Take-all patch	Davidson and Goss (1972)
Potassium (K)	Brown patch	Fidanza and Dernoeden (1996)
	Dollar spot	Goss and Gould (1968)
	Microdochium patch	Kowalewski et al. (2018)
	Spring dead spot	Dernoeden et al. (1991)
	Take-all patch	Goss and Gould (1968)
Calcium (Ca)	Pythium blight	Muse and Couch (1965)
	Red thread	Couch (1980)
Sulfur (S)	Microdochium patch	Brauen et al. (1975)
	Take-all patch	Dernoeden (1987)

Table 5.6. Documented cases of minor and non-essential elements imbalance on nominated turfgrass diseases.

Plant element	Fungal disease	References
Iron (Fe)	Dollar spot	Hill et al. (1999); Brecht et al. (2004, 2009); Ervin et al. (2016); Mattox et al. (2017)
	Microdochium patch	
Copper (Cu)	Take-all patch	
Manganese (Mn)	Take-all patch	
Silicon (Si)	Grey leaf spot	
	Leaf spot, melting-out	

pathological mechanism of silicon on turfgrass diseases is unclear.

Acidity/alkalinity (pH) level has been shown to impact the occurrence and severity of several turfgrass patch diseases (Table 5.7). Assessment of direct pH impacts on turfgrass diseases is problematic because of the indirect effects on nutrient availability of certain elements (e.g. phosphorus). Acidifying, soluble nitrogen forms (e.g. ammonium sulfate and ammonium chloride) assist to diminish take-all patch via acidification, as opposed to other forms (e.g. potassium nitrate, calcium nitrate and sodium nitrate) which may intensify the disease (Dernoeden, 2013).

Table 5.7. Specific patch diseases suppressed under acidic conditions and sulfur application.

Disease	References
Spring dead spot	Dernoeden and O'Neill (1983)
Summer patch	Tredway *et al.* (2023)
Take-all patch	Vargas (2005)

Rolling scheduling

Mechanized rolling, a fundamental surface-performance practice on cricket wickets, bowling and golf greens (Fig. 5.6), and increasingly on high-quality sportsgrounds, has been shown experimentally to reduce anthracnose and dollar spot over long periods on creeping bentgrass (Horvath *et al.*, 2009; Giordano *et al.*, 2012). The probable result is due to drier foliage and hence less disease.

Secondary Cultural Practices

Secondary or supplementary cultural practices are selectivity adopted on certain turfgrass systems (Box 5.2). The single and interactive effects on many cultural practices on common fungal diseases remain unclear.

Dew removal, syringing, and temporary placement of oscillating fans and synthetic

Fig. 5.6. Lightweight, multi-roll roller on a golf green. (Courtesy of Gary W. Beehag.)

growth cloths and covers are relatively common practices and technologies. Foliar application of certain chemical surfactants has been evaluated to minimize dew formation of closely mown turfgrass (Hooker *et al.*, 2000) with varying degrees of success. Oscillating fans are utilized on golf course sites to reduce humidity (Peacock and Lyford, 2022). Synthetic covers and growth cloths are widely used in cool and cold climates particularly in regions subject to snow coverage. Extended periods of use of synthetic covers and growth cloths may cause leaf etiolation and increase susceptibility to certain foliar diseases, depending on environmental conditions.

Temporary placement of portable lighting systems (Fig. 5.7) during months of low light intensity is a technology utilized on high-value turfgrass sites (Abelard and Galbrun, 2022). Illumination of the turfgrass has a potential risk of causing sporulation of certain fungal pathogens having high light requirements. Permanent installation of subsurface, electric heating cables and reverse-pumping/vacuum piping systems have been adopted for cricket wickets and golf greens and the impact on various diseases remains unclear.

Surface-aerification practices utilizing solid and/or hollow tines (Fig. 5.8) with surface application or subsurface injection of sand are routinely adopted as thatch-mat minimization

---

Box 5.2. Secondary turfgrass cultural practices.

- Dew removal
- Oscillating fans
- Syringing
- Portable lighting
- Growth cloths and cover
- Subsurface electric cables
- Aeration technologies
- Sand topdressing
- Fraze mowing
- Surface burning
- Overseeding

Fig. 5.7. Portable overhead lighting system on a sportsground in New Zealand. (Courtesy of Gary W. Beehag.)

Fig. 5.8. Combined hollow-tine aerification and core recycling operation on a golf fairway. (Courtesy of Gary W. Beehag.)

strategies on highly-quality turfgrass. Aerification practices assist to moderate the rootzone moisture–aeration balance. Sand topdressing on creeping bentgrass golf greens has been shown to reduce the incidence of anthracnose and dollar spot (Roberts and Murphy, 2014). Oversowing of dormant warm-season grasses by cool-season species is widely adopted in transition zones but the grasses may be prone to certain fungal diseases (e.g. damping-off).

Scarification (vertical mowing) practices alone (Fig. 5.9) or in combination with aerification and sand topdressing are routine practices on high-quality turfgrass surfaces. All the aforementioned practices moderate the microbial populations in turfgrass systems to lesser or greater degrees, depending on the intensity of the operations and environmental conditions.

The practice of 'fraze mowing' is commonplace on elite sportsgrounds and other turfgrass sites in certain countries. Adoption of fraze mowing to a depth of 8.0 mm on an experimental bermudagrass site demonstrated experimentally a lesser severity of spring dead spot (Miller *et al.*, 2017). Surface 'planing' of cotula bowling greens in New Zealand (Adams and Gibbs, 1994) and 'shaving' of couchgrass (syn. bermudagrass) bowling greens in Australia (Beehag, 1999) have been widely adopted since the 1970s as thatch-management practices (Fig. 5.10). Observations in Australia have shown a relationship between regular 'surface shaving' of bowling greens to a depth of around 5 mm every 2–4 years during the summer leading to a reduced incidence of spring dead spot (Beehag and Wong, 2021).

'Fraze mowing' using large-width machinery with rotating blades has been widely practised on sportsgrounds in Europe (Carson, 2015) and Australia and more recently introduced into North America (Munshaw *et al.*, 2017). Fraze mowing large turfgrass areas is adopted to partially remove the existing turfgrass coverage prior to oversowing, reuse the salvaged stolons

Fig. 5.9. Scarification (vertical mowing) operation on a golf fairway. (Courtesy of Gary W. Beehag.)

for re-establishment purposes or totally remove the existing surface.

Surfacing burning using gas-powered flame burners used in the northern United States as a thatch-management tool demonstrated experimentally the suppression of 'Fusarium blight' (now known as two distinct fungal diseases) and snow mould on golf course fairways (Meszaros, 1967). Surface burning of bermudagrass bowling greens in Australia using pressurized gas-powered burners was also adopted as a thatch-management tool.

Consumable Chemical Products

Multiple chemical products may be applied to turfgrass systems (Box 5.3). The products have the potential to impact negatively or positively turfgrass diseases in ways not fully understood.

Anti-transpirant products are simple compounds (e.g. potassium silicate), metabolic inhibitors (e.g. abscisic acid, flurprimidol and

---

Box 5.3. Consumable turfgrass products.

- Anti-transpirants
- Turfgrass colourants
- Plant-growth regulators
- Soil-wetting agents
- Biostimulants

---

melfuidide) or certain surfactants (alcohol- and ester-based). Proven efficacy of anti-transpirants on turfgrass diseases remains unclear (Huang and Fry, 2000). Turfgrass colourants contain a solvent (i.e. water) and chemical pigments and binders (e.g. carbon, iron oxide or titanium dioxide) and are applied to mask disease symptoms (Miller and Pinnix, 2018).

Plant-growth regulators (PGRs) are either natural plant substances (i.e. phytohormones) or their synthetic analogues. Synthetic plant-growth regulators have significance given their similarity in chemistry and biochemical mode of action to triazole fungicides. Several turfgrass growth regulators, at low concentrations in

Fig. 5.10. 'Surface shaving' a 'Tifdwarf' couchgrass bowling green in Australia. (Courtesy of Gary W. Beehag.)

Table 5.8. Documented examples of turfgrass growth regulators and fungal diseases.

Disease	Growth regulator	References
Anthracnose	Melfuidide, trinexapac-ethyl	Inguagiato et al. (2008)
Dollar spot	Flurprimidol, paclobutrazol, trinexapac-ethyl	Allan-Perkins et al. (2017)
Grey leaf spot	Ethephon, trinexapac-ethyl	Uddin and Soika (2000)
Microdochium patch	Trinexapac-ethyl	Aamlid and Petterson (2013)
Rhizoctonia blight	Flurprimidol, paclobutrazol, trinexapac-ethyl	Burpee (1998); Miller (2016)

culture agar, have been demonstrated experimentally to have variable suppression against certain foliar pathogens (Table 5.8).

Effects of turfgrass growth regulators on infected turfgrasses have ranged from beneficial, minimal to neutral when applied alone, following or tank-mixed with turfgrass fungicides, depending on circumstances (Fidanza et al., 2006; Latin, 2021).

Circumstantially, certain turfgrass growth regulators appear to possess dual fungistatic and growth-regulatory action in specific turfgrass–pathogen combinations. Caution needs to be taken when using turfgrass growth regulators as their overuse may result in reduced fungicide sensitivity or disease resistance.

Soil-wetting agents are chemically classified as surfactants generally grouped as anionic, cationic or non-ionic, dependent on how they dissociate in water. Wetting agents have been widely promoted against hydrophobicity and to assist the vertical movement of rootzone-targeted pesticides (Zontek and Kostka, 2012; Martin et al., 2022).

An ever-increasing range of products are marketed worldwide as plant biostimulants, plant-growth promotants or metabolic enhancers (Acuna *et al.*, 2022). The precise function of many biochemicals synthesized by plants remains unclear. Continued development of biostimulants for turfgrass application is set against the background of a vast suite of complex biocompounds synthesized in minute concentrations by plants growing under stressed and non-stressed conditions to perform various roles and functions.

The technical challenge for product manufacturers is attempting to mimic the complex biochemical, metabolic and physiological effects of biocompounds. Any measurable change from biostimulant application is ultimately influenced by complex interactions (Box 5.4). Many claimed benefits of biostimulant products remain to be universally demonstrated in turfgrass systems.

Many biostimulant formulations are based on more than one compound and may contain either soluble nitrogen and/or iron, thus making objective evaluations problematic. Biostimulants are difficult to categorize because of the multitude of specified and non-specified ingredients (Box 5.5). Biostimulants can be categorized according to their chemical or biochemical composition, biosynthetic pathway and effect on plant function, depending on their source origin (du Jardin, 2015).

The ongoing state of confusion about the roles and benefits of biostimulants emanates partly from the ever-increasing diversity of formulations and partly from unvalidated claims purported about their efficacy. Understanding and predicting the outcome of multi-component biostimulants becomes difficult in the absence of knowing their precise mode of action and which is the most common active compound. Turfgrass managers are strongly advised to consult the product label, request any technical literature and initiate their own on-site trials to evaluate the benefit or otherwise of any biostimulant product.

Biological Turfgrass Disease Management Options

Biological disease management attempts to minimize disease severity by biological manipulation of the microenvironment. Proven non-pathogenic microbial species or strains may displace or compete for a pathogen's food source. The microbes may then inhibit or suppress the pathogen's growth, spore germination or directly parasitize the pathogen (i.e. mycoparasitism), triggering host defence mechanisms from the biosynthesis of metabolites (i.e. phytoalexins).

Biological management of turfgrass diseases can be attempted through either the introduction of microbial inoculants or processed organic amendments and composts. Knowledge of the potential benefits afforded by microbial inoculants has been commercially exploited for turfgrass application as components of many biofertilizers, biostimulants and biofumigation-based products.

Box 5.4. Governing factors of biostimulant impacts.

- Intervening environmental conditions
- Chemical and/or biological constituents
- Microbial composition of the rhizosphere
- Physiological activity of turfgrass foliage and roots

Box 5.5. Main categories of biostimulants.

- Humic and fulvic acids
- Proteins and nitrogen compounds
- Plant-based extracts (e.g. seaweeds)
- Chitosan (arthropod exoskeletons)
- Inorganic compounds
- Beneficial bacteria and fungi

Microbial inoculants

Cultured strains of non-pathogenic bacteria and fungi have been demonstrated experimentally in culture to have suppression against nominated fungal pathogens to varying degrees (Fig. 5.11) depending on the population level.

Bacteria (i.e. *Bacillus* spp.) or fungi (i.e. *Trichoderma* spp.) represent the most widely evaluated microbes (Table 5.9) and are the most common, commercially available fungal antagonists used in turfgrass systems.

Symbiotic fungal endophytes (*Epichloe* spp. and *Neotyphodium* spp.) have been intentionally

Fig. 5.11. Experimental suppression of the fungal pathogen *Phialocephala bamuru* in culture using various concentrations of a *Trichoderma* sp. (Courtesy of Percy T. Wong.)

Table 5.9. Non-pathogenic bacterial and fungal genera evaluated against turfgrass fungal pathogens.

Group	Genera	Fungal disease	References
Actinomyces	*Streptomyces*	Brown patch, dollar spot, leaf spot, microdochium patch, Pythium root rot	Nelson (1997); Heydari (2008); Dernoeden (2013); Koppenhofer *et al.* (2013)
Bacteria	*Bacillus*	Anthracnose, dollar spot, grey leaf spot, Pythium blight, summer patch	
	Enterobacter	Dollar spot, Pythium blight, Pythium root rot, summer patch	
	Flavobacterium	Dollar spot	
	Paenibacillus	Brown patch	
	Pseudomonas	Dollar spot, leaf spot, Pythium blight, summer patch, take-all patch	
	Serratia	Summer patch	
	Stenotrophomonas	Brown patch, leaf spot	
	Xanthomonas	Summer patch	
Fungi	*Acremonium*	Dollar spot	
	Epichloe	Dollar spot	
	Fusarium	Dollar spot	
	Gaeumannomyces	Take-all patch	
	Gliocladium	Brown patch, dollar spot, microdochium patch	
	Laetisaria	Brown patch	
	Neotyphodium	Dollar spot	
	Phialophora	Take-all patch	
	Rhizoctonia	Brown patch	
	Trichoderma	Brown patch, dollar spot, Pythium blight, southern blight, Typhula blight	
	Typhula	Grey leaf spot, Typhula blight	

inoculated into seeds of many cool-season grass cultivars with a focus on suppressing dollar spot, red thread and pink patch (Funk *et al.*, 1994; Meyer *et al.*, 2013). Certain mycorrhizal fungi (e.g. *Glomus* spp.) artificially inoculated into seedlings have been demonstrated experimentally to enhance their establishment and subsequent plant drought tolerance (e.g. Nikbakht *et al.*,

2014; Elhindi et al., 2018). Most microbial inoculates have demonstrated acceptable outcomes in laboratory and greenhouse investigations but not consistently in field turfgrass sites (Stewart et al., 2022). Endophyte-infected seed must be stored for short periods only under dry and cold conditions to maintain viability.

Organic amendments and suppressive composts

Ecological benefits afforded by certain organic amendments and composts (e.g. farmyard manures, municipal green waste and industrial wastes) in turfgrass systems have been extensively documented (e.g. Nelson et al., 1994; Wong, 1997; Heydari, 2008). Organic amendments and composts may comprise more than one organic source; thus offer the opportunity of introducing a suite of non-pathogenic microbes.

'Mature' composts, aside from containing antagonistic microbes, also contain bulking agents; hence not all amendments and composts are equally suppressive against any one turfgrass disease. Members of the bacterial genera *Arthrobacter*, *Bacillus* and *Pseudomonas* and the fungal genera *Aspergillus*, *Fusarium*, *Penicillium* and *Trichoderma* are among the most common microbial antagonists in composts (Day and Shaw, 2000).

Different types of organic amendments and composts have been evaluated in numerous countries against specific fungal diseases on common turfgrasses (Table 5.10). Most amendment and compost applications are applied as a 'topdressing' on existing turfgrass and others as a rootzone amendment or both. Composts are well suited for amending rootzone sands during the turfgrass establishment phase.

The precise modes of biological activity of non-pathogenic microbes applied to turfgrass systems as inoculants or as organic amendments and composts are far from understood. Numerous technical and practical challenges remain for effective and predictive biological management of turfgrass diseases (Box 5.6).

Biofumigation

Biofumigation describes the suppression of soil-borne pests and pathogens by inhibitory biochemical compounds, generally utilizing glucosinolates synthesized by *Brassica* plants (Box 5.7). The biochemical mechanism of biofumigation is relatively well understood (dos Santos et al., 2021). Glucosinolates are a class of secondary

---

Box 5.6. Issues associated with biological management options.

- Understanding of antagonist biology and ecology
- Stable formulations and ease of application
- Predictability and measurement of outcomes
- Establishing and maintaining viable microbial populations
- Acceptance of a level of suppression
- Compatibility with turfgrass management practices

Table 5.10. Disease–turfgrass combinations used for amendments/composts.

Disease	Turfgrass	References
Brown patch	Creeping bentgrass, Kentucky bluegrass, perennial ryegrass, tall fescue	Nelson (1997); Heydari (2008); Koppenhofer et al. (2013)
Dollar spot		
Microdochium patch		
Necrotic ring spot		
Pythium blight		
Pythium root rot		
Red thread		
Southern blight		
Summer patch		
Take-all patch		
Typhula blight		

> **Box 5.7.** Reported biofumigant crop plants.
>
> - Indian mustard (*Brassica juncea*)
> - Black mustard (*Brassica nigra*)
> - White mustard (*Sinapis alba*)
> - Radish (*Raphanus sativus*)

plant metabolites containing nitrogen and sulfur compounds.

The concept of biofumigation has been commercially exploited by the production of purified extracts from *Brassica* residues in the form of liquid and pelletized products. Once in moistened rootzones, glucosinolates are hydrolysed into their biologically active, degraded compounds collectively known as isothiocyanates. Varying degrees of sensitivity to various concentrations of pure isothiocyanates among numerous cereal fungal pathogens, some of which are associated with turfgrasses, have been demonstrated under laboratory-controlled conditions (Matthiessen and Kirkegaard, 2006). Isothiocyanates may enhance populations of certain species of antagonistic microflora (Wong, 1997).

Published results of biofumigation efficacy on turfgrass are extremely limited (Pan *et al.*, 2017). Isothiocyanates are the biologically active ingredient in the synthetic fumigants dazomet and metham-sodium. Biofumigation should be viewed as a possible component in an overall integrated disease management system when establishing turfgrass.

Putting It All Together – Compiling an Integrated Disease Management Plan

The foregoing information provides an overview for the compilation of a customized, integrated turfgrass disease management plan of possible genetic, cultural and biological strategies. Customization of any integrated management programme against specific diseases is governed by expectations of overall turfgrass quality and the tolerance level to each disease in terms of occurrence, intensity and duration.

References

Aamlid, T.S. and Petterson, T. (2013) Effect of the plant growth regulator trinexapac-ethyl on turf quality, concentration of total non-structural carbohydrates, and infection of *Micodochium nivale* in greens-type *Poa annua* in Scandinavia. *International Turfgrass Society Research Journal* 12, 801–803.

Abelard, E. and Galbrun, C. (2022) The effects of artificial lighting on sports turf. *International Turfgrass Society Research Journal* 14, 1016–1021. DOI: 10.1002/its2.115

Acuna, A., Gardner, D., Villalobos, L. and Danneberger, K. (2022) Effects of plant biostimulants on seedling root and shoot growth of three cool-season turfgrass species in a controlled environment. *International Turfgrass Society Research Journal* 14, 416–421. DOI: 10.1002/its2.97

Adams, W.A. and Gibbs, R.J. (1994) *Natural Turf for Sport and Amenity: Science and Practice*. CAB International, Wallingford, UK.

Allan-Perkins, E., Campbell-Nelson, K., Popko, J.T., Sang, H. and Jung, G. (2017) Investigating selection of demethylation inhibitor fungicide-insensitive *Sclerotinia homeocarpa* isolates by boscalid, flurprimidol, and paclobutrazol. *Crop Science* 57, S-301–S-309.

Beehag, G.W. (1999) Prescription surface development: lawn bowling greens, croquet lawns and tennis courts. In: Aldous, D.E. (ed.) *International Turf Management Handbook*. Inkata Press, Melbourne, Australia, pp. 265–280.

Beehag, G.W. and Wong, P.T. (2021) Widespread doughnut-shaped patch disease symptoms require diagnosis and solutions. *The Bowling Greenkeeper* 80, 10–14.

Bonos, S.A. and Huff, D.A. (2013) Cool-season grasses: biology and breeding. In: Stier, J.C., Horgan, B.P. and Bonos, S.A. (eds) *Turfgrass: Biology, Use and Management*. Agronomy Monograph No. 56. American Society of Agronomy, Inc., Crop Science Society of America, Inc. and Soil Science Society of America, Inc. Madison, Wisconsin, pp. 591–660.

Bonos, S.A., Clarke, B.B. and Meyer, W.A. (2006) Breeding for diseases resistance in the major cool-season turfgrasses. *Annual Review of Phytopathology* 44, 213–234.

Brauen, S.E., Goss, R.L., Gould, C.J. and Orton, S.P. (1975) The effects of sulphur in combinations with nitrogen, phosphorus and potassium on colour and Fusarium patch disease on Agrostis putting green turf. *Journal of the Sports Research Institute* 51, 83–91.

Brecht, M., Dartnoff, L.E., Kucharaek, T.A. and Nagata, R.T. (2004) Influence of silicon and chlorothalonil on suppression of gray leaf spot and increase growth in St. Augustinegrass. *Plant Disease* 88, 338–344.

Brecht, M., Stiles, C. and Datnoff, L. (2009) Effect of high temperature stress and silicon fertilisation on pathogenicity of *Bipolaris cynodontis* and *Curvularia lunata* on Floradwarf bermudagrass. *International Turfgrass Society Research Journal* 11, 165–180.

Burpee, L.L. (1998) Effects of plant growth regulators and fungicides on Rhizoctonia blight of tall fescue. *Crop Protection* 17, 503–507.

Carson, T. (2015) Fraze (frase, fraize, fraise) mowing. *Golf Course Management* 83, 32.

Couch, H.B. (1980) Relationship of management practices to the incidence and severity of turfgrass diseases. In: Joyner, B.G. and Larsen, P.O. (eds) *Advances in Turfgrass Pathology*. Harcourt Brace Jovanovich, Inc., Duluth, Minnesota, pp. 65–72.

Davidson, R.M. and Goss, R.L. (1972) Effects of P, S and N, lime, chlordane and fungicides on Ophiobolus patch disease on turf. *Plant Disease Reporter* 56, 565–567.

Day, M. and Shaw, K. (2000) Biological, chemical, and physical processes of composting. In: Stoffell, P.J. and Kahn, B.A. (eds) *Compost Utilization in Horticultural Cropping Systems*. CRC Press, Boca Raton, Florida, pp. 17–50.

Dernoeden, P.H. (1987) Management of take-all patch on creeping bentgrass with nitrogen, sulphur, and phenyl mercuric acetate. *Plant Disease* 71, 226–229.

Dernoeden, P.H. (2013) *Creeping Bentgrass Management*, 2nd ed. CRC Press, Boca Raton, Florida.

Dernoeden, P.H. and O'Neill, N.R. (1983) Occurrence of *Gaeumannomyces* patch disease in Maryland and growth and pathogenicity of the pathogen. *Plant Disease* 67, 528–532. DOI: 10.1094/PD-67-528

Dernoeden, P.H., Crahay, J.N. and Davis, D.B. (1991) Spring dead spot and bermudagrass quality as influenced by nitrogen source and potassium. *Crop Science* 31, 1674–1680.

dos Santos, C.A., de Souza Abboud, A.C. and do Carmo, M.G.F. (2021) Biofumigation with species of Brassicaceae family: a review. *Ciência Rural* 51, e20200440. DOI: 10.1590/0103-8478cr2020040

du Jardin, P. (2015) Plant biostimulants: definition, concept, main categories and regulation. *Scientia Horticulturae* 196, 3–14.

Elhindi, K., Al-Suhalbani, N., El-Hendawy, S. and Al-Mana, F. (2018) Effects of arbuscular mycorrhizal fungi on the growth of two turfgrasses grown under greenhouse conditions. *Soil Science and Plant Nutrition* 64, 238–243. DOI: 10.1080/00380768.2017.1417694

Endo, R.M. (1972) The turfgrass community as an environment for the development of facultative fungal parasites. In: Younger, V. (ed.) *The Biology and Utilization of Grasses*. Elsevier, Amsterdam, the Netherlands, pp. 171–202.

Ervin, E.H., Shelton, C., McCall, D., Reams, N. and Askew, S. (2016) Influence of ferrous sulfate and its elemental components on dollar spot suppression. In: *Proceedings of the 3rd European Turfgrass Society Conference, Salgados, Portugal, 5–8 June 2016*. European Turfgrass Society, Livorno, Italy, pp. 129–130.

Espevig, T., Aamlid, T.S., Peterson, T.O. and Kvalbein, A. (2018) Effect of nitrogen in late autumn on Microdochium patch on Nordic golf courses. In: *Proceedings of the 6th European Turfgrass Society Conference, Manchester, UK, 2–4 July 2018*. European Turfgrass Society, Livorno, Italy, pp. 16–17.

Fidanza, M.A. and Dernoeden, P.H. (1996) Brown patch severity in perennial ryegrass as influenced by irrigation, fungicide, and fertiliser. *Crop Science* 36, 1620–1630.

Fidanza, M.A., Wetzel, H.C., Agnew, M.L. and Kaminski, J.E. (2006) Evaluation of fungicide and plant growth regulator tank-mix programme on dollar spot severity of creeping bentgrass. *Crop Protection* 25, 1032–1038.

Frank, K.W. and Guertal, E.A. (2013) Nitrogen research in turfgrass. In: Stier, J.C., Horgan, B.P. and Bonos, S.A. (eds) *Turfgrass: Biology, Use and Management*. Agronomy Monograph No. 56. American Society of Agronomy, Inc., Crop Science Society of America, Inc. and Soil Science Society of America, Inc., Madison, Wisconsin, pp. 457–492.

Fulkerson, W.J., Jennings, N.R., Callow, M., Harper, K.J., Wong, P.T.W. and Martin, P.M. (2021) Selection for resistance to fungal diseases and other desirable traits in kikuyu grass (*Cenchrus clandestinus*). *Tropical Grasslands* 9, 60–69. DOI: 10.17138/tgft(9)60-69

Funk, C.R., Belanger, F.C. and Murphy, J.A. (1994) Role of endophytes in grasses used for turf and soil conservation. In: Bacon, C.W. and White, J.F. (eds) *Biotechnology of Endophyte Fungi in Grasses*. CRC Press, Boca Raton, Florida, pp. 201–209.

Giordano, P.R., Nikola, T.A., Hammerschmidt, R. and Vargas, J.M. (2012) Timing and frequency of lightweight rolling on dollar spot disease in creeping bentgrass putting greens. *Crop Science* 52, 1371–1378.

Goss, R.L. and Gould, C.J. (1968) Turfgrass diseases: the relationship of potassium. *USGA Green Section Record* 5, 10–12.

Hanna, W., Raymer, P. and Schwartz, B. (2013) Warm-season grasses: biology and breeding. In: Stier, J.C., Horgan, B.P. and Bonos, S.A. (eds) *Turfgrass: Biology, Use and Management*. Agronomy Monograph No. 56. American Society of Agronomy, Inc., Crop Science Society of America, Inc. and Soil Science Society of America, Inc., Madison, Wisconsin, pp. 543–590.

Heydari, A. (2008) Biological control of turfgrass diseases. In: Pessarakli, M. (ed.) *Handbook of Turfgrass Management and Physiology*. CRC Press, Boca Raton, Florida, pp. 223–236.

Hill, W.J., Heckman, J.R., Clarke, B.B. and Murphy, J.A. (1999) Take-all patch suppression in creeping bentgrass with manganese and copper. *HortScience* 34, 891–892.

Hooker, M., Gibbs, R. and Wrigley, M. (2000) Prevention of dew formation using wetting agents. *Golf & Sports Turf* 3, 6–11.

Horvath, B.J., Nichols, A.E. and Cutulle, M.A. (2009) The effects of mowing height and rolling on golf green speed, quality and disease-severity of creeping bentgrass. *USGA Turfgrass and Environmental Research Online* 8, 1–5.

Huang, B. and Fry, J.D. (2000) Turfgrass evapotranspiration. *Journal of Crop Production* 2, 317–333.

Huber, D., Romheld, V. and Weinmann, M. (2012) Relationship between nutrition, plant diseases and pests. In: Marschner, P. (ed.) *Marschner's Mineral Nutrition of Higher Plants*, 3rd edn. Academic Press, London, pp. 283–298.

Inguagiato, J.C., Murphy, J.A. and Clarke, B.B. (2008) Anthracnose severity on annual bluegrass influenced by nitrogen fertilisation, growth regulators, and verticutting. *Crop Science* 48, 1595–1607.

Kackley, K.E., Grybauskas, A.P., Dernoeden, P.H. and Hill, R.L. (1990) Role of drought stress in the development of summer patch in field-inoculated Kentucky bluegrass. *Phytopathology* 80, 655–658.

Khan, A. and Hsiang, T. (2003) The infection process of *Colletotrichum graminicola* and relative aggressiveness on four turfgrass species. *Canadian Journal of Microbiology* 49, 433–442.

Koppenhofer, A.M., Latin, R., McGraw, B.A. and Crow, W.T. (2013) Integrated pest management. In: Stier, J.C., Horgan, B.P. and Bonos, S.A. (eds) *Turfgrass: Biology, Use and Management*. Agronomy Monograph No. 56. American Society of Agronomy, Inc., Crop Science Society of America, Inc. and Soil Science Society of America, Inc., Madison, Wisconsin, pp. 933–1006.

Kowalewski, A., McDonald, B., Mattocks, C. and Braithwaite, E. (2018) Effects of winter nitrogen, phosphate and potassium rates on Microdochium patch. In: *Proceedings of the 6th European Turfgrass Society Conference, Manchester, UK, 2–4 July 2018*. European Turfgrass Society, Livorno, Italy, pp. 26–37.

Latin, R. (2021) *A Practical Guide to Turfgrass Fungicides*, 2nd ed. The American Phytopathological Society, St, Paul, Minnesota.

McCarty, L.B. (2010) *Best Golf Course Management Practices: Construction, Watering, Fertilizing, Cultural Practices and Pest Management Strategies to Maintain Golf Course Turf with Minimal Environmental Impact*, 3rd edn. Prentice-Hall, Upper Saddle River, New Jersey.

Madison, J.H. (1966) Brown patch of turf grass caused by *Rhizoctonia solani* Kuhn. *California Turfgrass Culture* 16, 9–13.

Martin, T., Rothwell, S. and Stevens, C. (2022) An investigation into the principal modes of action of surfactants and how a novel formulation may improve turfgrass quality by increasing the dominance of *Agrostis* spp. in golf greens. *International Turfgrass Society Research Journal* 14, 1010–1015. DOI: 10.1002/its2.126

Martin, D.L., Bell, G.E., Baird, J.H., Taliaferro, C.M., Tisserat, N.A., et al. (2001) Spring dead spot resistance and quality of seeded bermudagrass under different mowing regimes. *Crop Science* 41, 451–456.

Matthiessen, J.N. and Kirkegaard, J.A. (2006) Biofumigation and enhanced biodegradation: opportunity and challenge in soilborne pest and disease management. *Critical Reviews in Plant Sciences* 25, 235–265. DOI: 10.1080/07352680600611543

Mattox, C.M., Kowalewski, A.R., McDonald, B.W., Lambrinos, J.G., Davidson, B.L. and Pscheidt, J.W. (2017) Nitrogen and iron sulphate affect Microdochium patch severity and turf quality on annual bluegrass putting greens. *Crop Science* 57, 1–8.

Melvin, B.P. and Vargas, J.M. (1994) Irrigation frequency, and fertiliser type influence necrotic ring spot of Kentucky bluegrass. *HortScience* 29, 1028–1030.

Meszaros, J.P. (1967) Burn thatch out. *The Golf Superintendent* 35, 51–57.

Meyer, W.A., Torres, M.S. and White, J.F. (2013) Biology and applications of fungal endophytes in turfgrasses. In: Stier, J.C., Horgan, B.P. and Bonos, S.A. (eds) *Turfgrass: Biology, Use and Management*. Agronomy Monograph No. 56. American Society of Agronomy, Inc., Crop Science Society of America, Inc. and Soil Science Society of America, Inc., Madison, Wisconsin, pp. 713–732.

Meyer, W.A., Hoffman, L. and Bonos, S.A. (2017) Breeding cool-season turfgrass cultivars for stress tolerance and sustainability in a changing environment. *International Turfgrass Society Research Journal* 13, 3–10. DOI: 10.2134/itsrj2016.0806

Miller, G.L. (2016) Effect of watered-in demethylation inhibitor fungicide and paclobutrazol applications on foliar disease severity and turfgrass quality on creeping bentgrass. *Crop Protection* 79, 64–69.

Miller, G. and Pinnix, D. (2018) *Guide to Using Turf Colorants*. North Carolina State University Extension, Raleigh, North Carolina.

Miller, G.L., Earlywine, D.T. and Fresenburg, B.S. (2017) Effect of fraze mowing on spring dead spot caused by *Ophiosphaerella herpotricha* of bermudagrass. *International Turfgrass Society Research Journal* 13, 225–228.

Morris, K.N. (2000) Guidelines for using NTEP trial data. *Golf Course Management* 4, 64–70.

Muchovej, J.J. and Couch, H.B. (1987) Colonisation of bentgrass by *Curvularia lunata* after clipping and heat stress. *Plant Disease* 71, 873–875.

Munshaw, G.C., Dickson, K.H., Cropper, K.L. and Sorochan, J.C. (2017) The effect of fraze mowing on overseed establishment in *Cynodon dactylon* turf. *International Turfgrass Society Research Journal* 13, 380–382. DOI: 10.2134/itsrj2016.10.0844

Muse, R.R. and Couch, H.B. (1965) Influence of environment on diseases of turfgrasses. 1V. Effects of nutrition and soil moisture on Corticium red thread of creeping red fescue. *Phytopathology* 55, 507–510.

Nelson, E.B. (1997) Biological control of turfgrass diseases. *Golf Course Management* 65, 60–69.

Nelson, E.B., Burpee, L.L. and Lawson, M.B. (1994) Biological control of turfgrass diseases. In: Leslie, A.R. (ed.) *Handbook of Integrated Pest Management for Turf and Ornamentals*. CRC Press, Boca Raton, Florida, pp. 409–428.

Nikbakht, A., Hakim-Meybodi, N.D. and Pessarakli, M. (2014) New approaches to turfgrass nutrition – humic substances and mycorrhizal inoculation. In: Pessarakli, M. (ed.) *Handbook of Plant & Crop Physiology*. CRC Press, Boca Raton, Florida, pp. 1141–1160.

Pan, X., Earlywine, D.T., Smeda, R.J., Teuton, T.C., English, J.T., et al. (2017) Effect of oriental mustard (*Brassica juncea*) seed meal for control of dollar spot on creeping bentgrass (*Agrostis stolonifera*) turf. *International Turfgrass Society Research Journal* 13, 166–174. DOI: 10.2134/itsrj2016.06.0455

Peacock, C.H. and Lyford, P.R. (2022) Use of fans and their effects on the turf microenvironment and disease development. *International Turfgrass Society Research Journal* 14, 1088–1091. DOI: 10.1002/its2.113

Putman, A.L. and Kaminski, J.E. (2011) Mowing frequency and plant growth regulator effects on dollar spot severity and duration of dollar spot control by fungicides. *Plant Disease* 95, 1433–1442. DOI: 10.1094/PDIS-04-11-0278

Roberts, J.A. and Murphy, J.A. (2014) Anthracnose disease on annual bluegrass as affected by foot traffic and sand topdressing. *Plant Disease* 98, 1321–1325. DOI: 10.1094/PDIS-08-13-0877-RE

Settle, D.M., Fry, J.D. and Tisserat, N.A. (2001) Development of brown patch and Pythium blight in tall fescue as affected by irrigation frequencies, clipping removal and fungicide application. *Plant Disease* 85, 543–546.

Smith, J.D., Jackson, N. and Woolhouse, A.R. (1989) *Fungal Diseases of Amenity Turf Grasses*, 3rd edn. E. & F.N. Spon, London.

Stewart, A.V., Barcellos, G. and Brilman, L. (2022) Use of endophyte fungi in turfgrasses: difficulties in delivery to the market. *International Turfgrass Society Research Journal* 14, 1070–1073. DOI: 10.1002/its2.131

Tredway, L.P., Tomaso-Peterson, M., Kerns, J.P. and Clarke, B.B. (2023) *Compendium of Turfgrass Diseases*, 4th edn. The American Phytopathological Society, St. Paul, Minnesota.

Turner, T.R. and Hummel, N.W. (1992) Nutritional requirements and fertilization. In: Waddington, D.V., Carrow, R.N. and Shearman, R.C. (eds) *Turfgrass*. Agronomy Monograph No. 32. American Society of Agronomy, Inc., Crop Science Society of America, Inc. and Soil Science Society of America, Inc., Madison, Wisconsin, pp. 385–440.

Uddin, W. and Soika, M.D. (2000) Effects of plant growth regulators, herbicides, and fungicides on development of blast disease (gray leaf spot) of perennial ryegrass turf. *Phytopathology* 90, S78.

Vargas, J.M. (2005) *Management of Turfgrass Diseases*, 3rd edn. Wiley, Hoboken, New Jersey.

Williams, D.W., Powell, A.J., Vincelli, P. and Dougherty, C.T. (1996) Dollar spot on bentgrass influenced by displacement of leaf surface moisture, nitrogen and clipping removal. *Agronomy Journal* 36, 1304–1309.

Wong, P.T.W. (1997) Organic amendments for the control of pests and diseases of turfgrasses. *International Turfgrass Society Research Journal* 8, 813–822.

Zontek, S.J. and Kostka, S.J. (2012) Understanding the different wetting agent chemistries. *USGA Green Section Record* 50, 1–6.

6

Cool-Weather Fungal Diseases of Mown Grasses

Abstract

This chapter covers the multitude of fungal pathogens that occur primarily during the cool and colder months of the year when conditions are highly conducive to their growth and development to cause various diseases on mown grasses. However, certain diseases caused by fungal pathogens may persist at any time of year, depending on regional environmental conditions of favourable temperature and moisture. Some fungal diseases are associated with the foliage and crown regions, others primarily with the stems and roots of host plants.

Many cool-weather diseases are common and readily recognizable while other diseases share common symptoms often leading to a misdiagnosis. Many fungal diseases result in disfiguring symptoms and once established can be problematic and challenging to manage. Effective management of many fungal diseases requires an integrated approach encompassing plant genetic resistance, proven cultural practices, and possible biological and chemical inputs.

Cool-weather fungal diseases of mown grasses are largely incited by mesophilic species and strains capable of growing and surviving moderate temperatures and to a lesser extent psychrophiles that are able to survive at very low or near freezing temperatures. These fungal pathogens can infect and colonize host tissues any time from early autumn through winter to late spring and early summer under environmental conditions favouring fungal growth and disease at the expense of their host. Cool-weather fungal pathogens may infect all host parts causing disease of seedlings and established plants, depending on the specific host–pathogen relationship.

Many cool-weather fungal diseases share common environmental and management-induced causes as covered in Chapter 2 (this volume). Numerous fungicides and certain biofungicides have registration in many countries for turfgrass application against the most common fungal diseases. Details of the fungicide groups and their characteristics are provided in Chapter 11 (this volume). Readers are strongly advised to continually consult fungicide and biofungicide labels for current information about specific product registrations.

Foliage-infecting fungal diseases

Foliage-infecting fungi colonize the foliage or crown tissues of their hosts. Other fungal pathogens can be seed-borne. Very-early-stage symptoms of many foliage-borne diseases may be difficult to accurately diagnose. Late-stage symptoms may range from small-sized, randomly scattered spots with some coalescing to form larger regions of chlorotic and necrotic areas with no definitive shape or boundary. Fungal infection typically begins on individual leaves then spreads to adjacent leaves of other plants. Certain foliar disease symptoms, under highly favourable conditions, may be the result of more than one fungal species, complicating the overall symptoms and a precise diagnosis.

Downy Mildew

Downy mildew, also called yellow tuft in the United States, causes largely cosmetic symptoms with the occasional presence of fungal mycelium and spores on infected leaves under conducive conditions.

Causal agent

Sclerophthora macrospora is the causal agent of yellow tuft (Dernoeden, 2013). *Sclerophthora macrospora* is an obligate pathogen, an oomycete, and not a true fungus.

Geographic distribution

Yellow tuft is widely distributed throughout North America and Europe (Smith *et al.*, 1989; Couch, 1995) and occurs in East Asia and Australia (Bransgrove, 2005; Han *et al.*, 2016). The global distribution of yellow tuft remains unclear.

Host range and susceptibility

Yellow tuft has a wide host range of cool-season and warm-season grasses. Bentgrasses, bluegrasses, fescues, ryegrasses, St. Augustinegrass and zoysiagrass are known hosts (Smith *et al.*, 1989; Couch, 1995). Cool-season grasses are generally more susceptible to the disease.

Signs and symptoms

Signs and early-stage symptoms of yellow tuft can be problematic to diagnose, depending on the host species. Slightly stunted leaf growth and widened leaf blades without significant discoloration are characteristic general symptoms of the disease. Early-stage symptoms on closely mown, cool-season species are small-sized, random chlorotic spots up to 3.0 cm in diameter (Fig. 6.1) and tend to be larger on higher-mown swards. Individual infected plants display clustered, yellow-coloured shoots and are easily removed due to shortened roots.

Late-stage symptoms on cool-season species may appear as coalescing spots that form a mosaic of patches up to 10 cm. Symptoms of yellow tuft on warm-season turfgrasses are

Fig. 6.1. Downy mildew (yellow tuft) symptoms on a creeping bentgrass golf green. (Courtesy of Nadeem Zreikat.)

generally expressed as linear white streaks or diffuse chlorosis. Co-infection of foliage by other fungal pathogens (e.g. *Bipolaris* spp. and *Ustilago* spp.) will disguise late-stage symptoms.

Disease profile

Yellow tuft is a disease of cool, wet weather that is most severe on immature cool-season grasses. Knowledge of the biology and epidemiology of yellow tuft remains limited (Dernoeden, 2013; Tredway *et al.*, 2023). The microbe survives as dormant zoospores in the upper thatch-mat region or inside infected plants. The pathogen invades the meristematic tissues and will grow systemically through the plant and can survive within the infected plant for a number of years.

Yellow tuft occurs most commonly during late spring and autumn and infection of leaves occurs between 10 and 25°C (Bruton and Toler, 1980). The pathogen grows outwardly through open stomates releasing white-coloured spores (sporangia) which freely germinate in the presence of moisture causing the 'downy appearance' on infected leaves. Movement by water and machinery and infection by released zoospores partly explain why the disease is commonly associated with low-lying areas.

Integrated management

Effective management of yellow tuft can be problematic on bowling and golf greens because of the ephemeral nature of the disease and limited cultural options (Box 6.1). Application of soluble nitrogen sources in excess of seasonal requirements may increase zoospore germination (Dernoeden and Jackson, 1980).

Box 6.1. Integrated management factors of downy mildew (yellow tuft).

- Disease-resistant varieties
- Monitor plants for signs of key fungal structures
- Free-draining rootzone and turfgrass sites
- Maintain consistent growth with use of a balanced NPK nutrition
- Application of soluble iron assists to mask symptoms
- Judicious irrigation water applications
- Fungicide application between 10 and 18°C

Disease-resistant varieties

Slight genetic resistance against yellow tuft has been documented among early-released St. Augustinegrass varieties (Couch, 1995).

Limonomyces-Incited Diseases

Limonomyces species are facultative pathogens that infect host foliage. *Limonomyces* species known to be associated with grasses are *Limonomyces cultigens* and *Limonomyces roseipellis* (Stalpers and Loerakker, 1982). *L. roseipellis* has different biotypes and is recognized as the causal agent of two distinct diseases: pink patch that primarily occurs on cool-season species and cream leaf blight that occurs on dwarf bermudagrasses.

An undescribed *Limonomyces* species causing light pinkish-coloured, small-sized patches has been isolated on bermudagrass (couchgrass) in Australia (Fig. 6.2) (P. Wong, New South Wales, 2021, personal communication).

Pink patch

Pink patch is a disease previously considered to be the same as another foliage disease, red thread. The name pink patch was ascribed to distinguish the causal agent and symptoms of the associated disease red thread, both of which may occur simultaneously on the same infected turfgrass. Awareness of pink patch as a distinct turfgrass disease has increased in recent years in countries where the disease occurs.

Causal agent

L. roseipellis (formerly *Laetisaria fuciformis*) is the causal agent of pink patch (Tredway *et al.*, 2023). The fungus has light pink-coloured mycelium and does not form the anther-like sclerotia associated with red thread.

Geographic distribution

Pink patch is distributed throughout the cool regions of Western Europe, the United Kingdom, North America and East Asia (Maccaroni *et al.*, 2002; Fermanian *et al.*, 2003; Zhang *et al.*, 2015a). The global distribution of pink patch remains unclear.

Fig. 6.2. Light pink-coloured patches caused by a *Limonomyces* sp. on couchgrass (bermudagrass) in Australia. (Courtesy of Gary W. Beehag.)

Host range and susceptibility

Bentgrass, bermudagrass, bluegrass, fescue, ryegrass, seashore paspalum and zoysiagrass are reported hosts of pink patch (Zhang *et al.*, 2015a; Tredway *et al.*, 2023).

Signs and symptoms

Appearance of light red- to pinkish-coloured mycelium in the canopy without the anther-like sclerotia produced by the red thread pathogen characterizes pink patch. The mycelium is readily observed by eye in the early morning. Symptoms of pink patch are similar to those of red thread, varying depending on level of cultural management and environmental conditions.

Initial symptoms may first appear as individual, light brown- to tan-coloured leaves that can have water-soaked lesions and result in necrotic, small-sized, irregular patches. Under highly favourable moisture conditions, masses of mycelium bind multiple leaves together. Late-stage symptoms on infrequently mown turfgrass present an unsightly appearance due to contrasting coloration between green foliage and pink mycelium. Individual patches may gradually enlarge to coalesce forming irregular shapes of blighted grass up to 60 cm.

Disease profile

Pink patch is a foliage disease generally considered to be more common on relatively slow-growing grass, either as a result of insufficient nitrogen and/or cold temperatures. The disease is favoured by cool, moist conditions when temperatures range from 18 to 24°C, coinciding with spring and autumn, but may occur outside these seasons provided moisture is adequate. The disease cycle and contributing environmental factors of pink patch are generally considered to mirror those of red thread (Couch, 1995; Tredway *et al.*, 2023).

Environmental factors

Key environmental and management-induced factors associated with pink patch (i.e. low temperatures, moderate moisture and humidity, and nitrogen deficiency) are similar to those of red thread. The interaction of moderate temperature, nitrogen deficiency, high humidity, moderate moisture and light is favourable for the disease.

Integrated management

Acceptable long-term management of pink patch necessitates an integrated approach (Box 6.2). Frequent monitoring for signs of

> **Box 6.2.** Integrated management factors of pink patch.
>
> - Monitor plants for signs of key fungal structures
> - Maintain consistent growth with use of a balanced NPK nutrition
> - Judicious irrigation water applications
> - Use of registered fungicides if required

pink- to reddish-coloured mycelium, disease symptoms and conducive weather patterns, cultural practices and fungicide inputs are required to effectively manage the disease.

Cream leaf blight

Cream leaf blight is a recently named disease of hybrid bermudagrass golf greens in the transition zone in the United States. Specific knowledge of the biology of the fungal biotype and epidemiology of the disease remains restricted to largely what is known about pink patch.

Causal agent

A biotype of *L. roseipellis* is the causal agent of cream leaf blight (Kerns and Butler, 2018).

Signs and symptoms

Aerial mycelium and leaf lesions are not obvious on infected plants. Symptoms of cream leaf blight are characterized by hundreds of randomly scattered spots which may be difficult to observe given their light tan colour on dormant bermudagrass. Symptoms of cream leaf blight on infected bermudagrass golf greens are largely cosmetic, suggestive of low fungal virulence. Early-stage symptoms commence as small-sized, off-white- to cream-coloured spots 8–20 cm in diameter. Spots may coalesce forming larger spots or small patches.

Geographic distribution and host range

Cream leaf blight has been observed on the partly dormant bermudagrass cultivars Champion and MiniVerde in the United States and possibly a reflection of the popularity of these two cultivars in the southern regions of the country. Global distribution of cream leaf blight on warm-season grasses remains unknown but undoubtedly occurs on bermudagrass bowling and golf greens in warm temperate and subtropical regions. In southern China, the pathogen has been found to be active during the spring on bermudagrass and seashore paspalum cultivars on golf courses and lawns (Zhang *et al.*, 2015) but without symptoms described.

Integrated management

Predictive management using cultural, biological and chemical measures against cream leaf blight remains unclear.

Powdery Mildew

Powdery mildew can be problematic on less-adapted grasses when managed in shaded and humid environments. What makes powdery mildew somewhat unique is the fungal spores do not require free moisture for germination. Powdery mildew usually causes an unsightly appearance and minor damage.

Causal agent

Blumeria graminis (formerly *Erysiphe graminis*) is the causal agent of powdery mildew (Tredway *et al.*, 2023). *B. graminis* is an obligate pathogen consisting of many fungal strains.

Geographic distribution and host range

Powdery mildew occurs on cultivated grasses worldwide (Smith *et al.*, 1989; Fermanian *et al.*, 2003). Cool-season grasses are hosts of powdery mildew. Kentucky bluegrass is a primary host.

Signs and symptoms

The presence of individual pustules of light-coloured fungal growth on foliage is characteristic of powdery mildew (Fig. 6.3). Disease progression under highly favourable conditions results in enlargement and coalescing of the fungal mycelium over multiple leaves partly or fully covered with white- to grey-coloured growth. Infected leaves have a dust-like appearance

Fig. 6.3. Light-coloured mycelium of powdery mildew on infected foliage. (Courtesy of Ben Evans.)

when rubbed. Advanced stages of the disease can result in a mosaic of mycelium-covered, chlorotic and dead leaves that become dark in colour due to spore-bearing structures.

Disease profile

Powdery mildew occurs primarily under cloudy, cool and humid conditions during spring and autumn. The fungus survives unfavourable conditions as spores or inside infected tissues. Masses of fungal conidia are produced in spring during favourable conditions and are the principal inoculum. Infection occurs directly through the leaf epidermis at temperatures from 15 to 22°C and during conditions of low light intensity (Fermanian et al., 2003).

Environmental influences

Several environmental factors are associated with powdery mildew (Box 6.3). Excess nitrogen has been implicated in promoting disease development.

Integrated management

Effective management of powdery mildew is problematic particularly under shaded environments given cool-season turfgrasses' greater adaptability to lower light intensity. A balanced approach of nitrogen, phosphorus and potassium (NPK) nutrition is required to avoid luxuriant growth (Box 6.4).

Box 6.3. Key environmental factors of powdery mildew.

- Low light intensity
- Moderate temperature
- Humidity
- Excess nitrogen

Box 6.4. Integrated management factors of powdery mildew.

- Disease-resistant and shade-tolerant varieties (e.g. fescues)
- Minimize shading and increase air circulation
- Monitor plants for signs of key fungal structures
- Maintain consistent growth with use of a balanced NPK nutrition
- Judicious irrigation water applications
- Use of registered fungicides if required

Disease-resistant varieties

Genetic resistance against powdery mildew is known to exist among Kentucky bluegrass, fine fescue, and tall fescue and perennial ryegrass cultivars (Fermanian et al., 2003).

Pythium-Incited Diseases

An extensive suite of described *Pythium* species and species complexes are associated with different disease symptoms on turfgrasses worldwide. Association between turfgrass disease and *Pythium* was reported on creeping bentgrass from the early 1900s throughout the eastern regions of the United States (Monteith and Dahl, 1932). Multiple *Pythium* species can occur simultaneously on all plant parts (i.e. foliage, crown and roots) depending on the host and environmental conditions, notably moisture and temperature.

Pythium species are not true fungi, but oomycetes often called 'water moulds' and can produce multiple types of spore-like structures or oospores. The pathogenicity or virulence among *Pythium* species varies widely and certain diseases caused by highly pathogenic species are highly problematic to manage once established on high-quality turfgrass (Fig. 6.4). All turfgrass species and their cultivars are potential *Pythium* hosts.

Many *Pythium* species have the ability to survive in water storages (e.g. recycled water and storage dams) potentially providing a constant source of inoculum when present in a closed irrigation system. *Pythium* species are only susceptible to certain fungicides or biofungicides formulated on specific active ingredients that have efficacy against this group of microbes.

Pythium nomenclature

Pythium is a complex genus of closely related species, some of whose taxonomy remains unclear. Certain fungal isolates have morphology intermediate between *Pythium* and *Phytopthora* resulting in taxonomic reclassification based on genetic methodologies reassigning certain species to the new genus *Phytopythium* and others to *Globisporangium* (Tkaczyk, 2020). Further reclassification of species complexes will undoubtedly occur in the future with the isolation of additional species and improved genetic-based methodologies.

Pythium and related pathogens and their diseases

The identity and description of *Pythium* species associated with cultivated turfgrasses have been widely documented (Schroeder *et al.*, 2013; Mahendra *et al.*, 2020). *Pythium*-incited diseases on turfgrasses occur in tropical, subtropical and temperate regions worldwide and may be known under more than one common name (Table 6.1) depending in which country they occur. Each name reflects the regions of infection and damage caused. Multiple *Pythium* and related species are associated with more than one turfgrass disease condition. *Pythium*-incited diseases are

Fig. 6.4. Extensive damage attributed primarily to *Pythium* spp. on creeping bentgrass. (Courtesy of Ben Evans.)

undoubtedly under-reported given the ubiquitous distribution of the genus.

Certain *Pythium* diseases are known worldwide (e.g. damping-off and Pythium blight), others restricted to certain countries (e.g. Pythium root dysfunction and Pythium spring dead spot) or in only one country.

Virulence among Pythium and related species

The potential among *Pythium* species and strains to infect and ultimately cause disease on seedling and mature turfgrass has been extensively investigated primarily on creeping bentgrass and bermudagrass cultivars in North America (e.g. Nelson and Craft, 1991; Abad *et al.*, 1994; Hodges and Campbell, 1994; Hsiang *et al.*, 1995; Kerns and Tredway, 2008) and on additional turfgrass species in in Asia (e.g. Ichitani *et al.*, 1986; Kim and Park, 1999; Guan *et al.*, 2009; Chang and Lee, 2013) and the Middle East (Marvasti and Banihashemi, 2011).

Results of these extensive investigations clearly indicate virulence varies widely among *Pythium* species and strains on a broad range of cool-season and warm-season turfgrasses. Not all *Pythium* species are highly pathogenic (Table 6.2) under all environmental conditions, with temperature and moisture most important. Age of turfgrass cultivars, whether seedling or mature, host–pathogen interaction and laboratory conditions under which pathogenicity is evaluated, potentially all influence results.

Most studies have indicated the most common *Pythium* species were not necessarily the most pathogenic and the degree of pathogenicity (virulence) varies between species at different temperatures. *Pythium graminicola* possesses variable pathogenicity and *Pythium torulosum* is non-pathogenic or weakly pathogenic (Nelson and Craft, 1991; Abad *et al.*, 1994). However, among certain *Pythium* species the optimum temperature for growth, as measured under laboratory-controlled conditions, and for infection of hosts may be similar. For other species, there

Table 6.1. *Pythium* diseases and common pathogenic fungal species.

Disease name	*Pythium* species	References
Damping-off	*P. aphanidermatum, P. arrhenomanes, P. graminicola*	Tani and Beard (1997); Dernoeden (2013); Tredway *et al.* (2023)
Pythium blight (cottony blight)	*P. aphanidermatum, P. graminicola, P. myriotylum*	
Pythium root rot	*P. aristosporum, P. catenulatum, P. torulosum*	
Pythium root dysfunction	*P. aphanidermatum, P. arrhenomanes, P. ultimatum*	
Pythium spring dead spot	*P. graminicola, P. vanterpooli*	
Snow blight	*P. graminicola, P. iwayamai, P. paddicum*	

Table 6.2. Comparative pathogenicity of selected *Pythium* and related species.

Genus/species	Pathogenicity rating	Reference
Pythium afertile	Low	Abad *et al.* (1994)
P. aphanidermatum	High	
P. arrhenomanes	High	
P. dissotocum	Moderate	
P. graminicola	High	
P. multisporum	Low	
P. pulchrum	Low	
P. splendens	Moderate	
Globisporangium irregulare	Moderate	
G. volutum	High	

is no relationship. Several studies have shown that *P. graminicola* possesses variable pathogenicity and *P. torulosum* is non-pathogenic or weakly pathogenic.

Temperature response among Pythium and related species

Pythium, *Phytopythium* and *Globisporangium* species grow and can survive over a very wide range of temperatures and each species has a minimum, optimum and maximum temperature for growth (Table 6.3). Relatively low temperatures of 11–21°C favour certain species while higher temperatures of 23–34°C favour others (Tredway *et al.*, 2023). Certain *Pythium* species (e.g. *Pythium ultimatum* and *Pythium vanterpooli*) can tolerate temperatures below 10°C while others (e.g. *Pythium catenulatum* and *Pythium aphanidermatum*) can grow at temperatures greater than 30°C (van der Plaats-Niterink, 1981). The overlap of temperatures among some *Pythium* and related species explains partly the isolation of more than one species or strain causing disease from the same infected plants.

Cool-weather *Pythium* diseases

Damping-off, Pythium crown and root rot and Pythium spring dead spot are recognized as cool-weather diseases.

Damping-off

Damping-off or seedling blight remains potentially problematic worldwide whenever grasses are established or oversown using seed. Damping-off is a generalized, historical term associated with fungal infection causing death of emergent seedlings (Fig. 6.5). Damping-off should be viewed as a disease syndrome of primary and secondary pathogens before, during and following

Table 6.3. Cardinal temperature ranges of nominated *Pythium* species.

Pythium species	Minimum (°C)	Optimum (°C)	Maximum (°C)	Reference
P. aphanidermatum	10	35–40	> 40	van der Plaats-Niterink (1981)
P. catenulatum	10	30–35	40	
P. graminicola	5	30	38	
P. myriotylum	5	37	> 40	
P. ultimatum	5	25–30	35	
P. vanterpooli	5	25	30	

Fig. 6.5. 'Damping-off' symptoms on seedling creeping bentgrass. (Courtesy of Gary W. Beehag.)

seed germination involving one or more *Pythium* species and other fungi (e.g. *Curvularia* spp. and *Rhizoctonia* spp.) (Smith *et al.*, 1989). Damping-off caused by *Pythium* species is favoured more so when establishing seed during cool and moist conditions under suboptimal environmental conditions for germination.

Signs and symptoms

Damping-off symptoms are preceded by a sequence of pre-emergent seed decay, post-emergence root necrosis and stem collapse. Early-stage symptoms on emergent seedlings may reveal a water-soaked appearance at the rootzone surface and lesions girdling individual leaves resulting in seedling collapse and foliage becoming chlorotic and turning brown before death. Advanced symptoms on seeded grasses can appear as irregular patches of bare soil surface and chlorotic and necrotic plants. A profusion of mycelium may be present during highly favourable conditions of temperature and moisture.

Integrated management of damping-off

Several specific management strategies apply to damping-off when establishing swards by seed (Box 6.5). Attention to rootzone temperature is critical and possible use of pre-treated seed and temporary covers should be considered.

Pythium crown and root rot

Pythium crown and root rot is a well-known disease throughout North America (Hsiang *et al.*, 1995; Inguagiato and Martin, 2015). The term Pythium root rot was assigned in the early 1960s to describe foliage, crown and root regions of cool-season turfgrasses infected by *Pythium* species in southern California (Endo, 1961). The name has also been designated to a *Pythium*-incited disease in New Zealand (Hood, 2006) and China (Guan *et al.*, 2009).

Pythium crown and root rot is associated primarily with highly managed golf greens mostly during cool and wet periods of spring. However, the disease may also occur during warmer periods, depending on environmental conditions. As the name suggests, Pythium root rot is associated primarily with severe infection, discoloration and rotting of the crown and roots. Pythium crown and root rot is a problematic disease once established and probably misdiagnosed.

Signs and symptoms

The presence of oospores embedded in infected crown tissues may be required to confirm the diagnosis of Pythium crown and root rot. Mycelium is typically absent on infected leaves. Initial infection and colonization by different *Pythium* species may occur simultaneously to the crown and root regions at any time of the year. Advanced symptoms of Pythium crown and root rot are a chlorosis and general decline rather than very specific patterns, making a precise diagnosis based on field symptoms challenging.

Early-stage symptoms typically appear in spring and autumn as relatively small (50–75 mm diameter), randomly scattered diffuse spots. Infected foliage ranges from yellow to reddish-brown to bronze-coloured, depending on circumstances. Disease progression at any time of the year under favourable conditions results in greater numbers of individual spots enlarging and coalescing forming an irregular mosaic of declining and dying grass. Severely damaged crowns may appear water-soaked and discoloured. Infected roots are discoloured and can have reduced length and mass.

Pythium spring dead spot

Pythium spring dead spot and an associated disease called Irregular Pythium patch occur as

---

Box 6.5. Integrated management factors of damping-off.

- Establish seed during periods of temperature favourable for the turfgrass species, i.e. temperate (16–30°C) and tropical (21–35°C)
- Consider using pre-treated seed or apply by hydro-seeding methods if practical
- Adequate NPK nutrition
- Judicious irrigation water applications
- Monitor plants for signs of key fungal structures
- Use of registered fungicides as a preventive application

roughly circular or irregular-shaped patch symptoms on ecotypes of manilla zoysiagrass (*Zoysia matrella*) in Japan (Tani and Beard, 1997). The ecology and epidemiology of this disease remain unclear.

Red Thread

Red thread causes an unsightly appearance on infected foliage and has the distinction of being the first-identified foliar disease of managed turfgrass. The characteristic signs of pink-reddish-coloured mycelium and sclerotia among symptomatic infected foliage were considered a component of a malady now known as two distinct diseases: pink patch and red thread. The name red thread was proposed to distinguish the causal agent and symptoms from those of pink patch. Awareness of red thread as a distinct turfgrass disease has increased in recent years in certain countries.

Causal agent

L. fuciformis (formerly *Corticium fuciforme*) is the causal agent of red thread (Smiley *et al.*, 2005). The fungus is a facultative pathogen possessing pale-coloured mycelium.

Geographic distribution

Red thread is widely distributed throughout the cooler temperate and subtropical regions worldwide (Smith *et al.*, 1989; Fermanian *et al.*, 2003). The fungus was initially identified under the name *Isaria fuciformis* in cereals in southern Australia then isolated from a bowling green in Victoria in 1904 (McAlpine, 1906). Red thread was recognized as a turfgrass disease in the 1930s in New Zealand (Brien, 1935) and became more widespread in the United Kingdom during the 1990s (Entwistle, 1999). Red thread is undoubtedly more widespread in cooler regions than currently reported.

Host range and susceptibility

Bentgrass, bermudagrasses, bluegrasses, fescues, ryegrasses, seashore paspalum and zoysiagrasses are hosts of red thread (Zhang *et al.*, 2015a; Tredway *et al.*, 2023). Fine fescues are highly susceptible to the disease. Kikuyugrass is a host of red thread in Australia (P. Wong, New South Wales, 2023, personal communication).

Signs and symptoms

The appearance of light red- to pinkish-coloured, anther-like, fine mycelium and sclerotia spreading in the canopy is characteristic of red thread. Sclerotia are 1–6 mm in length and may extend some 10 mm beyond the infected leaf tip and are more readily observed in the early morning. Red thread sclerotia can be easily removed for close examination using a 10× magnifying glass.

Symptoms of red thread vary depending on level of turfgrass management and environmental conditions. Initial symptoms of red thread may first appear as individual, light brown- to tan-coloured leaves becoming water-soaked and necrotic in small-sized, irregular patches. Late-stage symptoms on infrequently mown swards present an unsightly appearance due to contrasting coloration between healthy green foliage, diseased plants and pink mycelium (Fig. 6.6).

Disease progression results in gradual enlargement of spots coalescing to form irregular shapes of blighted grass up to half a metre in diameter. Perimeters of coalescing patches form irregular shapes. Aerial sclerotia protrude from infected foliage partly covering the canopy, giving an overall light red to pink appearance to the stand when in late stages of disease development (Fig. 6.7). Under highly favourable moisture conditions, masses of mycelium can bind multiple leaves together. Diagnosis of field symptoms of red thread may be difficult on closely mown swards in the absence of reddish-coloured aerial mycelium and sclerotia, raising the possibly of it being another disease (e.g. pink patch).

Disease profile

Red thread is a disease generally considered common in relatively slow-growing, mature grass, as a result of reduced temperatures, effects of insufficient nitrogen or plant-growth regulators. The biology, ecology and epidemiology of red thread are well understood in certain regions of the United Kingdom, Western Europe and North America (Smith *et al.*, 1989; Smiley *et al.*, 2005).

Fig. 6.6. Late-stage, red thread symptoms on a fine fescue lawn. (Courtesy of Gary W. Beehag.)

Fig. 6.7. Red thread on kikuyugrass in Australia. (Courtesy of Ben Evans.)

Box 6.6. Key environmental factors of red thread.

- Low to moderate temperatures (18–24°C)
- Nitrogen deficiency and imbalanced nutrition
- High humidity (> 80%)
- Prolonged leaf wetness (minimum of 12 h)
- Lower mowing heights
- Reduced light intensity
- Host–fungicide interactions

The pathogen survives unfavourable periods as mycelium in decomposing infected leaf tissues and sclerotia in the canopy. Sclerotia may remain viable for more than 2 years and require moisture for germination. Fungal infection and colonization typically occur during spring and autumn. Outward-growing mycelium may enter host plants indirectly through intact leaves via stomates or directly in open wounds and cut leaves.

The mycelium grows intercellularly within tissues. Diseased foliage may undergo necrosis within 2 days of infection. Physical spread of the disease occurs by dissemination of sclerotia and infected foliage or wind-borne over long distances via spore-type structures (arthroconidia). In geographic regions having a combination of moderate seasonal temperature and favourable moisture, the disease may occur year-round.

Environmental factors

Several key environmental and management-induced factors have been shown to be associated with red thread (Box 6.6).

TEMPERATURE, MOISTURE AND LIGHT QUALITY The red thread fungus is a psychrophilic species capable of infection at temperatures anywhere between 4 and 27°C (Endo, 1963) depending on fungal isolate and environmental factors. Diurnal temperature ranges between daytime and night-time in combination with atmospheric moisture in the forms of high humidity and prolonged periods of leaf wetness can favour red thread development.

Prolonged events of light rain and fog associated with reduced light intensity have been stated to be favourable for development of red thread when temperatures are favourable for disease development (Smiley et al., 2005). Periods of alternating light intensity may be associated with greater production of red thread structures or arthroconidia (Zhang et al., 2015a).

HOST NUTRITION Low nitrogen generally favours the occurrence of red thread. Application of soluble nitrogen forms indirectly lessens disease incidence (Cahill *et al.*, 1983; Tredway *et al.*, 2001) probably by promoting grass recovery. Application of potassium, phosphorus and calcium appears to have minimal or inconsistent impacts on reducing red thread severity (Tredway *et al.*, 2001; Cagas *et al.*, 2010).

PLANT-GROWTH REGULATORS Application of plant-growth regulators has been shown to favour the occurrence of red thread (Chastagner and Vassey, 1979) and probably enhance pathogen infection; hence the need to avoid their application during periods of minimal or variable grass growth.

MOWING HEIGHT Relatively low mowing heights generally favour red thread (Goss and Gould, 1971). Hence, an optimal mowing height and frequency with clipping removal needs to be adopted to suit the specific turfgrass species.

Integrated management

Effective management of red thread on high-value swards necessitates an integrated approach (Box 6.7). Hence, frequent monitoring of disease symptoms, changing weather patterns and cultural, biological and/or fungicidal options are required for effective management of the disease. Application of soluble nitrogen requires attention in cool, temperate regions during spring and autumn to avoid stimulation of certain other diseases (e.g. snow moulds) in some regions.

Disease-resistant varieties

Genetic resistance against red thread has been documented among fine fescue, tall fescue, Kentucky bluegrass and perennial ryegrass cultivars (Aamlid *et al.*, 2012; Bonos and Huff, 2013).

Rust and Smut Diseases

Leaf rust and smut diseases occur on cultivated grasses worldwide. Collectively, leaf rust and smut diseases are generally considered relatively minor turfgrass diseases primarily causing cosmetic or aesthetic problems to the foliage or seedheads, depending on fungal species. Rusts primarily infect and their uredinia (fruiting bodies) discolour the foliage and stem region. Leaf smuts manifest in the seedheads of the host resulting in management problems for grass seed and sod producers. Rust and smut fungi are closely related, being obligate pathogens having certain biological differences as well as similarities; hence are covered together in this section.

Causal agent

Leaf rust and smut diseases are caused by many fungal pathogens (Table 6.4). Certain rust fungi have a wide host range and other rusts are host-specific. Association of the turfgrass host and rust symptoms is often sufficient to confidently

Box 6.7. Integrated management factors of red thread.

- Monitor plants for signs of key fungal structures
- Maintain consistent growth using balanced NPK nutrition
- Judicious irrigation water applications
- Use of registered fungicides if required

Table 6.4. Rust and smut genera and common species on turfgrasses.

Disease	Fungal genera	Common species	References
Rust	*Puccinia*	*P. brachypodii, P. coronata, P. graminis, P. striiformis*	Smith *et al.* (1989); Fermanian *et al.* (2003); Vargas (2005); Tredway *et al.* (2023)
	Physopella	*P. compressa*	
	Uromyces	*U. dactylidis, U. setariae-italicae*	
Smut	*Entyloma*	*E. camusianum, E. dactylidis*	
	Tilletia	*T. sterilis*	
	Urocystis	*U. agropyri, U. occulata*	
	Ustilago	*U. spegazzinii, U. striiformis*	

identify the specific rust fungi. Rust-causing fungi occur commonly on cool-season grasses but are less common on warm-season grasses.

Geographic distribution and host range

Rust and smut diseases occur on wild and cultivated grasses worldwide. Most if not all cool-season and warm-season turfgrass species are hosts for certain rust and smut fungi. Bermudagrass smut (*Ustilago cynodontis*) is widespread on bermudagrasses (Tran *et al.*, 2020) and flag smut (*Urocystis agropyri*) is widespread on mature Kentucky bluegrass (Vargas, 2005).

Signs and symptoms

Signs and symptoms of leaf smut and rust diseases share many similarities depending on host–pathogen combination and environmental factors. The presence of 'rust-coloured' uredinia bursting out on leaves typically in the absence of lesions visible to the unaided eye is common for rusts. Misshapen, dark-coloured leaves or seedheads characterize smut diseases.

Rusts

Orange-, yellow- or brick-red-coloured flecks or steaks on leaves are early symptoms of rust diseases (Fig. 6.8). Disease progression results in coloured flecks enlarging and elongating, showing definitive orientation in rows parallel with leaf veins down towards leaf bases and stems. Late-stage symptoms appear as entire areas of light red- to yellow-coloured, blighted leaves (Fig. 6.9) depending on host and fungal species. Masses of spores are readily removed by hand and observed by eye on infected leaves.

Smuts

Most smut diseases are generally characterized by twisted and partly shredded and split seedheads ranging in colour and presence of dark-coloured masses of spores. Disease progression results in severely chlorotic and/or misshapen foliage or seedheads if present (Fig. 6.10). Smut-infected plants may be stunted, chlorotic with reduced plant densities and have twisted, malformed seedheads.

Fig. 6.8. Uredinia structures of *Puccinia graminis* on perennial ryegrass. (Courtesy of Ben Evans.)

Fig. 6.9. Leaf rust on a zoysiagrass domestic lawn. (Courtesy of Gary W. Beehag.)

Fig. 6.10. Misshapen, dark-coloured leaves caused by smut fungi. (Courtesy of Ben Evans.)

Rust and smut disease profile

Much of the biology, ecology and epidemiology of most rust and smut diseases on turfgrass remains limited. Disease development of certain rust species (e.g. *Puccinia graminis*) on cool-season grasses and smut (e.g. *U. cynodontis* on bermudagrass) is relatively well understood. Rust and smut pathogens overwinter in infected swards either as dormant mycelium inside the hosts or externally as spores. Spores germinate and infect foliage during favourable conditions. Germinating spores may penetrate leaves directly through the cuticle or indirectly through open stomates, depending on rust species. Rusts require two different host plant species to complete their life cycle. Rust and smut spores are disseminated easily by wind.

Disease development among rust fungi is complex because of the stages of spore development which may involve numerous, alternative grasses and other plant species, depending on rust species. Smut fungi require only one host to complete their life cycle. Additional stem fungi may further parasitize host plants weakened by either disease.

Most rust diseases of grasses occur during summer to autumn primarily on slow-growing plants under stressful conditions, most notably low fertility and drought. Certain rust diseases are most common during warmer conditions (e.g. *P. graminis*) while others (e.g. *Puccinia coronata*) during cooler temperatures. Rust species associated with warm-season grasses typically have temperature optima for growth and sporulation of between 20 and 30°C, as opposed to many rust species associated with cool-season turfgrasses which may have temperature optima between 10 and 20°C (Couch, 1995). Light intensity and leaf moisture are co-dependent factors further governing infection and severity of rust diseases (Smiley *et al.*, 2005).

Smut diseases may occur any time during spring and autumn under highly favourable conditions. *Ustilago* and *Uromyces* smut fungi are considered to be favoured between 10 and 20°C coinciding with low soil moisture. Certain smut fungi (e.g. *Ustilago* and *Uromyces*) may grow systemically throughout infected hosts as opposed to others (e.g. *Entyloma* spp.) which grow on the foliage. The smut fungus *U. cynodontis* is highly prevalent on common bermudagrass ecotypes having prolific seedhead production; hence utilization of hybrid cultivars offers an effective means of smut management.

Integrated management

Effective management of rust and smut diseases requires an integrated approach (Box 6.8). Monitoring of disease signs and changing weather patterns, utilization of disease-resistant cultivars and proven cultural practices and chemical options are required.

Disease-resistant varieties

Limited genetic resistance against rust and smut diseases has been documented among Kentucky bluegrass and perennial ryegrass cultivars (Bonos *et al.*, 2006; Bonos and Huff, 2013).

Stem- and root-infecting fungal diseases

Soil-borne fungi infect and colonize stem and root tissues of the hosts. Symptoms are often displayed

---

Box 6.8. Integrated management factors of rust and smut diseases.

- Monitor plants and weather patterns
- Use of disease-resistant varieties
- Maintain consistent growth using balanced NPK nutrition
- Use of registered fungicides if required

as ill-defined mosaic areas or large spots or patches, depending on the prevailing epidemiological conditions. Symptoms can coalesce to form larger-sized, roughly circular-shaped patches or circular patches to complete circles with unaffected centres often called a 'frog eye' or 'doughnut'-shaped symptom, depending on the host–pathogen combination.

Fungal infection of stems or roots may begin weeks or months prior to symptom expression. Certain stem and root disease symptoms may be the result of more than one active pathogen species under favourable conditions, often complicating disease symptoms and a precise diagnosis. Effective management of many soil-borne fungal diseases can only be attained by preventive cultural practices.

Fairy Ring Formations

Strictly speaking, fairy ring formation is not an infectious fungal disease but a unique, biological phenomenon in grassland and turfgrass systems worldwide. Fairy ring formations are the result caused by numerous genera and species of basidiomycete or mushroom-forming fungi eliciting various responses from host plants by altering the rootzone environment. Certain aspects of the ecology of fairy ring formations remain unknown despite the long-known existence of the disorders.

The English folklore name fairy ring or 'fairy circles' describes what was considered to be the mystical appearance and disappearance of mushrooms, toadstools or puffballs (fruiting bodies) and roughly circular, green-coloured rings of grass (Fig. 6.11). Symptomatology of fairy ring formation varies widely depending on environmental conditions far from fully understood.

Fairy ring formations can have longevity and be very large in size, as a result of continual radial expansion that can in some cases occur over many decades, even centuries. Antagonistic microbial competition or food exhaustion occurs when two or more rings intersect, making fairy rings unique in plant ecosystems. Multiple fairy rings are a challenging and problematic malady to effectively manage once established in turfgrass systems. Association of water repellence on sand-based rootzones with plant death adds another dimension to their management.

Causal agent

Fairy rings are associated with a multitude of known (Table 6.5) and unknown basidiomycete fungi. The actual basidiomycete species associated with a specific fairy ring formation is seldom identified or documented. *Marasmius oreades* is a well-known, classic basidiomycete fungus known to be associated with fairy ring formations from the 1700s on English grasslands (Wollaston, 1807).

The basidiomycete *Calvatia cyathiformis* (formerly *Lycoperdon cyathiforme*) was identified from fairy rings, some intersecting, of varying sizes some up to 12 ft (3.6 m) in diameter on a bowling green in the late 1800s in southern Australia (McAlpine, 1898). The basidiomycetes *Bovista dermoxantha* and *Vascellum curtisii* have been isolated on golf greens in the United States (Miller *et al.*, 2011). Basidiomycete fungi associated with lectophilic fairy rings are members in the genera *Clavaria*, *Clitocybe*, *Coprinus*, *Cristella*, *Hygrophorus*, *Psilocybe* and *Trechispora* (Smith *et al.*, 1989; Fermanian *et al.*, 2003). The absence of fungal fruiting bodies for some species makes precise identification of the associated fungus to species level problematic.

Geographic distribution and host range

Fairy ring formations have been documented on turfgrass systems worldwide (Smith *et al.*, 1989; Charbonneau and Hsiang, 2015). Fairy rings of one sort or another undoubtedly occur wherever turfgrass is managed.

Signs and symptoms

The presence of fruiting bodies and proliferation of fungal mycelium in the thatch-mat and rootzone regions, and reduced soil water content, are characteristic of fairy rings. Fruiting bodies may emerge in a roughly circular arrangement coinciding with consistent rainfall events or frequent irrigation.

Fig. 6.11. Fruiting bodies around the perimeter of a fairy ring formation. (Courtesy of Ben Evans.)

Table 6.5. Common fairy-ring-forming fungal genera.

Fungal genera	References
Agaricus, Agrocybe, Calocybe, Calvatia, Cantharellus, Chlorophyllum, Clitocybe, Coprinopsis, Cortinarius, Gymnopus, Hydnellum, Hydrocybe, Lepista, Lycoperdon, Marasmius, Melanoleuca, Suillus, Tricholoma	Couch (1995); Keighley (2017); Tredway *et al.* (2023)

The presence of fairy ring fungi can be confirmed by placing turfgrass core samples in a humid environment for a few days to allow for the proliferation of fungal mycelium on the outside of the sample (Fig. 6.12). A slight 'mushroom' odour may be detected when samples are examined.

Fig. 6.12. Basidiomycete fungal mycelium growth around a core sample. (Courtesy of Gary W. Beehag.)

> **Box 6.9.** Categorization of edaphic fairy ring symptoms.
>
> **Type 1** – turfgrass is ultimately killed or badly damaged in rings.
> **Type 2** – turfgrass growth is only stimulated and is often darker-coloured.
> **Type 3** – turfgrass remains unchanged with the presence of mushrooms.

Fairy ring formations are broadly categorized as either edaphic (soil) or lectophilic (thatch-mat) borne.

Edaphic fairy rings

Individual edaphic fairy rings can vary in size, shape, rate of outward growth and overall pattern, encompassing stimulated plant growth and possible regions of plant death. Multiple fairy rings of varying diameters and shapes may be present at one time in the same location. Initial symptoms may appear as small regions of stimulated green-coloured turfgrass or simply a random cluster of fruiting bodies. Late-stage symptoms of individual fairy rings may range from large, 'banana'-shaped arcs to roughly circular rings of luxuriously growing turfgrass or dead plants due to soil hydrophobicity.

Edaphic fairy rings are characterized according to stimulation or retarding effects on plant growth based on a descriptive system of three types (Box 6.9) developed for natural grasslands (Shantz and Piemeisel, 1917). The designation generally applies for turfgrass systems. More than one symptom type may occur simultaneously which may alter in size and even disappear over time with changing environmental conditions. In all categories, fruiting bodies may emerge for a limited time following favourable rainfall events.

Type-1 fairy ring symptoms vary, but generally comprise roughly circular, complete rings or banana-shaped arcs of dead plants with a central zone of dead or stunted turfgrass and an outer zone of slightly stimulated turfgrass (Fig. 6.13). These fairy ring formations are more common during moist periods often following rainfall events. Type-1 fairy rings are associated with rootzone water repellence due to prolific growth of fungal mycelium in the thatch-mat region. Incomplete, mature rings may indicate some form of subsurface obstruction (e.g. subsurface rocks).

Affected grass foliage may display a distinct yellow-orange colour. The width of the dead zone can be highly variable from centimetres to metres depending on environmental conditions, fungal species and age of the ring. Type-1 fairy rings are commonly associated with the fungal genera *Agaricus*, *Agrocybe*, *Chlorophyllum*, *Coprinus*, *Lepiota*, *Lycerpodon* and *Marasmius* (Dernoeden, 2013).

Type-2 fairy ring symptoms are typically expressed as roughly circular rings of a single, darker-coloured, stimulated turfgrass (Fig. 6.14). The inner region often remains unaffected. Type-2 symptoms are particularly common on domestic lawns, golf fairways and parklands in rootzones with high levels of organic matter. Type-2 symptoms may also be similar to Type-1 symptoms under highly favourable conditions.

Fig. 6.13. Type-1 fairy ring formation on a semi-dormant 'Tifdwarf' couchgrass (bermudagrass) bowling green. (Courtesy of Gary W. Beehag.)

Fig. 6.14. Type-2 fairy ring formation on a semi-dormant, seashore paspalum bowling green in southern Australia. (Courtesy of Gary W. Beehag.)

The stimulated bands occur in roughly circular bands immediately above the rootzone region where the fungus is most active.

Type-2 symptoms become apparent during and immediately following extended periods of warm and dry weather and become less severe in cooler months. Type-3 symptoms are categorized on the basis of appearance of fruiting bodies in incomplete or complete circular patterns (Fig. 6.15) often with minimal apparent effect to turfgrass.

Lectophilic fairy rings

Lectophilic fairy rings, or superficial as the name suggests, commonly appear as roughly circular, indiscrete blighted patches. Symptoms of lectophilic fairy ring vary and have been characterized according to effects on turfgrass growth (Smith *et al.*, 1989). The descriptive system categorizes three types (Box 6.10).

Lectophilic fairy rings are common on bowling and golf greens expressing roughly circular, whitish-coloured patches measuring up to 100 cm diameter or more with white mycelium around the perimeter (Fig. 6.16) during highly favourable conditions.

Fairy ring profile

Fairy ring formation is an ecologically complex malady whereby the fungi colonize the thatch-mat region or organic matter to obtain nutrition as saprophytes. It may be possible certain basidiomycete fungi are root endophytes or possibly very weak pathogens, depending on the turfgrass health and the fungal species involved. Ecological understanding of edaphic fairy rings is based largely on the species *Agaricus campestris* and *M. oreades* and other common basidiomycete fungi present in natural grasslands (e.g. Smith *et al.*, 1989; Keighley, 2017). Thus, only a generalized disease profile can be outlined for fairy rings on turfgrass given an absence of detailed understanding of the fungal biology and ecology.

New fairy ring formation may occur from germination of fungal spores and mycelium growth in the thatch-mat or rootzone region.

Box 6.10. Categorization of lectophilic fairy ring symptoms.

Type A – produces sparse to abundant mycelium with or without mushrooms, and in the thatch with little or no apparent effect on grass growth.

Type B – produces stimulated grass growth and/or turfgrass discoloration. Thatch degradation is apparent, but the plants are not severely injured and will eventually recover.

Type C – produces severe injury to the grass. During the initial stages of formation, the growth of the turf may or may not be stimulated.

Fig. 6.15. Type-3 fairy ring formation with *Agaricus* sp. fruiting bodies. (Courtesy of Lawrie Greenup.)

Fig. 6.16. Lectophilic fairy ring on a creeping bentgrass bowling green. (Courtesy of Gary W. Beehag.)

The relative importance of mycelium growth and likelihood of spore germination to initiate a new fairy ring remains unclear. In order to grow, the basidiomycete fungi must overcome microbial competition from existing microflora and individuals of the same or different fungal species. The outward progression of fungal mycelium is associated with long-term physicochemical and biological changes of the thatch-mat and rootzone environment (Smith, 1980).

Microbial antagonism and suppression among *M. oreades* individuals and non-related fungi or bacterial species for Type-1 fairy rings has been documented (Smith and Rupps, 1978). However, species distribution and possible biological effects of other fungi and bacteria in the inner and outer zones remain unclear. Physicochemical changes have been widely investigated which together with the possible evolution of hydrogen cyanide (cyanogenesis) by the fungi, an increase in soil nitrogen level and development of soil hydrophobicity (Whiteley and Baldwin, 1990; Fidanza, 2007) have all been reported for various fairy-ring-causing fungi.

Water repellence (hydrophobicity) widely associated with fairy rings in natural grasslands and turfgrass systems (York and Canaway, 2000; Hallett *et al.*, 2001) has been shown to be associated with fungal hydrophobins (Rillig, 2005). Knowledge of changes in microbial composition associated with fairy ring formation is lacking. Numerous species of bacteria and fungi isolated from fairy ring formation have been experimentally shown in culture to be antagonistic towards *M. oreades* (Smith and Rupps, 1978).

Luxuriant turfgrass growth in stimulated zones has been suggested to be the result of liberation of nitrogen from organic matter by the fungus (Smith *et al.*, 1989). Certain basidiomycete fungi are capable of biodecomposing lignin, a material associated with turfgrass thatch (Baetsen-Young *et al.*, 2016).

Thatch collapse

The rapid degradation of thatch is a condition generally known as thatch collapse and is similar to what occurs in fairy ring formation. Thatch collapse is generally associated with the fungus *Sphaerobolus stellatus*, an unusually named basidiomycete called the artillery fungus (Dernoeden *et al.*, 2011; Baetsen-Young *et al.*, 2016). *Sphaerobolus* is Greek, meaning 'spear thrower', describing the unique ability of the fungus of discharging spores a considerable distance and hence its name. Symptoms of thatch collapse on golf greens (Fig. 6.17) result from the degradation of the thatch-mat layer as assessed by pressing down on the dark-coloured, infected patches. Published information about the distribution, ecology and management of thatch collapse on turfgrass is limited.

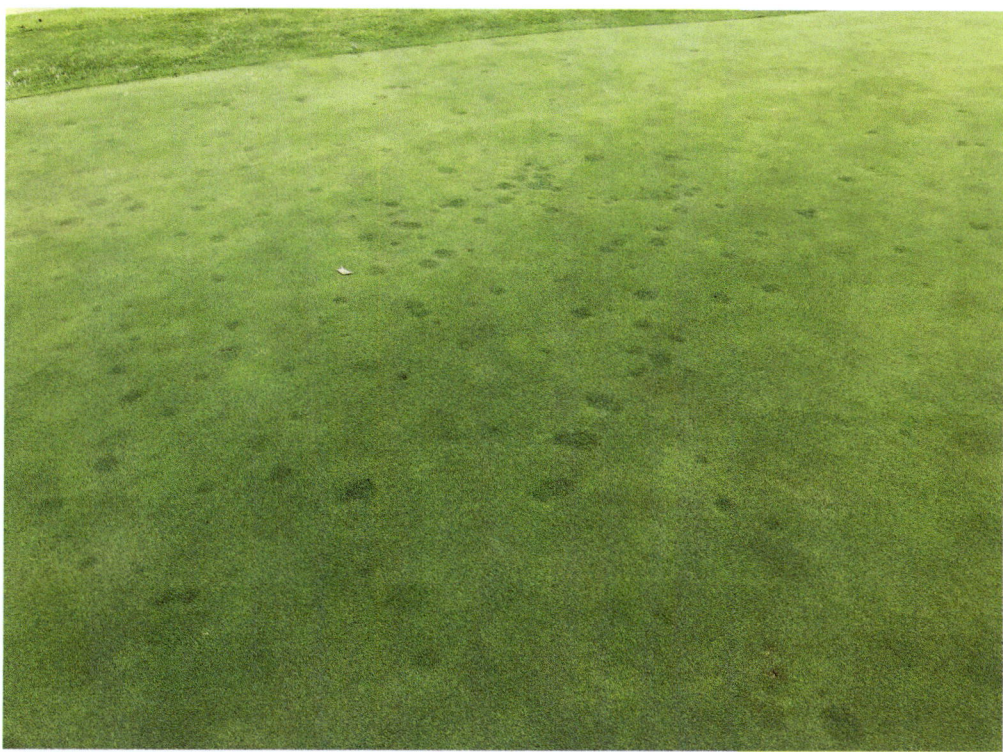

Fig. 6.17. Thatch collapse symptoms on a creeping bentgrass golf green in Australia. (Courtesy of John Forrest.)

Biology of intersecting edaphic fairy rings

The unique biology of fairy rings is no better exemplified than when two or more enlarging edaphic fairy rings expand and contact each other (Fig. 6.18). Cases of intersecting fairy rings caused by *M. oreades* have long been reported in natural grasslands (e.g. Wollaston, 1807; Shantz and Piemeisel, 1917) but less frequently in turfgrass (Smith *et al.*, 1989). Explanation of the biological behaviour of intersecting fairy rings is that the fungi have exhausted the substrate nutrient supply or produce unknown compounds resulting in mutual antagonism between the same and different species (Smith *et al.*, 1989; Dernoeden, 2013).

When two enlarging fairy rings caused by the same fungus and eventually meet, one of three outcomes occurs at the points of intersection (Box 6.11). Rarely is one fairy ring found inside another but occasionally a small-sized fairy ring will be found inside an incomplete larger fairy ring.

Environmental factors

Several key environmental and management-induced factors are associated with edaphic fairy ring formation (Box 6.12). Temperature and moisture are key determinant factors of fairy ring formation.

ROOTZONE TEMPERATURE AND MOISTURE   Fairy ring formation can occur over a wide temperature range as observed by the growth and presence throughout all seasons. Moisture in association with temperature is another key determinant factor of fairy ring formation. Relatively dry and warmer conditions generally favour the onset of Type-1 ring symptoms while relatively moist and cooler conditions favour onset of Type-2 symptoms (Fidanza, 2010; Mann, 2011).

ROOTZONE STRUCTURE   Sandy rootzones, as opposed to fine-textured clay soils, generally favour faster lateral and vertical growth of edaphic fairy rings (York, 1998). Penetration depth of

Fig. 6.18. Multiple, intersecting Type-2 fairy rings on a hybrid couchgrass (bermudagrass) golf green in Australia. (Courtesy of Gary W. Beehag.)

Box 6.11. Categorization of edaphic fairy ring competition.

A. **Indifference** – each fairy ring continues to radially grow producing fruiting bodies.
B. **Bilateral extinction** – surface regions of both intersecting fairy ring portions are obliterated.
C. **Unilateral extinction** – one entire fairy ring remains unaffected, the other fairy ring is obliterated.

Box 6.12. Key environmental factors of fairy ring formation.

- Rootzone temperature and moisture
- Rootzone structure
- Nutrient status and organic matter

---

fungal mycelium has been recorded to depths of 300–500 mm in sandy loams and limited to 25–30 mm in high-clay-content soils (Smith *et al.*, 1989). The impact of organic matter, soil pH and nutrition appears to differ between fungal species and fairy ring formation (Smith *et al.*, 1989; Watschke *et al.*, 2013).

Integrated management

Effective management of multiple fairy ring formation, once established on highly managed turfgrass systems, is problematic (Box 6.13). The cryptic nature of basidiomycete fungi and frequent absence of fruiting bodies on closely mown turfgrass, uncertainty of the identity and nomenclature of associated species, and non-specification on many pesticide labels contribute to unpredictable fairy ring management. Applied fungicides must target the rootzone. High-pressure water injection has been shown experimentally to assist management of Type-1 fairy rings (Fidanza *et al.*, 2005).

Possible antagonistic mechanisms of other fungal and bacterial species in *M. oreades*-based fairy rings have raised the potential of biological management (Smith, 1980). However, effective management of fairy ring formation using microbial antagonistic products for turfgrass application has yet to be realized.

> **Box 6.13.** Integrated management factors of fairy ring formation.
>
> - Monitoring turfgrass health and early disease signs
> - Maintain consistent growth with use of a balanced NPK nutrition
> - Minimization of thatch-mat thickness
> - Judicious irrigation water applications
> - Targeted application of registered fungicides and biofungicides
> - Use of proven soil-wetting agents

Gaeumannomyces-Incited Diseases

Gaeumannomyces is a complex genus of fungal species and strains associated with cool-season and warm-season grasses. Differentiation between various *Gaeumannomyces* species and variants has been traditionally based on ascospore length and width (Hernández-Restrepo *et al.*, 2016) but more recently using genetic methodologies (e.g. DNA sequencing).

The *Gaeumannomyces* species *Gaeumannomyces avenae* and *Gaeumannomyces tritici* (previously classified as distinct strains of *Gaeumannomyces graminis*) are closely related and well-known pathogens of cool-season grasses worldwide. *G. graminis* is a well-known pathogen of warm-season grasses (Freeman and Ward, 2004). The genus *Gaeumannomyces* (formerly *Ophiobolus*), now reclassified, embraces previously known and undescribed fungal species continually being isolated from warm-season grasses worldwide. *Gaeumannomyces* are mesophilic fungi that grow well between 20 and 30°C depending on fungal species (Weste, 1970; Plumley *et al.*, 1997).

However, not all *Gaeumannomyces* species and variants isolated from warm-season grasses are managed as formal turfgrass pathogens. Pathogenicity (virulence) among known *Gaeumannomyces* species is highly variable and remains unproven among others. The *Gaeumannomyces*-incited diseases of bermudagrass decline (USA), zoysia decline and others are grouped as a collective. Root decline of warm-season turfgrasses is covered elsewhere in this chapter.

Take-all patch of cool-season grasses

Take-all patch of cool-season grasses is among the most extensively studied of turfgrass diseases, but many key aspects of the biology and epidemiology of the disease remain uncertain. Take-all patch is a problematic disease once established on closely mown, cool-season swards (e.g. bowling and golf greens and tennis courts) and classically characterized by roughly circular rings with unaffected centres (Fig. 6.19) often displaying a frog eye appearance. The disease name take-all patch originated from take-all of cereal grains, referring to the host's symptoms following death of entire plants. Initial symptoms may result in take-all patch being misdiagnosed for other patch diseases of cool-season turfgrasses.

What distinguishes take-all patch on cool-season grasses (formerly called Ophiobolus patch) from other soil-borne turfgrass diseases is the disease typically occurs on newer swards and often will undergo gradual decline over time, possibly mitigated by suppressive microflora in affected areas. However, disease symptoms may occur on mature swards of highly managed, cool-season grasses associated with elevated soil pH following application of alkaline-containing materials or possibly highly alkaline irrigation water. The disease is challenging to effectively manage once it becomes well established.

Causal agents and pathogenicity

G. avenae (previously classified as *G. graminis* var. *avenae*) and *G. tritici* (previously classified as *G. graminis* var. *tritici*) are the causal agents of take-all patch on cool-season grasses (Wong and Siviour, 1979; Charbonneau and Hsiang, 2015). *G. avenae* generally causes more severe damage to cool-season grasses than *G. tritici* (Smith *et al.*, 1989).

Geographic distribution

Take-all patch on cool-season grasses occurs worldwide primarily in cool temperate regions (Table 6.6). The disease probably occurs in temperate and subtropical regions wherever susceptible, cool-season grasses are managed.

UNITED KINGDOM AND WESTERN EUROPE Take-all of patch was first reported to be caused by *G. graminis* in Holland (Europe) in the early 1930s (Schoevers, 1937). Take-all patch occurs widely throughout Western European and Scandinavian

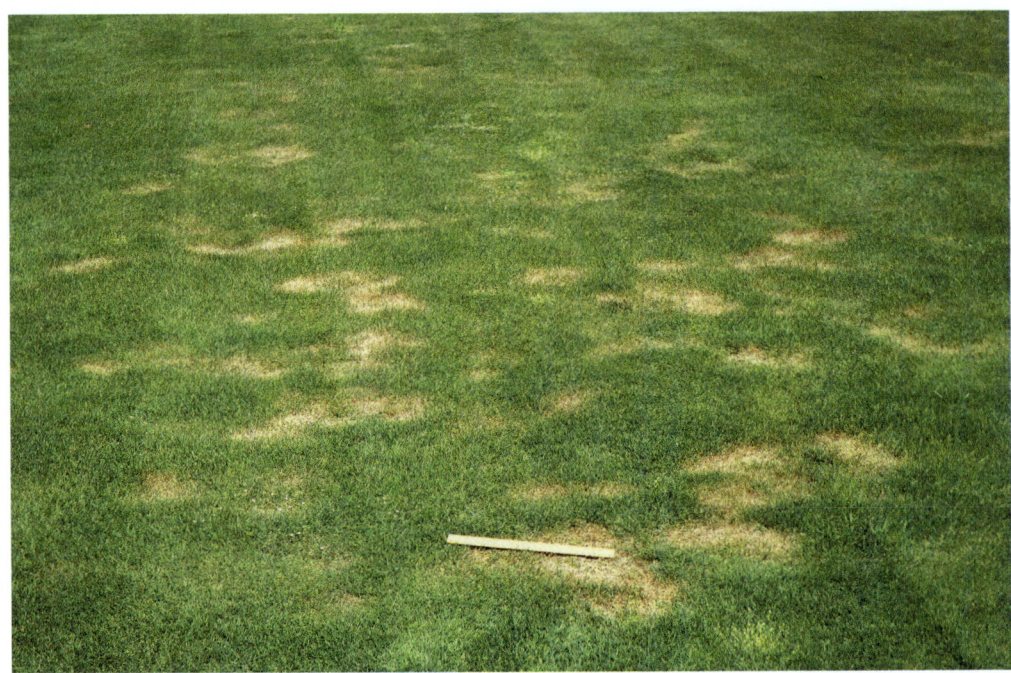

Fig. 6.19. Take-all patch on newly seeded, cool-season turfgrass. (Courtesy of Gary W. Beehag.)

Table 6.6. Documented global distribution of take-all patch on cool-season grasses.

Region/country	References
Western Europe	Smith et al. (1989); Stuttard et al. (2005)
United Kingdom	Smith et al. (1989)
Canada	Bucher and Wilkinson (2007); Charbonneau and Hsiang (2015)
United States	Dernoeden (2013); Turgeon and Kaminski (2019)
Australia	Walker (1975); Woodcock (1983)
New Zealand	Christensen (1985); Ormsby (2021)

countries, and the United Kingdom (Nilsson and Smith, 1981; Smith et al., 1989).

NORTH AMERICA Take-all of patch was not confirmed in the United States until the 1960s in Washington State (Jackson, 1981) although undiagnosed symptoms were initially observed during the early 1900s (Monteith and Dahl, 1932). Take-all patch occurs from the Pacific northwestern states and central west across into the upper central and north-eastern regions. Take-all patch was reported in Canada in the mid-1980s and now occurs in southern British Columbia, southern Ontario, Saskatchewan and the Atlantic provinces (Charbonneau, 1999).

AUSTRALASIA Take-all patch was first reported in Australia in the mid-1960s (Smith, 1969) and can occur on bowling and golf greens throughout southern Australia. In New Zealand, take-all patch was described during the 1980s and confirmed in the late 1990s on an experimental green at Palmerston North, North Island (Anon., 2020). The incidence and distribution of the disease in New Zealand remain unclear (Ormsby, 2021).

Host range and susceptibility

G. avenae has a relatively wide host range of cool-season grass species, many of which are managed as turfgrass (Table 6.7). The degree of susceptibility as demonstrated experimentally varies widely between grass species (Smith *et al.*, 1989; Couch, 1995). *G. tritici* generally has a more restricted host range (Smith *et al.*, 1989; Chng *et al.*, 2005). The degree of host susceptibility probably varies depending on the fungal isolate present and existing environmental conditions.

In mixed cool-season grass swards of bentgrass and fine fescue, damage by take-all patch can cause a succession over time in the absence of intervention and may result in damaged areas being slowly invaded by lesser susceptible or non-symptomatic plant species.

Signs and symptoms

The presence of dark-coloured runner hyphae and flask-shaped fungal structures (perithecia) examined microscopically along the stems of infected cool-season grasses can be indicative of take-all patch. Early- and late-stage symptoms of take-all patch are highly variable in patch distribution, size, shape and coloration, governed partly by host–pathogen combination and environmental conditions. Symptoms of take-all patch generally are most obvious during cooler and moist periods coinciding with late summer, spring and autumn, depending on seasonal environmental conditions.

Seedling and immature grasses may succumb rapidly to infection under highly favourable environmental conditions. Early-stage symptoms may occur as relatively small-sized, 5–8 cm diameter, light to reddish-brown-coloured, randomly scattered patches gradually changing in shape and size (Fig. 6.20). Advancing margins of individual,

Table 6.7. Hosts of take-all patch of cool-season grasses.

Grass species	References
Annual bluegrass (annual meadow grass), colonial bentgrass, creeping bentgrass and velvet bentgrass, rough bluegrass, Kentucky bluegrass, creeping dog's tooth, creeping red fescue, tall fescue, annual and perennial ryegrasses, timothy grass and tufted hair grass	Smith *et al.* (1989); Couch (1995); Tredway *et al.* (2023)

Fig. 6.20. 'Classic' early-stage symptom of take-all patch on newly seeded cool-season turfgrass. (Courtesy of the NZSTI.)

actively growing patches may be reddish-brown- to bronze-coloured eventually increasing in overall diameter up to 0.75 m or more as roughly circular patches.

Late-stage symptoms of individual patches on closely mown turfgrass may appear as irregularly shaped, declining areas that may display the frog eye or doughnut-ring symptom with unaffected centres. Progression of symptom expression results from gradual growth of the fungi on plants. Patches may appear in clusters that eventually coalesce to form larger areas of damaged grass. Patches may persist to lesser or greater degrees throughout the entire growing season, while other affected areas may disappear or even reappear the following season. Symptoms tend to lessen in overall severity towards the end of the year and recur the following year when fungal growth resumes. On higher-mown swards, late-stage patches may appear as roughly circular declining patches.

Disease profile

Take-all patch of cool-season grasses is an ecologically complex disease and its frequency and severity are influenced partly by rhizosphere chemistry and microbial antagonists. The disease cycle and epidemiology of take-all patch have been extensively documented in the United Kingdom and the northern United States (e.g. Smith *et al.*, 1989; Smiley *et al.*, 2005; Kerns and Tredway, 2013) partly based on the ecology of take-all of cereal grains (e.g. Freeman and Ward, 2004).

Take-all of cereal grains and take-all patch of cool-season grasses share many ecological characteristics assumed to be similar for both diseases. Gradual decline of take-all patch over time is believed to be the result of an increase in antagonistic microflora (Wong and Baker, 1986; Wong and Worrad, 1989) in accordance with suppressive mechanisms associated with take-all decline of monocultured cereal grains (Cook, 2003). Key aspects of the epidemiology of take-all patch of cool-season grasses remain unknown.

The fungi causing take-all patch survive unfavourable conditions saprophytically on tissues of previously infected plants or as spores. Primary fungal infection and colonization on mown grasses results from lateral growth of runner hyphae along plant organs in the soil. Fungal attachment and infection of plants emanates from specialized fungal structures (hypopodia) or flattened mycelial structures during highly favourable cool temperatures and moisture coinciding with spring and autumn. Secondary infection follows development of lesions and lateral growth of mycelium from infected to non-infected plant organs. Non-susceptible grass species may recolonize the centres of infected patches (Fig. 6.21).

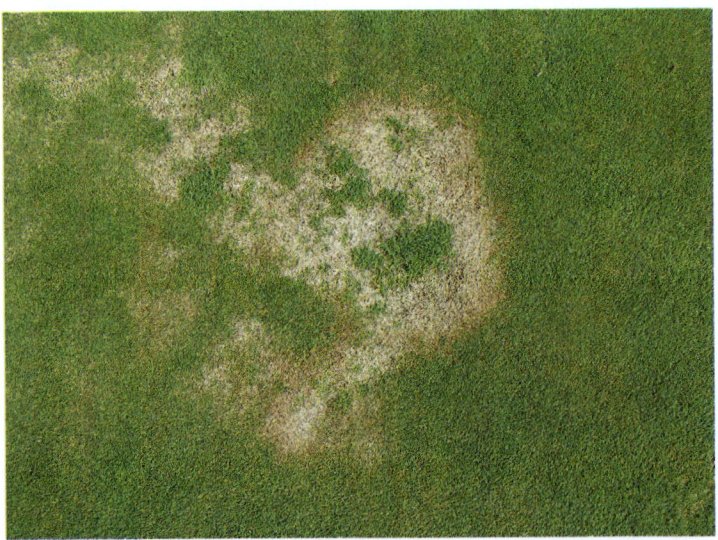

Fig. 6.21. Late-stage symptoms of take-all patch with recolonization in the centre by non-susceptible grass species. (Courtesy of the NZSTI.)

Seasonal occurrence of take-all patch of cool-season grasses varies somewhat between countries. In Great Britain and Ireland, take-all patch occurs throughout the summer months and into late autumn but may persist during mild winters (Smith *et al.*, 1989; Stuttard *et al.*, 2005). In the United States, where the disease occurs, symptoms are observed any time from late spring throughout summer and may reoccur in autumn (Dernoeden, 2013). In southern Canada, symptoms occur from June to September (Charbonneau and Hsiang, 2015). In temperate Australia, take-all patch occurs during early spring, late autumn and following cool, moist conditions in late summer (Siviour, 1972). In New Zealand, take-all patch symptoms occur throughout the growing season and are enhanced when plants suffer stress (Ormsby, 2021).

Necrosis of infected plant tissues occurs slowly. Host tissues become progressively damaged by colonizing fungal hyphae, resulting in brown-coloured, necrotic, plant soil organs. The precise biological mechanisms by which take-all pathogens cause stunting and tissue necrosis in susceptible grasses are unclear. Take-all fungi produce a range of cell-wall-degrading enzymes which may have a role in disease development. Dissemination of the pathogen can occur through direct plant contact, colonized organic material, infected sod and turfgrass mechanical equipment.

Environmental influences

Several key environmental and management-induced factors are associated with take-all patch on cool-season grasses (Box 6.14). Manipulation of certain rhizosphere properties by proven cultural practices in combination with efficacious fungicides during spring and autumn is crucial for effective prevention and long-term management of the disease.

Box 6.14. Key environmental factors of take-all patch of cool-season grasses.

- Cool temperatures
- Soil alkalinity
- Nitrogen source
- Soil moisture
- Microbial antagonists

TEMPERATURE AND MOISTURE *G. avenae* and *G. tritici* are mesophilic fungi which grow well and infect hosts between 10 and 19°C (Grose *et al.*, 1984; Couch, 1995), which partly explains their infection and colonization during moist and moderate conditions. Higher temperatures and drier conditions increase host stress often resulting in plant death.

ROOTZONE PROPERTIES Development of take-all patch is generally more prevalent in sand-based rootzones having low organic matter content, neutral to alkaline pH, and imbalances of nitrogen, phosphorus, potassium and manganese (Dernoeden, 2013; Kerns and Tredway, 2013). *Gaeumannomyces* species are capable of growing over a wide pH range but grow faster in alkaline or near-alkaline soils more so than acidic soils, which partly explains why the disease is more frequently seen if slight increases in rhizosphere pH occur from higher than required amounts of finely graded, alkaline materials (e.g. lime).

Alkaline sands used for construction and topdressing ('dusting') and irrigation water may promote the disease. Hence the need to regularly monitor irrigation water and rootzone pH under such circumstances.

The impact on the severity and occurrence of take-all patch from a minimal content of organic material generally associated with newly constructed greens and its gradual accumulation in ageing greens is unclear. A gradual increase in organic content in maturing greens may encourage populations of antagonistic microbes and may partly explain disease decline over time.

Imbalances in nitrogen, phosphorus and potassium to varying degrees have all been linked to the incidence of take-all patch (Dernoeden, 2013). Acidifying synthetic fertilizers will lessen but not eliminate take-all patch, as opposed to alkaline ones, which may intensify the disease (Latin, 2005). Numerous acidifying or alkaline fertilizers are available as soluble nitrogen-based fertilizers alone or in mixtures (Box 6.15).

Ammonium-nitrogen fertilizers, as opposed to nitrate-nitrogen forms, generally reduce take-all in cereals, which is attributed to antagonistic microflora (Colbach *et al.*, 1997). The source and ratios of soluble nitrogen are considered to be factors in the severity of take-all patch (Vargas, 2005). However, because ammonium-nitrate is biologically converted to nitrate-nitrogen

> **Box 6.15.** Common acidifying and alkaline fertilizers.
>
> | Acidifying | Alkaline |
> |---|---|
> | Ammonium sulfate | Calcium nitrate |
> | Aluminium sulfate | Calcium cyanamide |
> | Di-ammonium phosphate | Potassium nitrate |
> | Mono-ammonium phosphate | Sodium nitrate |

by bacteria (e.g. *Nitrosomonas* spp.) in nitrate-deficient soil, precise comparisons between nitrogen source and the effects on take-all can be problematic. None the less, most authors generally conclude that the beneficial effects on increased root growth partly by ammonium-nitrate outweigh any detrimental impact of increased infection caused by nitrate-nitrogen. Thus, a balanced amount, ratio and rate of ammonium-nitrate and nitrate-nitrogen are required to optimize growth of the host and management of take-all patch.

Rhizosphere acidification is generally believed to be a primary factor for mitigating take-all. Application of elemental sulfur has been shown to suppress take-all patch (Dernoeden, 2013). Whether the effect is due to rootzone acidification or because of its fungicidal properties remains unclear. Manganese, a minor essential element, has been associated with take-all patch (Dernoeden, 2013). Manganese oxidation by soil-borne microbes is linked to take-all in cereals and acidification of the rhizosphere limits the ability of microbes to oxidize manganese resulting in increased availability for root absorption; thus aiding defence against the disease. Very small applications of soluble manganese sulfate generally have shown experimentally to reduce the severity of take-all patch on bentgrass over 3 years but only when combined with ammonium sulfate (Heckman *et al.*, 2003).

ANTAGONISTIC MICROBES The close association between take-all pathogens and antagonistic, soil-borne microbes causing suppression of take-all diseases has been demonstrated experimentally using certain bacteria (e.g. fluorescent pseudomonads) and non-pathogenic fungi (e.g. *Trichoderma* sp. and avirulent isolates of *G. graminis* and *Phialophora* sp.) with varying degrees of success (Wong *et al.*, 1988; Sarniguet and Lucas, 1992). Commercialization of microbial antagonists providing effective, long-term suppression against take-all patch pathogens has yet to be realized.

ASSOCIATION OF FUMIGATION Pre-establishment incorporation of methyl bromide in sand-based rootzones followed by a rapid emergence of take-all patch on newly seeded bentgrass bowling and golf greens was widely documented in Australia (Siviour, 1972), England (Smith *et al.*, 1989) and the United States (Gould, 1973). The assumption is significant populations of antagonistic or competitive microbes were selectively reduced, even eliminated, immediately following fumigation. The impact of other fumigant-like compounds (e.g. metham-sodium, methyl isothiocynate) on the occurrence of take-all patch of cool-season grasses is unclear.

Emergence of take-all patch following fumigation of high-sand-content rootzones raises questions of the source and type of fungal inoculum; and, importantly, how to minimize infection and colonization on establishing swards. Fumigation causes a temporary 'biological vacuum' from the death of organisms and the release of organic nutrients from organism decomposition enables reinvading pathogens having an ecological advantage in the absence of beneficial microbes. Ecologically, *Gaeumannomyces* are considered poor competitors which may partly explain the apparent ability of the take-all pathogens to colonize fumigated soils in the absence of strong microbial competition.

Integrated management

Preventive and long-term management of take-all patch on cool-season grasses necessitates cultural manipulation of the rhizosphere and promoting a highly functional root system (Box 6.16). Consideration of rootzone and topdressing sands, frequent monitoring of early symptoms and changing weather patterns combined with proven cultural and fungicidal applications are required for effective management of take-all patch in the long term.

Disease-resistant cultivars

Colonial bentgrass and creeping bentgrass cultivars have been demonstrated experimentally to possess reduced susceptibility to take-all patch (Weibel *et al.*, 2002). Establishing and maintaining a mixed bentgrass–fine fescue composition may assist to minimize the disease in certain regions. Kentucky bluegrass–fescue mixes may provide better disease-tolerant turfgrass in other situations (Smith *et al.*, 1989).

Box 6.16. Integrated management factors of take-all patch on cool-season grass.

- Avoid using alkaline and calcareous sands for construction
- Minimization of thatch-mat thickness
- Minimize application of alkaline-based fertilizers and products
- Monitoring of pH levels of irrigation water and rootzone
- Possible use of acid injection to irrigation water
- Maintain consistent growth with use of a balanced NPK nutrition
- Supplemented application of soluble manganese-based fertilizer
- Judicious irrigation water applications
- Use of registered fungicides

Microdochium Patch

Microdochium patch is a common and disfiguring turfgrass disease in temperate regions subject to prolonged periods of cool and wet conditions (Fig. 6.22) and partial snow coverage. Microdochium patch remains problematic to effectively manage where it occurs, especially as cases of fungicide resistance have been documented in certain countries.

Causal agent

Monographella nivalis (formerly *Microdochium nivale* and *Fusarium nivale* then later changed to *Gerlachia nivalis*) is the causal agent, hence the

Fig. 6.22. Severe infestation of microdochium patch on an aged golf green in southern Australia. (Courtesy of Gary W. Beehag.)

disease name. *M. nivalis* is an opportunistic, cold-tolerant (psychrotolerant–mesophilic) fungus capable of growth at near freezing temperatures. The fungus is characterized by light pink-coloured mycelium and banana-shaped conidia which is why it was first thought to be a species of *Fusarium*.

M. nivalis comprises various fungal strains each differing in temperature tolerance, morphology and virulence. Confirming the identity and proving pathogenicity of individual fungal isolates from swards expressing symptoms consistent with microdochium patch is challenging. *M. nivalis* can occur singly or simultaneously with other cool-temperature pathogens (e.g. *Rhizoctonia* spp. and *Typhula* sp.) raising the possibility of a patch–disease complex (Fig. 6.23) depending on fungal isolates and environmental conditions.

M. nivalis appears to survive from year to year generally within the same locations. Conidia have been shown to remain viable in soil for extended periods under highly favourable temperatures (Tronsmo *et al.*, 2001). The fungus may survive and be spread on infected seed and is thus possible to be also seed-borne.

M. nivalis should not be confused with the cool-weather fungus *Microdochium bolleyi* nor two relatively newly described, warm-weather pathogens: *Microdochium paspali* (sparse leaf patch) on seashore paspalum (Zhang *et al.*, 2015b) or *Microdochium poae* (leaf blight) on bentgrass and bluegrass (Liang *et al.*, 2019).

Geographic distribution

Microdochium patch is widely distributed throughout the subarctic and temperate regions in the northern hemisphere and the subalpine and temperate regions in the southern hemisphere (Table 6.8). Microdochium patch came to prominence as an increasingly common and problematic turfgrass disease in the early 1900s (Smith *et al.*, 1989).

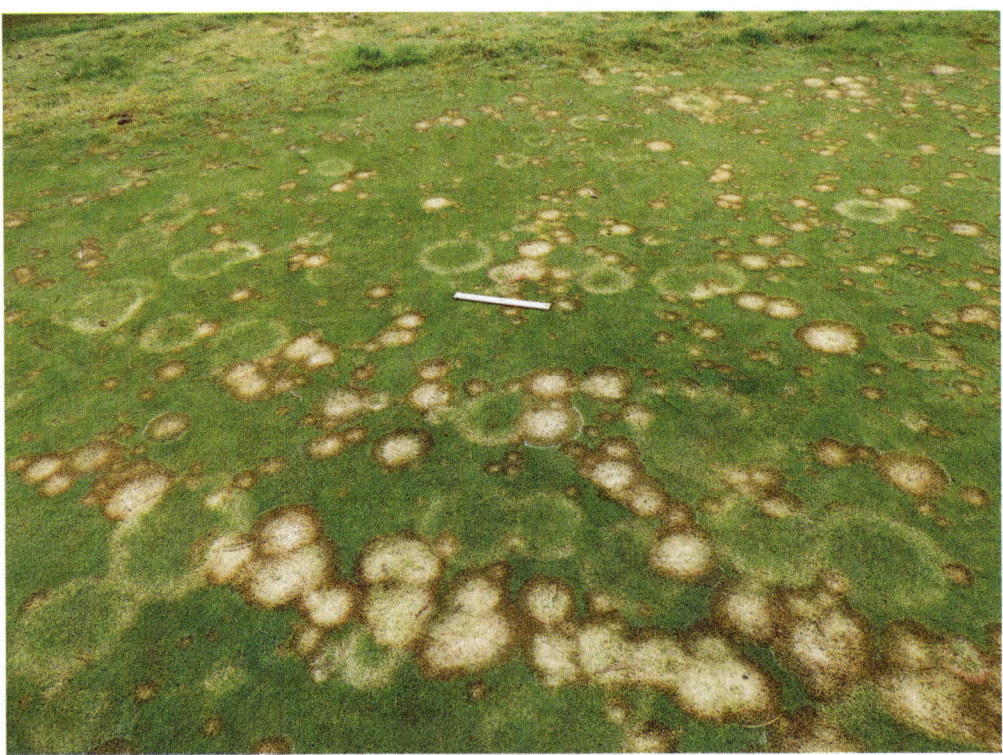

Fig. 6.23. Microdochium patch and *Rhizoctonia* sp. symptoms on a mixed cool-season grass golf green in southern Australia. (Courtesy of Gary W. Beehag.)

Host range and susceptibility

M. nivalis has an extensive host range of common cool-season and certain warm-season grasses. Annual bluegrass can be severely affected due to the loss of coverage during highly favourable conditions for the disease (Mann, 2011; Tredway *et al.*, 2023).

Signs and symptoms

The presence of specialized fruiting bodies (sporodochia) protruding from infected leaves and white- to salmon-pink-coloured, cobweb-like growth of fungal mycelium can be characteristic of microdochium patch (Fig. 6.24) depending on environmental conditions.

Symptoms of microdochium patch vary widely in the size, distribution and coloration of patches and their frequency and severity, depending on host species and environmental conditions, particularly the duration and frequency of wet and cool, freely available moisture or snow events. In the absence of prolonged snow, early-stage symptoms may appear as a limited number of relatively small, roughly circular spots 2.5–5.0 cm in diameter (Fig. 6.25). Individual, randomly scattered spots initially may contain water-soaked leaves that change colour from yellow, orange-brown to dark reddish-brown, possibly turning light grey or tan, and may eventually increase in size, typically no greater than 20 cm. In other cases, microdochium patch

Table 6.8. Documented global distribution of microdochium patch.

Region/country	References
Canada	Charbonneau and Hsiang (2015)
United States	Tredway *et al.* (2023)
Scandinavia	Kvalbein *et al.* (2017)
Western Europe	Mann and Newell (2005)
United Kingdom	Mann (2004)
East Asia	Tani and Beard (1997)
Australia	Woodcock (1983)
New Zealand	Howard (1986)
South Africa	Machielse (1968)

Fig. 6.24. Pink-coloured mycelium of *Monographella nivalis* (*Microdochium nivale*) on turfgrass in Canada. (Courtesy of Tom Hsiang.)

Fig. 6.25. Microdochium patch on a colonial bentgrass golf green in New Zealand. (Courtesy of Gary W. Beehag.)

symptoms may be more widespread (Fig. 6.26) depending on environmental conditions.

Disease progression results in enlarging patches that can coalesce to form relatively larger-sized unsightly and disfiguring patches. Late-stage patches may develop in 48–72 h from onset of initial symptoms depending on conditions and susceptibility of the host species. The presence of pinkish-coloured mycelium (Fig. 6.27) may be temporary and most noticeable during early morning. Infected foliage at the margins of late-stage patches may appear matted.

On closely mown turfgrass (e.g. bowling and golf greens) certain patches may display a concentric-ring or frog eye appearance indicative of recovering foliage behind active fungal growth (Fig. 6.28). Left unmanaged, patches may be colonized by plant species not affected by the disease.

Disease profile

Microdochium patch is the most common fungal disease in cold regions and the biology, ecology and management have been widely documented in the United Kingdom, Western Europe and North America (Smith *et al.*, 1989; Couch, 1995; Tredway *et al.*, 2023).

The fungus has been described as opportunistic and widely reported surviving unfavourable summer conditions on decomposing plant tissues in the thatch-mat region or as dormant mycelium inside infected hosts. Reports of the fungus (mycelia, conidia and ascospores) present on decomposing plant tissues and the presence of haustoria (fungal structures that are used to absorb plant cell nutrients) on cereals circumstantially support the possibility that the fungus may assume an alternative life stage as a saprophytic to slightly parasitic or a hemibiotrophic lifestyle (Hofgaard *et al.*, 2008).

Microdochium patch epidemics in the northern hemisphere occur under highly favourable conditions during spring and autumn. Infection by the fungus is generally considered to occur by germinating conidia and mycelium disseminated from infected soil and plant tissues. Symptoms may occur within 2 days following fungal penetration. Massed mycelium during prolonged overcast and cool conditions may occur but may gradually disappear under drying conditions of direct sunlight and wind. Disease outbreaks outside spring and autumn are triggered by snowfall and thawing events.

Environmental factors

Several key environmental and management-induced factors have been shown to specifically influence development of microdochium patch (Box 6.17).

Fig. 6.26. Microdochium patch symptoms on a partly shaded, creeping bentgrass golf green apron in southern Australia. (Courtesy of Gary W. Beehag.)

TEMPERATURE AND MOISTURE Temperature and moisture in combination are the key factors that influence the occurrence and frequency of microdochium patch dependent on the country in which the disease occurs. Host infection and disease progression occurs most rapidly around

Fig. 6.27. Microdochium patch symptoms on creeping bentgrass golf green showing 'white-coloured' and 'pink-coloured' mycelium. (Courtesy of Gary W. Beehag.)

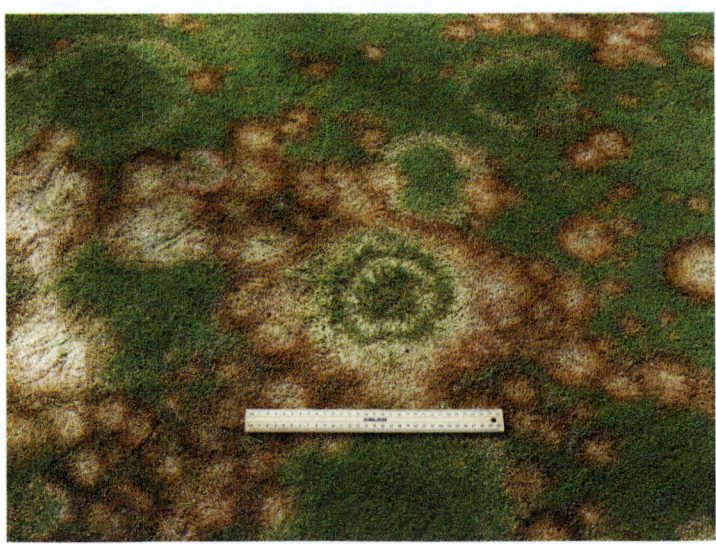

Fig. 6.28. Spectacular microdochium patch 'frog eye' symptoms on an aged creeping bentgrass golf green in southern Australia. (Courtesy of Gary W. Beehag.)

Box 6.17. Key environmental factors of microdochium patch.

- Cool to cold temperatures (8–20°C)
- Humid conditions (> 90% for more than 24 h)
- Long periods of leaf wetness
- Excessive nitrogen

0–8°C but may occur up to 18°C with increased humidity levels and prolonged leaf wetness (Smith, 1987; Couch, 1995; Smiley et al., 2005) caused by frequent, light rainfall events and prolonged foggy conditions. Under glasshouse-controlled conditions, infection on creeping bentgrass by an American isolate of *M. nivalis* (*M. nivale*) ceased at 32°C (Endo, 1963).

Tolerance of fungal strains at near-freezing conditions and a relatively wide infection temperature range largely explain the widespread distribution of the disease in temperate and subarctic regions subject to variations in snow cover. The biological reason explaining how *M. nivalis* has activity at near freezing temperatures allowing for its growth remains unclear. Physiological changes to the host at lower temperatures may also be involved.

LIGHT   Exposure to sunlight induces the production of *M. nivalis* spores in a pink-coloured matrix of mycelium and sporodochia in the turfgrass canopy (Fig. 6.29).

HOST NUTRITION   The relationship between turfgrass nutrition and host susceptibility to microdochium patch has been extensively documented. Generally, most studies have shown that imbalanced fertilizer or soluble nitrogen applications at more than normal requirements coinciding with falling temperatures in autumn increase the disease. Insufficient cold acclimatization (hardening) because of carbohydrate modification in excessively fertilized turfgrass has been proposed as an explanation of lower tolerance to the disease (Smith *et al.*, 1989).

Conversely, most studies have shown application of potassium, phosphorus and iron sulfate decreased microdochium patch (Mann, 2011; Mattox *et al.*, 2017). Direct association between rootzone pH and incidence of microdochium patch is unclear. Most studies have generally shown alkaline rootzones resultant of alkaline fertilizers or liming materials cause an increase in microdochium patch (Handoll, 1966) and application of sulfur-containing products reduces the incidence of the disease (Mattox *et al.*, 2017). However, it remains unclear if suppression of microdochium patch was because of the fungicidal or acidifying effects of sulfur.

Integrated management

The complex epidemiology of microdochium patch on high-value turfgrass necessitates an integrated approach (Box 6.18). Regular monitoring of turfgrass health and weather conditions and the use of disease-resistant varieties in conjunction with proven cultural, biological and fungicidal inputs are required for effective management of the disease. Appropriate and careful programming of registered fungicides to minimize potential of fungicide resistance is a necessity given the high risk of fungicide resistance.

Fig. 6.29. *Monographella nivalis* spores in a turfgrass canopy in Canada. (Courtesy of Tom Hsiang.)

Disease-resistant varieties

Genetic resistance against microdochium patch has been documented among creeping bentgrass, colonial bentgrass and fine fescue cultivars (Gregos *et al.*, 2011; Aamlid *et al.*, 2012).

Ophiosphaerella-Incited Diseases

Ophiosphaerella spp. are slow-growing, facultative pathogens characterized by dark brown- to black-coloured mycelium that infect the stems and roots of cool-season and warm-season grasses worldwide. *Ophiosphaerella*-incited diseases are problematic to effectively manage once established, requiring a concerted, integrated approach combining cultural practices and timely application of registered fungicides. Diseases such as spring dead spot, necrotic ring spot, dead spot and others are all caused by species in the genus *Ophiosphaerella*. Each *Ophiosphaerella* species and strain varies in virulence depending on the host–pathogen combination and environmental conditions. *Ophiosphaerella* spp. survive adverse environmental conditions in colonized tissues or by spores.

The genus *Ophiosphaerella* (some were formerly *Leptosphaeria*) was reclassified based on DNA sequences that compare numerous isolates collected from various countries (Tredway *et al.*, 2009). Several different *Ophiosphaerella* species cause disease in temperate and tropical grasses (Table 6.9). As a group, *Ophiosphaerella* species grow over a relatively wide temperature range, but each species has an optimum temperature range for host infection. Certain *Ophiosphaerella* species and other root-infecting fungi may form a root–rot disease complex of warm-season grasses, as listed separately in this chapter.

Spring dead spot

Spring dead spot is a destructive and problematic disease once established on closely mown bermudagrass (couchgrass) managed in locations where environmental conditions induce cold-temperature dormancy of the grass. What partly distinguishes spring dead spot from other soil-borne diseases is there are three causal agents, and the disease may be observed during autumn with symptoms obvious in spring and damage persisting into the summer season. Classic symptoms of spring dead spot may appear as large dead spots 5–15 cm in diameter or, more commonly, as patches, 20–100 cm diameter (Fig. 6.30) depending on environmental conditions not fully understood.

Spring dead spot represents one of the longest challenges undertaken by turfgrass pathologists. The disease name was originally ascribed in the United States to advanced symptoms observed in spring on a bermudagrass as non-diseased grass resumed growth. In southern Australia, where the disease is also called spring dead patch, early reports described symptoms

---

Box 6.18. Integrated management factors of microdochium patch.

- Disease-resistant cultivars
- Monitoring turfgrass health and early disease symptoms
- Maintain consistent growth using balanced NPK nutrition
- Minimize leaf wetness duration
- Minimization of thatch-mat thickness
- Removal of fallen leaves in autumn
- Avoid maintaining high mowing height late in the season
- Judicious irrigation water applications
- Foliage application of registered fungicides and biofungicides

---

Table 6.9. *Ophiosphaerella*-incited diseases and their causal agents by host group.

Ophiosphaerella species	Disease name	Host group	References
O. agrostidis	Dead spot	C3/C4	Tredway *et al.* (2009);
O. korrae	Necrotic ring spot	C3	Kerns and Tredway
O. herpotricha	Spring dead spot	C4	(2013)
O. korrae	Spring dead spot	C4	
O. narmari	Spring dead spot	C4	

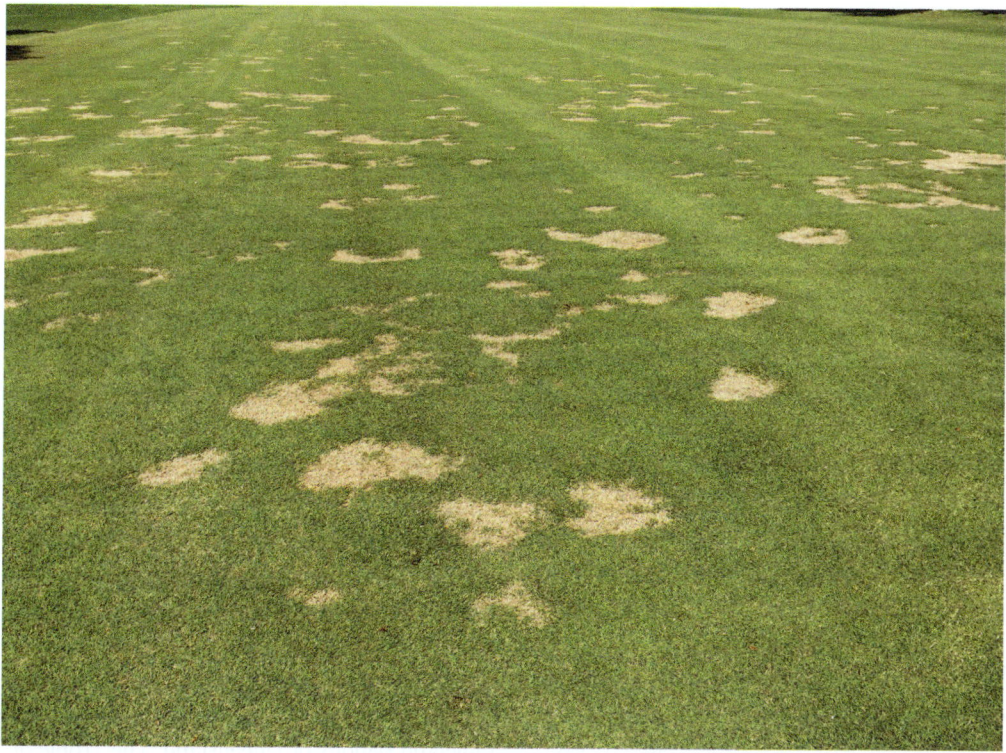

Fig. 6.30. Spring dead spot on a bermudagrass fairway in the United States. (Courtesy of Nathan R. Walker.)

appearing as early as autumn and pronounced in spring as well-defined, circular dead patches. Effective management of spring dead spot requires an integrated programme of proven cultural practices and correctly timed applications of efficacious fungicides to minimize the disease.

Causal agent

Spring dead spot is a disease complex caused by one of three fungal species in the genus *Ophiosphaerella* (formerly *Leptosphaeria*): *Ophiosphaerella herpotricha*, *Ophiosphaerella korrae* and *Ophiosphaerella narmari*. Isolation and identification of the causal agents of spring dead spot were resolved following considerable research effort and debate over many decades among American and Australian researchers. Reclassification from *Leptosphaeria* to *Ophiosphaerella* was based on morphological differences and DNA sequencing of numerous fungal isolates collected throughout the United States and from Australia. The genus name *Ophiosphaerella* might change in the future once the type species for the genus is rediscovered and genetically studied.

The fact that different *Ophiosphaerella* fungal species cause spring dead spot has practical implications in the management of the disease because of differences in virulence between fungal species and given the fact more than one *Ophiosphaerella* sp. has been shown to be present in the same turfgrass site.

Geographic distribution

Verified distribution of spring dead spot on bermudagrass continues to widen globally (Table 6.10). Spring dead spot has long been recognized throughout the transition zone of the United States and southern mainland Australia. The disease has been more recently reported in the North Island of New Zealand and in certain East Asian, South American and southern European countries. Due to climate change and water restrictions, global distribution of spring dead spot will inevitably

Table 6.10. Documented global distribution of spring dead spot and their causal agents.

Region/country	*Ophiosphaerella* species	References
South America (Argentina)	*O. herpotricha*	Luc et al. (2005)
Southern Europe (Italy)		Gullino et al. (2007)
United States		Tredway et al. (2023)
Australia	*O. korrae*	Walker and Smith (1972)
United States		Tredway et al. (2023)
Australia	*O. narmari*	Hawkes (1987)
New Zealand		Walker and Smith (1972)
United States		Tredway et al. (2023)
East Asia (China)		Geng et al. (2021)

increase given transport of *Ophiosphaerella*-infected, vegetatively propagated bermudagrass cultivars and their utilization into areas traditionally known to grow cool-season grasses.

THE AMERICAS Symptoms of the disease were first reported in the United States in the 1930s, but the name spring dead spot was first used to describe symptoms found in 1960 (Wadsworth and Young, 1960). In the United States, the disease is distributed from California in the west, throughout the mid-western states across and into the central and upper east regions of the country. However, distribution of the disease has in more recent decades has been extended further northwards in the central United States as utilization of bermudagrass expands northwards.

All three species of *Ophiosphaerella* have been isolated from infected bermudagrass in the United States but the distribution of the causal agents varies widely between regions. In the mid-west, the principal pathogens are *O. herpotricha* and *O. korrae*. *O. korrae* is predominant in California and the south-western United States.

O. narmari is less prevalent in the United States but is found in California, Kansas, Oklahoma and North Carolina (Tredway et al., 2009). The genetic diversity among *Ophiosphaerella* isolates suggests *O. herpotricha* is native to the United States (Iriarte et al., 2004).

AUSTRALASIA Unconfirmed symptoms were reported on couchgrass (bermudagrass) bowling greens in northern New South Wales in the mid-1940s (Ryan, 1986) but the accuracy of the diagnosis remains unknown. Symptoms consistent with spring dead spot in the United States became known in Australia from the early 1960s (Fig. 6.31). *O. herpotricha* (formerly reported as *Ophiobolus herpotricha*) was the first species reported to cause spring dead spot symptoms during the mid-1960s in Sydney (New South Wales) (Smith, 1965).

Observations and limited isolations indicate that spring dead spot occurs in south-west Western Australia, south-eastern South Australia, throughout Victoria up to the Australian Capital Territory throughout southern, central and coastal New South Wales, and into coastal and tableland regions of Queensland. The Tropic of Capricorn (approximately 24°S) delineates approximately the northern limit of couchgrass dormancy along coastal Queensland below which spring dead spot generally occurs. *O. korrae* and *O. narmari* have been isolated throughout southern mainland Australia (Walker and Smith, 1972; Plant Health Australia, 2001). The precise distribution of spring dead spot and the causal pathogens throughout Australia have not been documented.

In New Zealand, *O. korrae* was isolated at Palmerston North (North Island) by the late Dr Noel Jackson, USA (New Zealand Fungal and Plant Disease Collection). Characteristic symptoms of spring dead spot have been reported in recent years on golf courses in the North Island but verification of the pathogen(s) and the distribution of the disease in New Zealand remain uncertain (D. Ormsby, New Zealand, 2020, personal communication).

WESTERN EUROPE AND EAST ASIA The spring dead spot pathogen *O. korrae* has been identified from infected bermudagrass in central and southern Italy (Titone et al., 2004; Gullino et al., 2007). *O. narmari* has recently been isolated from symptomatic

Fig. 6.31. An early photograph of spring dead spot on a common couchgrass bowling green in southern Australia. (Courtesy of the late Dermott Reilly.)

bermudagrass in southern China (Geng *et al.*, 2021). The reasons accounting for the distribution of spring dead spot in Western Europe and East Asia on bermudagrass are unknown but possibly may be attributed to the utilization of vegetatively propagated infected grasses.

Host range and susceptibility

Common bermudagrass ecotypes (*Cynodon dactylon*), South African (*Cynodon tranvaalensis*) and interspecific hybrids of *C. dactylon* × *C. tranvaalensis* are the principal hosts of spring dead spot. Susceptibility among bermudagrass cultivars to spring dead spot and the causal species vary widely (Baird *et al.*, 1998; McCarty and Miller, 2002). Zoysiagrass is also a known symptomatic host that can be infected by *O. korrae* and typically occurs when the grass is extensively fertilized.

Centipede grass, kikuyugrass, seashore paspalum, St. Augustinegrass (also called buffalo grass) and tropical carpetgrass have been diseased to varying degrees by *Ophiosphaerella narmari* in several Australian mainland states (Walker and Smith, 1972). The full turfgrass host range of the spring dead spot pathogens is unknown.

Signs and symptoms

The presence of dead patches in the spring, dead plants that have darkened stems and black-coloured runner hyphae and fruiting bodies or pseudothecia (Fig. 6.32) are typically characteristic of the spring dead spot pathogens. Symptomatology of spring dead spot on bermudagrass varies widely depending on duration and severity of winter dormancy, management intensity, and differential fungus pathogenicity and cultivar susceptibility.

Under intensive, low mowing on bowling and golf greens, indiscrete symptoms may initially develop in the previous autumn or very early spring and persist through to late summer depending on environmental conditions. Individual blighted spots or patches are often roughly circular (Fig. 6.33) but may vary in size and shape ranging from 10 cm up to 1.0 m or more in diameter.

Late-stage symptoms of spring dead spot become most obvious and characteristic during spring green-up. Adjacent patches may coalesce forming extensive damaged areas of odd-shaped or doughnut-like shapes of straw-coloured and dead regions impacting on surface aesthetics and playing quality. Generally, individual patches reappear in the same location year after year, some may increase in size, produce frog eye symptoms or diminish.

Disease profile

Spring dead spot is an intensively studied turfgrass disease because of its impact on golf courses primarily in the United States and worldwide utilization of an ever-increasing number of bermudagrass cultivars. Scientific understanding of the spring dead spot pathosystem has been challenging because of the co-occurrence of three fungal species and a primary host that undergoes winter dormancy.

The biology, ecology and epidemiology of spring dead spot on bermudagrass have been extensively documented in the United States (e.g. Walker *et al.*, 2006; Tredway *et al.*, 2009) and to a much lesser degree in Australia (Smith, 1971; Hawkes, 1987; Robinson *et al.*, 1995). None the less, greater understanding of certain aspects of the biology of the causal agents and epidemiology of the disease continues to attract scientific investigation.

Spring dead spot is monocyclic (i.e. one disease cycle per year). Initial infection and early-stage

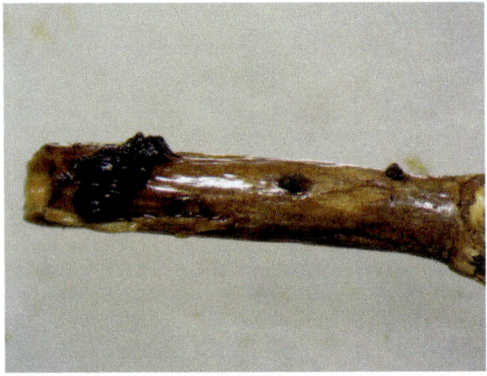

Fig. 6.32. Black-coloured fruiting bodies of an *Ophiosphaerella* sp. on a couchgrass (bermudagrass) stem. (Courtesy of Percy T. Wong.)

Fig. 6.33. An individual spring dead spot symptom on semi-dormant couchgrass in Australia. (Courtesy of Gary W. Beehag.)

symptoms may occur in the preceding autumn in certain geographic regions coinciding with falling temperatures. Advancing and late-stage symptoms of larger-sized, blighted spots and patches appear in late spring coinciding with resumption of healthy grass growth. Progression of spring dead spot is relatively slow in autumn on infected turfgrass and has been reported most severe on mature swards that are 3 years old or more.

Pathogen activity occurs during cool conditions that are not favourable for host growth and are favourable for fungal infection and colonization in late autumn or early spring; typically, once soil temperatures are below 23°C for several days at a depth of 4.0 cm. Commencing from colonization by *Ophiosphaerella*, the hyphae occur along stems and roots coinciding with a period of reduced host growth providing an ecological advantage to the invading pathogen.

Ophiosphaerella species produce finer hyphae reported to penetrate intact and damaged stem tissues. Different mechanisms of fungal colonization of roots and stems and subsequent infection of vascular root tissues among bermudagrass species may be associated with tolerance or susceptibility towards spring dead spot (Cassi *et al.*, 2011). Damage by spring dead spot pathogens becomes most apparent in late spring with necrotic or dead patches contrasting against green-coloured regions of active normal growth (Fig. 6.34).

Diseased patches are perennial and will recur, increasing in size each succeeding year, and new symptomatic areas may appear. Infected stems that survive cold-temperature dormancy can occur outside the margins of dead patches. Infected bermudagrasses are less tolerant to winter temperatures; thus undergoing greater damage and resulting in death of the plant. Diseased patches do not increase in diameter through summer. This widely reported comment gave rise to the view for the presence of a fungal toxin (Fermanian *et al.*, 1981). Fungal metabolites have been shown to be synthesized by *Ophiosphaerella* spp. grown in culture media (Venkatasubbaiah *et al.*, 1994). However, the use of root-growth-inhibiting herbicides is most likely the result of field observations.

The precise biological mechanism by which the invading pathogens cause tissue necrosis of susceptible bermudagrasses remains unclear. However, results of one study (Flores *et al.*, 2017) suggest root necrosis could be caused by a combination of non-specific toxins produced by the fungi or reactive oxygen species (ROS) produced by the plant that results in independent cell death by the host in an effort to stop pathogens progressing.

Fig. 6.34. Late-stage symptoms of spring dead spot on semi-dormant couchgrass in Australia. (Courtesy of Percy T. Wong.)

Environmental influences

Several key environmental and management-induced factors are known to be associated with spring dead spot on closely mown bermudagrass (Box 6.19). Low atmospheric and rootzone temperatures are a key determinant factor of occurrence and severity of the disease.

TEMPERATURE The spring dead spot pathogens are able to cause disease across a relatively wide range of low temperatures as confirmed by glasshouse-controlled studies (Walker et al., 2006; Tredway et al., 2009). These temperature ranges typically occur during early autumn and late spring in temperate regions. Infection and colonization of bermudagrass roots by O. herpotricha was found to occur in the United States in spring and autumn (Walker et al., 2006) and suggests the pathogen has two distinct periods of activity. In Australia, infection and initial symptoms of spring dead spot caused by O. narmari are commonly observed in autumn (Walker, 1975).

The contrasting temperature optima of less than 21°C for fungal infection (Endo et al., 1985) and 24–35°C for optimal bermudagrass root growth (McCarty and Miller, 2002) support the view that bermudagrass root growth during warmer temperatures is faster than the ability of the pathogen to injure plant organs. The invading pathogen has a competitive advantage to injure plant roots at the lower temperature over the slow-growing host (Tredway et al., 2009).

Reduced soil temperatures causing root senescence or dieback of bermudagrass (Sifers et al., 1985) have been suggested as a factor in the occurrence of spring dead spot because of reduced normal root function (Smith et al., 1989). This scenario is consistent with a growth chamber study which showed plants infected by O. korrae and O. herpotricha recorded less tolerance to freezing temperatures (Nus and Shashikumar, 1993).

SOIL MOISTURE AND NUTRITION Maintaining adequate moisture enhances a functional root system and more efficient nutrient absorption. Highly fertilized bermudagrass has long been observed as a factor associated with spring dead spot in the United States and southern Australia (Kozelnicky, 1974; Hawkes, 1987). Adequately fertilized bermudagrass aids recovery from the disease during summer but excessive nitrogen may contribute to disease development in late autumn because it promotes plant growth during those cool temperatures favourable for disease when the plants should be entering cold-temperature-induced dormancy. Field studies conducted in the United States have shown that certain sources of soluble nitrogen applied throughout summer in excess of normal turfgrass requirements exacerbate the disease (Vincelli and Williams, 1998).

Applications of soluble potassium during the cooler periods have been indirectly associated with reduced spring dead spot. Most studies have shown soluble potassium sources directly increase bermudagrass cold tolerance (Vincelli and Williams, 1998), thus indirectly resulting in less spring dead spot. In contrast, other studies showed autumn applications of potassium sulfate resulted in no change or increased spring dead spot (Tredway et al., 2009).

THATCH-MAT MANAGEMENT Thatch-mat accumulation on bermudagrass has long been reported associated with spring dead spot on golf greens in the United States (Lucas, 1980) and on bowling greens in Australia (Martin, 2000). Excessive thatch-mat accumulation causes irreversible adsorption of soil-applied pesticides (Niemczyk and Krueger, 1980); thus reducing fungicide movement and concentration in the rootzone.

Integrated management

Effective management of spring dead spot requires a concerted, integrated approach (Box 6.20). Frequent monitoring of early disease symptoms in conjunction with proven cultural and biological technologies and timely, targeted applications of registered fungicides is required to minimize the intensity of the disease. There are no commercialized, microbial inoculants with proven or predictive long-term suppression against spring dead spot.

Box 6.19. Key environmental factors of spring dead spot.

- Cool to cold temperatures (around 23°C)
- Low soil moisture
- Potassium and nitrogen
- Thatch-mat accumulation

Disease-resistant cultivars

Variable disease resistance to spring dead spot among seeded and vegetatively propagated bermudagrass cultivars has been demonstrated experimentally in the United States (Baird *et al.*, 1998; Martin *et al.*, 2001). Observations in Australia have also indicated variable resistance to the disease among natural ecotypes and interspecific couchgrass hybrids. Research in North America has suggested that cultivars with greater cold-tolerance are more tolerant to the disease. Patriot and Riviera bermudagrass cultivars developed at Oklahoma State University (USA) with good cold tolerance are still very susceptible; however, south-western (Arizona) types with less cold tolerance are even more susceptible to the disease.

Necrotic ring spot

Necrotic ring spot is widespread throughout the northern United States and regions of southern Canada during summer on cool-season grasses. Necrotic ring spot became recognized as an economically important disease in North America in the early 1980s on Kentucky bluegrass cultivars. The disease was given the name necrotic ring spot when application of the fungicide triadimefon was unsuccessful in controlling another patch disease (Worf *et al.*, 1986). Symptoms of necrotic ring spot can resemble other similar, patch-type diseases (e.g. summer patch) on cool-season grasses except for the season of occurrence.

Causal agent

O. korrae is the causal agent of necrotic ring spot (Smiley *et al.*, 2005). Much of what is known about the identity and biology of *O. korrae* originates from what is known from spring dead spot on bermudagrass.

Geographic distribution and host range

Necrotic ring spot has been documented in North America and East Asia (Table 6.11). The global distribution of necrotic ring spot remains unknown.

In the United States, necrotic ring spot is widely distributed throughout the upper midwest, Pacific north-west and north-eastern regions. Necrotic ring spot occurs in southern Canada (Ontario and Quebec). Annual bluegrass, Kentucky bluegrass and fine fescues are hosts of necrotic ring spot (Turgeon and Kaminski, 2019).

Signs and symptoms

The presence of dead patches in summer with dark black stems and the possible presence of black-coloured fruiting bodies (pseudothecia) may indicate the presence of the necrotic ring spot pathogen. Extent and rate of symptom expression vary depending on level of cultural intensity of the turfgrass and seasonal weather conditions. Initial symptoms appear as relatively small-sized, chlorotic patches ranging 15–30 cm in diameter (Fig. 6.35).

Box 6.20. Integrated management factors of spring dead spot.

- Disease-resistant bermudagrass cultivars
- Monitor for signs of key fungal structures
- Minimization of thatch-mat thickness
- Encourage and maintain a highly functional root system
- Maintain consistent growth based on balanced NPK nutrition
- Maintain slightly acidic pH
- Judicious irrigation water applications
- Use of registered fungicides as soil temperatures approach 23°C
- Apply adequate post-application irrigation water
- Avoid or minimize pre-emergent herbicides in early spring

Table 6.11. Documented global distribution of necrotic ring spot.

Region/country	References
United States	Jackson (1993); Tredway *et al.* (2023)
Southern Canada	Charbonneau and Hsiang (2015)
East Asia (Japan)	Hayakawa *et al.* (2007a)

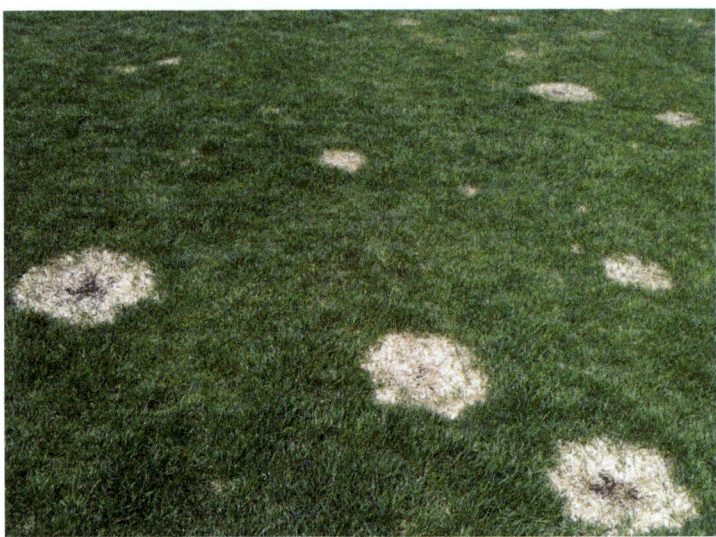

Fig. 6.35. Necrotic ring spot symptoms on Kentucky bluegrass in northern United States. (Courtesy of Nathan R. Walker.)

Disease profile

Necrotic ring spot is considered primarily problematic on relatively young turfgrass less than 5 years old. The pathogen survives as mycelium on host roots or as spores during unfavourable conditions. Necrotic ring spot may appear in spring, late summer and into autumn. The period from infection to initial disease expression may take a year or more probably due to the slow growth of the pathogen. Necrotic ring spot is most prominent during summer associated with heat and drought stress of the host grass.

Symptoms may become less obvious with onset of cooler temperatures and infected grass may recover. Infected leaves change colour to bronze or purple that become brown and bleached as plants die. Certain infected plants may partially recover forming a doughnut-ring or frog eye appearance. Late-stage symptoms may see a series of coalescing rings and patches over larger areas depending on seasonal temperatures.

Environmental influences

Several key environmental and management-induced factors appear to be strongly associated with necrotic ring spot on Kentucky bluegrass (i.e. cool temperatures, moisture and nutrition).

Necrotic ring spot has been demonstrated experimentally to develop severely on inoculated bluegrass turfgrass between 24 and 28°C (Worf *et al.*, 1986). Necrotic ring spot occurs over a relatively wide range of rootzone moisture conditions (Landschoot *et al.*, 1993).

Integrated management

Effective management of necrotic ring spot requires an integrated approach. Frequent monitoring of disease signs and changing weather patterns combined with proven cultural and fungicidal options are required (Box 6.21). Resistance to necrotic ring spot among Kentucky bluegrass cultivars has been reported (Melvin and Vargas, 1994).

Box 6.21. Integrated management factors of necrotic ring spot.

- Maintaining acceptable thatch-mat thickness and aeration
- Maintain consistent growth with use of a balanced NPK nutrition
- Judicious irrigation water application
- Use of registered fungicides
- Use of improved Kentucky bluegrass cultivars where appropriate

Dead spot

Dead spot emerged rapidly on creeping bentgrass golf greens during the late 1990s in the north-east United States. The disease was associated with relatively newly constructed greens of near-pure sands or with mature greens previously fumigated with methyl bromide. The name was changed from bentgrass dead spot (acknowledging the primary host) to dead spot, accounting for the pathogen being additionally isolated on symptomatic bermudagrass (Kerns and Tredway, 2013).

In recent years, the disease has declined in occurrence in the United States. Consequently, the precise epidemiological factors underlying its initial emergence and apparent decline and predictable cultural, biological and fungicidal management practices remain unknown given the relatively short history of the disease.

Causal agent

Ophiosphaerella agrostidis (formerly *Ophiosphaerella agrostis*) is the causal agent of dead spot (Turgeon and Kaminski, 2019). The fungus has multiple strains which grow over a wide temperature range.

Geographic distribution and host range

Dead spot has been reported in North America and East Asia (Table 6.12). The global distribution of dead spot is unknown. Distribution of dead spot became widespread during the early 2000s throughout the United States. The disease was reported in the north-east, mid-west and mid-Atlantic regions from Michigan in the north, Missouri to the west and along the eastern seaboard from Massachusetts to Georgia. Dead spot has occurred on bermudagrass in Florida, South Carolina and Texas (Kerns and Tredway, 2013).

Bentgrass, rough bluegrass and hybrid ultradwarf bermudagrass are considered to be the primary hosts. All creeping bentgrass cultivars and certain colonial and velvet bentgrasses are susceptible with varying degrees to dead spot. 'Champion' and 'Tifdwarf' bermudagrass are susceptible to dead spot (Kaminski and Dernoeden, 2005) under highly favourable conditions.

Signs and symptoms

The presence of dark-coloured, flask-shaped fruiting bodies (pseudothecia) embedded along stems together with runner hyphae and discoloured stems and roots characterize dead spot. Pseudothecia are generally absent on fungicide-treated turfgrass. Late-stage symptoms of dead spot are relatively small-sized patches as opposed to the symptoms caused by other *Ophiosphaerella* species. Individual symptoms on closely mown bentgrass greens are reddish-brown, roughly circular spots initially around 1 cm and can enlarge up to 6–8 cm in diameter. Symptoms may appear in locations subject to the disease the previous year.

Disease progression causes infected foliage to turn tan-coloured with bronzed to reddish-brown-coloured edges and may resemble other spot-type fungal diseases. Infected plants are randomly scattered or localized on individual greens. Infected spots generally do not coalesce unless there is a severe epidemic and unlike dollar spot, morning foliar fungal mycelium does not occur. On golf greens, infected areas may form slight depressions. On closely mown bermudagrass golf greens individual spots range 3–10 cm in diameter and are tan or reddish brown in colour.

Disease profile

Dead spot is a polyclinic disease due to the ability of the fungus to undergo multiple cycles of spore

Table 6.12. Documented global distribution of dead spot.

Region/country	Hosts	References
United States	Creeping bentgrass	Kerns and Tredway (2013)
	Hybrid bermudagrass	Krausz *et al.* (2001)
Southern Canada	Creeping bentgrass	Charbonneau and Hsiang (2015)
East Asia (Japan)	Creeping bentgrass	Hayakawa *et al.* (2007b)

production over short time periods. The disease declines in severity over time. Dead spot occurred in North America from mid- to late summer through to mid- to late autumn even into early winter depending on geographic location and seasonal conditions (Kaminski and Dernoeden, 2002; Kaminski and Hsiang, 2006).

The pathogen survives unfavourable conditions during winter in infected tissues and as spores. Favourable environmental conditions result in the pathogen resuming growth and development when soil temperatures attain 20°C and relative humidity is less than 80%. Discharge and dissemination of large numbers of ascospores by wind and water-splash is triggered at dawn and dusk or by rainfall events.

Fungal spores can germinate under favourable conditions aided by leaf exudates and light within 2 h of attachment to host foliage or stems. Fungal penetration can be direct into host tissues or indirectly through open stomates. The dead spot pathogen appears to first infect foliage then progressively enters stem bases and stolons. Symptoms typically occur on hosts between 4 and 10 days following infection.

Dead spot is most severe on golf greens in exposed locations and on mounds or slopes in full sunlight subject to heat and drought stress. The incidence of dead spot reduces with onset of cool conditions in autumn until occurrence of frosts coinciding with reduced temperatures in early winter. Dead spot undergoes a decline in symptom expression 1–3 years after the first year of occurrence. Antagonistic microbial competition has been suggested as an explanation in the disease decline (Kaminski, 2008) but remains unknown.

Environmental influences

Several environmental and management-induced factors have been shown to specifically influence the development of dead spot (Box 6.22).

TEMPERATURE AND MOISTURE Host infection and development of dead spot resumes when atmospheric temperatures increase to around 20–30°C coinciding with early summer. Symptoms generally do not occur at atmospheric and soil temperatures less than 15°C (Kaminski *et al.*, 2007). In the absence of limiting factors, the pathogen potentially has a physiological advantage over

---

Box 6.22. Key environmental factors of dead spot.

- Moderate temperatures (20–30°C)
- Low relative humidity (< 80%)
- Heat and drought stress
- Shortened leaf wetness (< 12 h)
- High solar radiation
- Nitrogen and pH

---

bentgrass, as opposed to bermudagrass, during summer given its temperature requirements of growth.

Other factors aside from temperature are implicated in the incidence and severity of dead spot. Increased development of pseudothecia in Maryland (mid-Atlantic) from June, in comparison to the observed occurrence on infected bermudagrass greens in Florida and Texas in March, suggests photoperiod and accumulation of heat may impact the development of pseudothecia depending on seasonal conditions. A multiparameter model based on atmospheric and soil temperature (greater than 27°C and 18°C respectively), low relative humidity (less than 80%), shortened periods of leaf wetness (less than 14 h) and high levels of solar radiation was developed for prediction of dead spot epidemics.

NUTRITION Soluble nitrogen sources and rootzone pH in the short term have been demonstrated experimentally to impact the incidence and recovery of plants infected by dead spot. Generally, ammonium-based (i.e. ammonium sulfate) as opposed to nitrate-based (i.e. calcium nitrate, sodium nitrate) fertilizers reduced dead spot severity (Kaminski and Dernoeden, 2005; Dernoeden, 2013). Long-term impacts of a combination of acidifying and alkaline-reacting products (e.g. lime) and elevated pH levels of irrigation water remain unknown.

Integrated management

Effective management of dead spot is complicated given the shortcomings in knowledge of the ecology and epidemiology of the disease. Regular monitoring of turfgrass health and weather variations, in conjunction with proven cultural and possible fungicidal options, is required for effective management of the disease (Box 6.23).

Disease-resistant varieties

Limited genetic resistance among creeping bentgrass cultivars against dead spot has been documented (Bonos and Huff, 2013).

Rhizoctonia- and Waitea-Incited Diseases

Rhizoctonia and *Waitea* are two closely related fungal genera displaying variable morphology, genetic diversity and virulence among species and strains, with wide host ranges. Multiple species and strains from both genera are associated with diseases on cool-season and warm-season grasses worldwide. Certain *Rhizoctonia* species are associated with cool and wet conditions and other species are commonly associated with warm to hot and humid conditions.

Symptom expression of *Rhizoctonia*- and *Waitea*-incited diseases shows many similarities and differences, depending on host–pathogen combination, nitrogen fertilization rates and environmental conditions (Fig. 6.36). Precise identification of certain *Rhizoctonia*- and *Waitea*-incited diseases can be problematic, complicating management efforts.

Box 6.23. Generalized management factors of dead spot.

- Disease-resistant cultivars
- Monitoring turfgrass health and early disease signs
- Maintain consistent growth using balanced NPK nutrition
- Minimize leaf wetness duration
- Minimization of thatch-mat thickness
- Judicious irrigation water applications
- Foliage application of registered fungicides and biofungicides

Rhizoctonia and Waitea nomenclature

Rhizoctonia solani was the first species of the genus recognized as a turfgrass pathogen. Laboratory identification of *Rhizoctonia* and *Waitea* species is initially based on morphological characteristics (i.e. mycelium growth rate and colour) and presence of right-angled branched mycelium. New *Rhizoctonia* and *Waitea* species and strains are now recognized as the causal agents, singly or in combination, under several taxonomic names

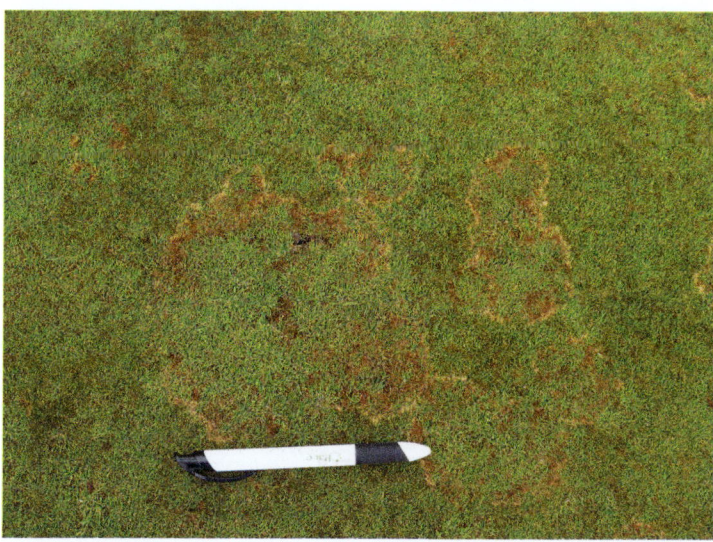

Fig. 6.36. Unusual patch disease symptoms attributed to a *Rhizoctonia* sp. on a bentgrass bowling green. (Courtesy of Nadeem Zreikat.)

(Stalpers et al., 2021). The binomial taxonomy of *Rhizoctonia* and *Waitea* species has undergone reclassification and may further change with refinements in genetic techniques (e.g. DNA sequencing).

Most *Rhizoctonia* species are categorized according to their anastomosis group (a laboratory assay, e.g. AG) and subgroups represented by genetically related isolates and the ability to self-fuse that occurs between fungal isolates of the same AG (Burpee and Martin, 1996).

Rhizoctonia and *Waitea* pathogens and their diseases

The identity and description of *Rhizoctonia* and *Waitea* species associated with turfgrasses have been extensively documented (Table 6.13) particularly in North America on creeping bentgrass. *Rhizoctonia*- and *Waitea*-incited diseases occur worldwide in tropical, subtropical and temperate regions on a wide range of cool-season and warm-season turfgrasses. *Rhizoctonia*- and *Waitea*-incited diseases are undoubtedly under-reported and misdiagnosed.

Temperature responses among *Rhizoctonia*/*Waitea* species

As a group, *Rhizoctonia* and *Waitea* species and strains have the ability to grow over a very wide range of temperatures and each species has an optimum temperature for growth and causing disease (Table 6.14). The highest risk of infection among almost all *Rhizoctonia* and *Waitea* species is at around 25–30°C (Kammerer and Harmon, 2008). Yellow patch is the exception, occurring at much cooler temperatures. The overlap of certain temperature ranges raises the possibility of more than one *Rhizoctonia* or *Waitea* pathogen occurring simultaneously, depending on environmental conditions, resulting in a disease complex.

Table 6.13. *Rhizoctonia* and *Waitea* pathogens and their diseases and host groups.

Rhizoctonia or *Waitea* species	Disease name	Host group	References
R. solani (AG-1-1A, 1-1B; AG-2-111B)	Brown patch	C3/C4	Wong and Kaminski (2007); Kammerer and Harmon (2008); Kammerer et al. (2011)
R. solani (AG-2-21V; AG-2-2LP)	Large patch	C4	
R. cerealis (AG-D)	Yellow patch	C3	
W. circinata	Brown ring patch	C3	
W. prodiga	Basal leaf blight	C3	
W. oryzae, W. zeae	Leaf and sheath spot	C3	

Table 6.14. Temperature optima of growth and infection of *Rhizoctonia* and *Waitea* species.

Disease name and pathogen	Optimal growth (°C)	Optimal infection (°C)	References
Brown patch *Rhizoctonia solani*	21–28	21–32	Burpee (1980); Burpee and Martin (1996); Green et al. (1993); Wong and Kaminski (2007); Kammerer and Harmon (2008); Kammerer et al. (2011)
Large patch *R. solani*	25	21–32	
Yellow patch *R. cerealis*	23	10–20	
Brown ring patch *Waitea circinata*	25–30	18–29	
Basal leaf blight *W. prodiga*	30	25–35	
Leaf and sheath spot *W. oryzae, W. zeae*	28–36	28–36	

Cool-weather *Rhizoctonia* diseases

Large patch

Large patch is a rapidly expanding disease once established in warm-season grasses (Fig. 6.37). The disease has become more widespread coinciding with increased utilization of warm-season grasses and recycled irrigation water in some countries. The disease is known under various names (e.g. zoysia patch, large brown patch or Rhizoctonia large patch) depending on the country in which the disease occurs.

Causal agent

R. solani (AG-2-2LP) is the causal agent of large patch (Hyakumachi and Hayakawa, 2008). Mature mycelium grown in culture varies in colour from off-white, grey to dark brown and sclerotia are coloured off-white to dark brown (Tredway *et al.*, 2023).

Geographic distribution

Large patch occurs throughout the transition zone of the United States (Kerns and Tredway, 2013) and in several countries in East Asia (Shim and Kim, 1995; Tani and Beard, 1997). Large patch has recently been confirmed on kikuyugrass on mainland Australia (J. Kappro, New South Wales, 2022, personal communication) (Fig. 6.38).

Host range

Zoysiagrass, hybrid bermudagrass, carpetgrass, centipede grass, St. Augustinegrass and seashore paspalum are hosts of large patch (Kerns and Tredway, 2013).

Signs and symptoms

The presence of soft, dark-coloured rotting to the lower sections of leaf sheaths can be an indication of large patch. Reddish-brown necrosis may be observed during relatively moist periods. Infected leaves are easily removed.

Symptoms of large patch caused by *R. solani* on warm-season grasses are distinctly different from those on cool-season grasses and may vary between turfgrass species and seasons (Tisserat *et al.*, 1994). Early-stage symptoms commencing occasionally in the autumn, but more often in spring, appear as individual, small-sized patches 50–75 mm in diameter, enlarging rapidly into crescent-shaped rings or fully formed,

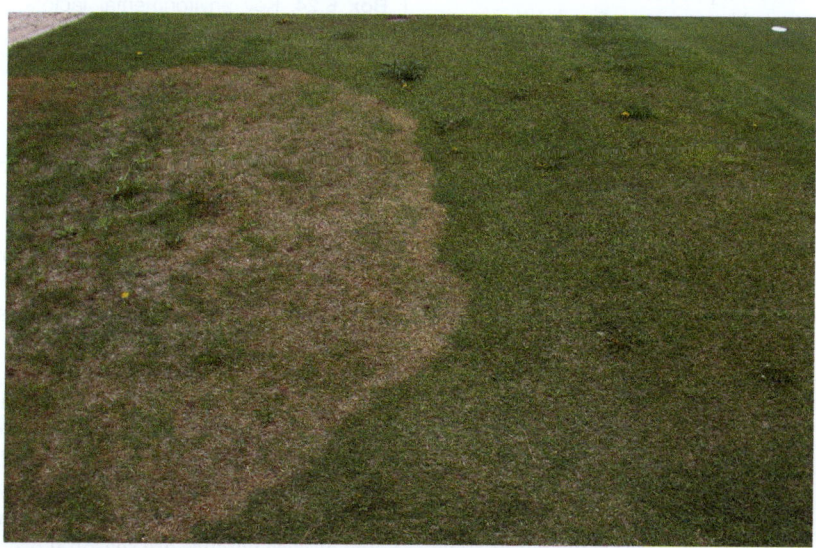

Fig. 6.37. Late-stage symptoms of large patch showing orange-fired perimeters on zoysiagrass in the United States. (Courtesy of Nathan R Walker.)

Fig. 6.38. Late-stage symptoms of large patch on a kikuyugrass sportsground in Australia. (Courtesy of Percy T. Wong.)

roughly circular-shaped patches, 1 m or more in diameter. Infected leaves range from chlorotic to necrotic. Leaf lesions and smoke rings are usually absent. Advancing patches may expand rapidly up to 15 cm per week under highly favourable conditions to coalesce forming large-sized, irregularly shaped areas.

Late-stage symptoms of large patch on infected zoysiagrass can be dramatic, displaying a bright yellow-orange- to orange-coloured perimeter around actively growing individual or coalesced patches. Coalesced patches may range up to several metres in overall diameter, with significant reduction in turfgrass density and remaining until the next autumn.

Environmental influences

Several key environmental and management-induced factors, similarly as for brown patch, have been shown to influence development of large patch (Box 6.24). Moderate temperatures and humidity are key factors.

Box 6.24. Key environmental factors of large patch.

- Moderate temperatures (20–25°C)
- High humidity (above 95%)
- Prolonged periods of leaf wetness

Disease profile

Large patch is a disease linked to the growth of warm-season grasses in regions subject to winter dormancy. Large patch disease occurs during cooler seasons when the host begins to enter cool-temperature-induced dormancy in the autumn or when growth resumes with increasing temperatures in spring. Dormant sclerotia germinate between 15 and 25°C (Green et al., 1993).

Development of large patch of zoysiagrass is suppressed at soil temperatures above 30°C. Some disease patches are seasonal occurring in either spring or autumn. Other disease patches are perennial developing in the same location year after year for several years.

Disease-resistant varieties

Genetic resistance against large patch has been documented among St. Augustinegrass, seashore paspalum and zoysiagrass cultivars (Canegallo *et al.*, 2006; Flor *et al.*, 2017; Patton *et al.*, 2017).

Yellow patch

Yellow patch is a distinctive *Rhizoctonia*-incited disease on cool-season grasses because the pathogen is capable of causing disease during cooler conditions. The disease was previously called cool-weather brown patch to differentiate the symptoms from the brown patch disease (*R. solani*) that occurs during warmer conditions. The name yellow patch describes roughly circular-shaped, chlorotic or yellow-appearing symptoms (Fig. 6.39). Alternative names in certain countries are Rhizoctonia patch and winter patch.

Causal agent

Rhizoctonia cerealis (AG-D1), a binucleate mesophilic fungus, is the causal agent of yellow patch (Dernoeden, 2013). *R. cerealis* also causes Rhizoctonia spring dead spot on zoysiagrass in Japan (Tani and Beard, 1997).

Geographic distribution

Yellow patch has been documented in certain regions in the Americas, East Asia, Western Europe and Australasia (Table 6.15). The pathogen was isolated and identified on infected bermudagrass in 1980 from Bermuda (Caribbean) and was given the name yellow patch (Burpee, 1980). Yellow patch undoubtedly occurs in many countries in temperate regions.

Host range and pathogenicity

Bluegrasses, creeping bentgrass, perennial ryegrass and tall fescue are common hosts of yellow patch, together with bermudagrass and zoysiagrass (Tredway *et al.*, 2023). *R. cerealis*, compared to *R. solani*, is less pathogenic on warm-season grasses.

Table 6.15. Documented global distribution of yellow patch.

Region/country	References
United States	Tredway *et al.* (2023)
Southern Canada	Charbonneau and Hsiang (2015)
East Asia	Tani and Beard (1997)
Western Europe	Smith *et al.* (1989)
Australasia	Christensen (1992)

Fig. 6.39. Multiple yellow patch symptoms on a creeping bentgrass golf green. (Courtesy of Gary W. Beehag.)

Symptoms

Symptom expression of yellow patch is generally superficial and diminishes with the onset of cloudless skies and warmer and drier conditions. Late-stage symptoms on greens occur as randomly scattered, single or multiple rings or concentric rings that can sometimes coalesce (Fig. 6.40). Rings range from 20 to 80 cm or more in diameter with a perimeter band generally having a width of 3–5 cm. Colour of the rings varies from yellow on creeping bentgrass to yellow or reddish-brown on annual bluegrass. Chlorosis begins at the leaf tip progressing downwards. Infected grass often in the centre section of the symptomatic area recovers or remains healthy.

Advanced symptoms may be associated with a partly sunken turfgrass surface, as indicated by touch, because of organic matter decomposition. Mycelium is not produced in abundance and whitish-yellow-coloured rings may be seen after snow melt in colder regions.

Integrated management

Effective management strategies of yellow patch are similar to those for other *Rhizoctonia* diseases. Proven cultural, biological and fungicide options combined with frequent monitoring of disease symptoms and changing weather patterns are required for effective management of the disease.

Root Decline of Warm-Season Grasses

The emergence and recurrence of previously undescribed, patch-like disease symptoms on warm-season turfgrasses over recent decades continue to be reported in several regions of the world. Confirmation of the causal agent(s) and proof of pathogenicity remain confined largely to the southern United States and Australasia.

These disfiguring patch diseases are characterized by the presence of multiple, randomly distributed chlorotic and necrotic patches of varying shapes and sizes; brittle, dysfunctional stems and roots; and the presence of root-infecting fungi. A range of symptoms have been observed most commonly on closely mown, warm-season grasses managed on bowling greens and golf courses (Fig. 6.41), croquet lawns, tennis courts and residential lawns, depending on the country in which the various disease symptoms occur.

Fig. 6.40. Concentric ring symptoms of yellow patch on a creeping bentgrass golf green. (Courtesy of Gary W. Beehag.)

Fig. 6.41. Disfiguring symptoms of a root decline disease on a couchgrass (bermudagrass) fairway in southern Australia. (Courtesy of Percy T. Wong.)

The real significance of these root declines of warm-season grasses lies in the fact multiple, ectotrophically growing fungal species from different genera can co-colonize and infect the stems and roots of the same symptomatic host, suggestive of a disease complex, and not be caused by one fungal pathogen. One fungus may be the primary pathogen, others may be secondary pathogens or be present but not pathogenic. Certain fungal combinations may be antagonistic or even synergistic with each other.

Spring dead spot and bermudagrass decline could both be considered the first-described root decline diseases of warm-season turfgrasses. Spring dead spot is widely accepted to be caused by one or more *Ophiosphaerella* species (Tredway *et al.*, 2009). *G. graminis* was previously isolated from hybrid bermudagrass golf greens in the United States expressing spring dead spot symptoms (McCarty and Lucas, 1989); however, this was later proven incorrect. Bermudagrass decline is attributed largely to *Gaeumannomyces* spp. being the principal causal agent (Tredway *et al.*, 2023); however, other fungi have been implicated as being involved.

Overviews of recent work conducted primarily in the United States (Bronzato-Badial *et al.*, 2020) but also in Australia (Beehag and Wong, 2023) have identified a suite of root-infecting fungal species from different genera on symptomatic grasses. These findings have resulted in the adoption of several terms describing the symptoms. The term 'take-all root rot' adopted in the United States describes root disease on warm-season turfgrasses caused by *G. graminis* which is distinctly different from the root disease 'take-all patch' of cool-season grasses caused by primarily *G. avenae*. The associated term 'root decline' describes 'roughly a continuum of non-distinct visual symptoms including thinned turf stands, irregular chlorotic patches and distinct circular patches of diseased turf' (Martin, 2017).

The term 'root decline of warm-season turfgrasses' is adopted in this book and covers root-infecting fungal species singly or in combination from several genera associated with a stress-related gradual decline and dysfunction of the stem and root system of warm-season grasses. Spring dead spot is covered as a separate root disease in this chapter.

Effective management of root decline diseases on warm-season turfgrasses is highly problematic once these diseases become established on high-quality hybrid bermudagrass bowling and golf greens. Minimizing the impact of root decline of warm-season turfgrasses requires an integrated approach of cultural practices and timely application of proven fungicides.

Characterization of the causal agents

Root-infecting fungi associated with root decline of warm-season grasses were initially reported on hybrid bermudagrass golf greens during late 1980s in the southern United States (Smiley, 1993) and later in the 1990s in eastern mainland Australia (Stirling, 2001). Fungal species associated with the root decline of warm-season grasses are classified as ascomycetes characterized as possessing flask-shaped, fruiting bodies (perithecia) containing sac-like asci producing ascospores (sexual spores). However, not all ascomycetes produce recognizable spores in nature or under induced laboratory conditions and are referred to as being sterile.

Numerous ascomycete fungal genera and species have been isolated from both symptomatic and non-symptomatic warm-season grasses (Table 6.16). Some fungal isolates are new species and other isolates remain to be identified using genetic-based methodologies (e.g. DNA sequencing and others).

Determination of the identity and basic biology of ectotrophic root-infecting fungi (ERI) on warm-season grasses emanates from investigations conducted primarily in the United States (Bronzato-Badial et al., 2020; Stephens et al., 2022) and to a lesser extent in Australia (Wong, 2019; Beehag and Wong, 2023). Improvements in genetic methodologies will undoubtedly confirm reports in other countries of patch-like symptoms to be caused by known and other ERI fungal species.

ERI fungi comprise multiple fungal genera and species characterized as growing laterally along the external surface of host organs. Precise diagnosis of the causal agents of root decline requires close observation for characteristic signs by highly skilled and trained personnel using laboratory-based microscopy and culturing techniques supported by genetic methodologies.

Table 6.16. Warm-season grass hosts and their documented ERI fungal genera and species.

Host species	Fungal genera/species	References
Bermudagrass (couchgrass)	Budhanggurabania cyndonticola Candidacolonium cynodontis Gaeumannomyces graminis Magnaporthiopsis cynodontis, M. dharug, M. gumbaynggirr, M. incrustans, M. taurocanis Penicillifer martini Phialocephala bamuru Wongia garrettii, W. griffinii	Scott (1989); Wong et al. (2000, 2003, 2012, 2014, 2015a,b, 2022); Elmore et al. (2002); Wong (2002); Cha et al. (2012); Wang (2015); Hernández-Restrepo et al. (2016); Khemmuk et al. (2016); Stephens and Kerns (2020); Vines et al. (2021); Zong et al. (2022)
Centipede grass Kikuyugrass	Gaeumannomyces graminicola G. graminis Magnaporthiopsis gadigal P. martini P. bamuru	
Seashore paspalum	G. graminis	
St. Augustinegrass (buffalo grass)	Gaeumannomyces arxii, G. californicus, G. floridanus, G. graminis, G. graminicola, G. wongoonoo	
Zoysiagrass	G. graminis	

The presence of root-infecting fungi may not always be readily visible by microscopic examination of infected organs. Extensive colonization by the fungi may be revealed using a microscope and may be observed as dark brown- to black-coloured, runner hyphae (thickened hyphal strands) on stems and primary roots (Fig. 6.42). Flask-shaped perithecia may also be seen embedded in host tissues. Fungal structures may be found on infected hosts but do not always develop in culture.

The occasional absence of fruiting bodies (e.g. perithecia) except under highly favourable conditions and the spore production among many of these newly described, root-infecting fungal species make identification based solely on morphological features problematic. When grown in culture, the mycelium or hyphae of different ERI species can vary in size, colour and growth rate, depending on moisture and temperature (Zidek et al., 2021; Goetze and Jo, 2022). Root decline pathogens grow optimally in culture around 25–30°C but can grow between 10 and 35°C (Evans et al., 2016; Zidek et al., 2021; Stephens et al., 2022) depending on the species present, which may partly explain damage levels to particular host grasses.

Certain ERI genera can be broadly differentiated based on the shape of their hyphopodia (hyphal end) if present. Hyphopodia may range between either slightly or highly lobed (e.g. *G. graminis* var. *graminis*) and simple or slightly swollen (e.g. *Magnaporthiopsis* spp.) with slight variations depending on the fungal isolate. However, not all ERI genera possess clearly defined hyphopodia. Ascospores (when present) vary widely in size, shape, colour and cell number depending on the fungal genus and species (Fig. 6.43).

Fig. 6.42. Ectotrophic fungal hyphae and fruiting bodies on an infected couchgrass (bermudagrass) stem. (Courtesy of the NZSTI.)

Global distribution and host range

The global distribution of root decline diseases remains unknown but will inevitably increase on warm-season turfgrasses in tropical and subtropical regions. The occurrence of root decline on warm-season grasses has been confirmed widely throughout the south-eastern United

Fig. 6.43. Ascospores of *Wongia griffinii*. (Courtesy of Roger Shivas.)

States and mainland Australia and to a lesser extent in New Zealand, East Asia and South Africa. Confirmed hosts of root decline pathogens of warm-season grasses are bermudagrass, centipede grass, kikuyugrass, seashore paspalum, St. Augustinegrass and zoysiagrass.

Hybrid bermudagrass (couchgrass) cultivars represent the most common and widespread host for root-infecting fungi. In the United States, highly symptomatic bermudagrasses are the older (i.e. Tifdwarf and Tifton 419) and the more common dwarf (i.e. Champion, MiniVerde, and TifEagle) cultivars. In Australia, ERI fungal species have been isolated from hybrid (i.e. Santa Ana, Tifdwarf and TifEagle) and Australian couchgrass selections (Legend, Greenlees Park, Wintergreen and WindsorGreen).

Champion bermudagrass has been shown to be highly susceptible to *Gaeumannomyces* species as opposed to *Candidacolonium cynodontis* and *Magnaporthiopsis cynodontis* (Stephens *et al.*, 2022). Non-symptomatic grasses (e.g. centipede grass) from which an ERI fungal species was isolated may not necessarily have been managed as a formal turfgrass; hence the level of susceptibility of certain grasses as hosts of many ERI species remains unclear.

G. graminis also occurs on lesser-known warm-season grasses (e.g. *Brachiaria* and *Digitaria*) in tropical regions (Peixoto *et al.*, 2013). Species of these two grass genera and others occur where managed as low-intensity turfgrass in certain countries, raising the possibility some of these grass species may be hosts for these root-infecting pathogens.

Global distribution of root decline of warm-season grasses will inevitably increase given the widespread adoption of newer bermudagrass, seashore paspalum and zoysiagrass cultivars particularly on golf courses and sportsgrounds with a potential risk from infected vegetatively propagated material shared within and between countries.

Specific root decline diseases of warm-season grasses

Several warm-season grass host–ERI fungal combinations have a specific disease name. Bermudagrass decline in the United States and fairway patch and summer decline in Australia represent the most well-known root decline diseases in their respective countries. Designation of a specific warm-season grass host–root pathogen combination is now generally considered inappropriate because of multiple causal agents occurring in disease complexes.

Bermudagrass decline

Bermudagrass decline was primarily attributed to *G. graminis* var. *graminis* (Elliott, 1991) but other root-infecting fungi have since been associated with the disease on bermudagrass golf greens (Stephens *et al.*, 2022; Tucker *et al.*, 2022). Symptoms on golf greens appear as irregular, chlorotic areas with thinning around the boundaries (Fig. 6.44). Bermudagrass decline occurs on hybrid bermudagrass golf greens throughout the south-eastern United States from the Carolinas, to Florida, and into Texas (Bronzato-Badial *et al.*, 2020).

Fairway patch

Fairway patch, caused primarily by *Phialocephala bamuru*, is a relatively recent confirmed disease throughout mainland Australia on closely mown couchgrass (bermudagrass) and kikuyugrass golf fairways (Fig. 6.45). The disease, having a distinctive patch symptom with a chlorotic centre, unexpectedly appeared in 2000 on couchgrass on a golf course in Sydney (New South Wales); hence the ascribed name (Wong *et al.*, 2015a).

Fairway patch is a highly disfiguring disease of national significance throughout mainland Australia because of widespread utilization of numerous couchgrass cultivars and kikuyugrass. The disease is highly problematic once established on high-quality turfgrass due to limited knowledge of its ecology and integrated management using cultural practices and efficacious fungicides. Another ERI fungal species, *Penicillifer martini* (Wong *et al.*, 2014), has been associated with fairway patch on some golf course fairways in mainland Australia. Fairway

Fig. 6.44. Bermudagrass decline on a hybrid bermudagrass golf green in the southern United States. (Courtesy of Maria Tomaso-Peterson.)

patch also occurs on couchgrass in the North Island of New Zealand (Fig. 6.46). The full extent of the distribution of fairway patch in Australia and in New Zealand remains unknown (Anon., 2016; P. Wong, New South Wales, 2022, personal communication).

Summer decline

In Australia, summer decline, caused by *Wongia griffinii* (formerly *Magnaporthe griffinii*), is a term coined by golf superintendents during the 1990s describing the progressive loss of putting surfaces and dieback primarily on hybrid couchgrass greens in south-eastern Queensland. The host range now encompasses kikuyugrass (Fig. 6.47). A similar term is adopted in the United States describing surface decline and stunted roots on hybrid bermudagrass golf greens (Tredway *et al.*, 2023).

Lesser-known root decline diseases

In regional Australia, several other root-infecting fungi causing named root decline diseases on warm-season grasses are recognized (Beehag and Wong, 2023). The diseases are known as Adelaide patch, Deniliquin patch, summer decline and wongoonoo patch (Table 6.17). Distribution of these ERI fungal species in Australia currently remains unclear.

Deniliquin patch (Fig. 6.48) has been diagnosed on golf and bowling greens while another disease, wongoonoo patch (Fig. 6.49), has been diagnosed on St. Augustinegrass (buffalo grass) residential lawns.

Diagnosis of root decline diseases of warm-season grasses

Different symptoms coupled with occasional lack of characteristic fungal structures on infected

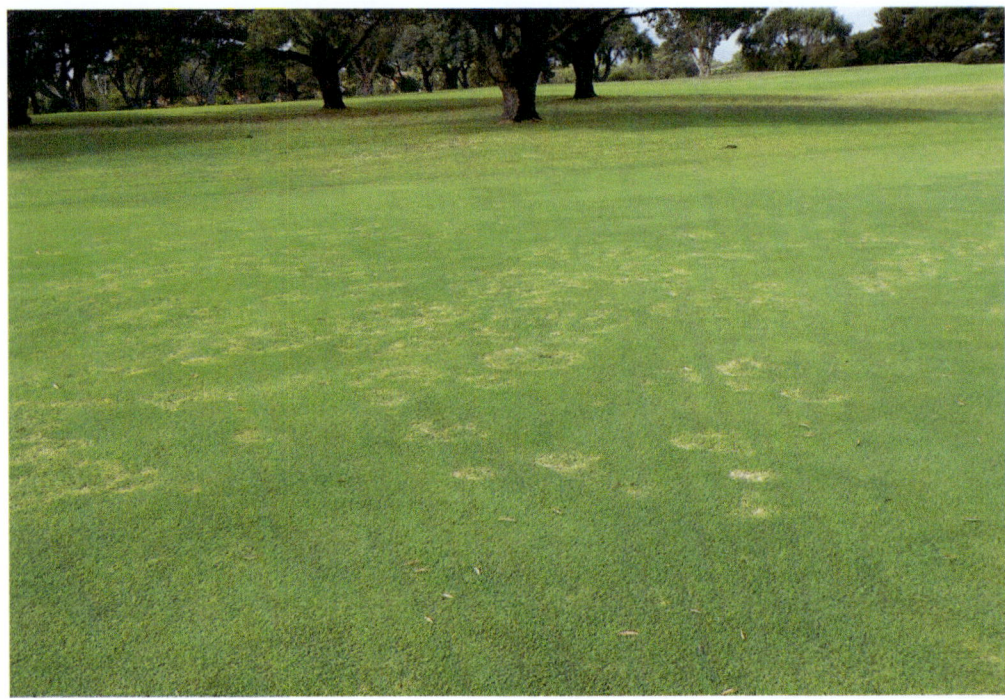

Fig. 6.45. Severe disfiguration caused by fairway patch on a couchgrass fairway in southern Australia. (Courtesy of Percy T. Wong.)

Fig. 6.46. Fairway patch on a highly-quality couchgrass fairway in New Zealand. (Courtesy of Mark Hooker.)

Fig. 6.47. Summer decline symptoms on a kikuyugrass fairway in southern Australia. (Courtesy of Percy T. Wong.)

Table 6.17. Lesser-known ERI fungi and root declines of warm-season grasses in Australasia.

Disease name	Causal agent	References
Adelaide patch	*Wongia garrettii*	Khemmuk *et al.* (2016)
Deniliquin patch	*Budhanggurabania cyndonticola*	Wong *et al.* (2015b)
Summer decline	*Wongia griffinii*	Khemmuk *et al.* (2016)
Wongoonoo patch	*Gaeumannomyces wongoonoo*	Wong (2002)

roots make a precise identification of the fungi challenging using morphological techniques. Field sampling revealing misshapen or abnormal stems displaying a darkened colour, reduced root length and decreased root branching may be indicative of root decline as observed on washed plant organs (Fig. 6.50). Extensive dieback commences on root tips progressing slowly to other plant organs.

Fungal structures characteristic of ERI fungi are not always present, tending only to exist under highly favourable conditions. The physical presence is not diagnostic of a specific fungus nor a root decline disease. Microscopic examination of mycelium actively growing in laboratory culture, under favourable conditions of temperature and moisture, and taken from suspect stem and root sections may reveal differences in colour, morphology and radial growth of the fungal isolates present.

Skilful observation of the shape of specialized cells (hyphopobia) located at the ends of fungal hyphae can assist in differentiating certain ERI fungal genera (Fig. 6.51). Hyphopobia are usually flattened but vary in shape ranging between finger-like or highly lobed (e.g. *G. graminis*) and simple or slightly swollen (e.g. *Magnaporthiopsis* and *Phialocephala* spp.) with slight morphological variations between both extremes among some isolates in the same species (e.g. *Gaeumannomyces* spp.).

Characteristic ascospores of certain non-sterile fungal may be found on infected hosts but do not always develop in laboratory culture. Furthermore, isolation of one or more root-infecting fungal isolates from symptomatic hosts

Fig. 6.48. Deniliquin patch symptoms on a common couchgrass fairway in southern Australia. (Courtesy of Terry Howe.)

does not necessarily mean a 'cause-and-effect' relationship exists. Pathogenicity (ability of a pathogen to cause disease) must be demonstrated using Koch's postulates.

Symptomatology

Symptomatology (i.e. pathogen signs and disease symptoms) of root declines of warm-season grasses bears certain similarities and differences, depending on host–pathogen(s) combination and environmental factors far from fully understood. Overall symptom expression can vary from general surface thinning, irregularly shaped blighted and chlorotic areas with ill-defined boundaries or roughly circular but variably shaped and sized chlorotic and necrotic patches to roughly circular doughnut-shaped rings often coalescing (Fig. 6.52). Whether certain patch-like symptoms are new or 'novel' diseases or diseases that have appeared in the past and their causal agents been misdiagnosed remains speculative.

The foregoing information highlights the highly variable expression, overall pattern and severity of root decline symptoms. Severity of root decline symptoms is undoubtedly dependent partly on the most virulent ERI fungal species present and partly on the intensity of cultural management. Symptoms are generally very pronounced often with discrete perimeters on extremely closely mown turfgrasses (i.e. less than 2.0 mm bench setting) on bowling and golf greens but often less pronounced on medium-height turfgrass (i.e. 5–10 mm).

Distinguishable symptoms may occur any time during autumn and again in mid- to late summer during wet conditions all depending on host–pathogen combination and environmental conditions. Early-stage symptoms may appear as small-sized patches or irregularly shaped chlorotic areas. Disease progression may result in patches gradually enlarging, some coalescing forming irregularly shaped regions of chlorotic patchiness or doughnut-shaped rings with unaffected centres.

Fig. 6.49. Wongoonoo patch symptoms on a buffalo grass domestic lawn in Western Australia. (Courtesy of Ken Johnston.)

Fig. 6.50. Dark-coloured and rotted stems of washed couchgrass sod infected with a *Magnaporthiopsis* sp. (Courtesy of Gary W. Beehag.)

New grass tillers may emerge recolonizing affected areas but may soon decline. Severe loss ultimately results from the inability of the turfgrass to recover from the stressful effects of dysfunctioning roots. Foliar symptoms on less-intensely managed, warm-season grasses appear as very irregular-shaped areas of chlorotic and necrotic plants and areas devoid of live plants.

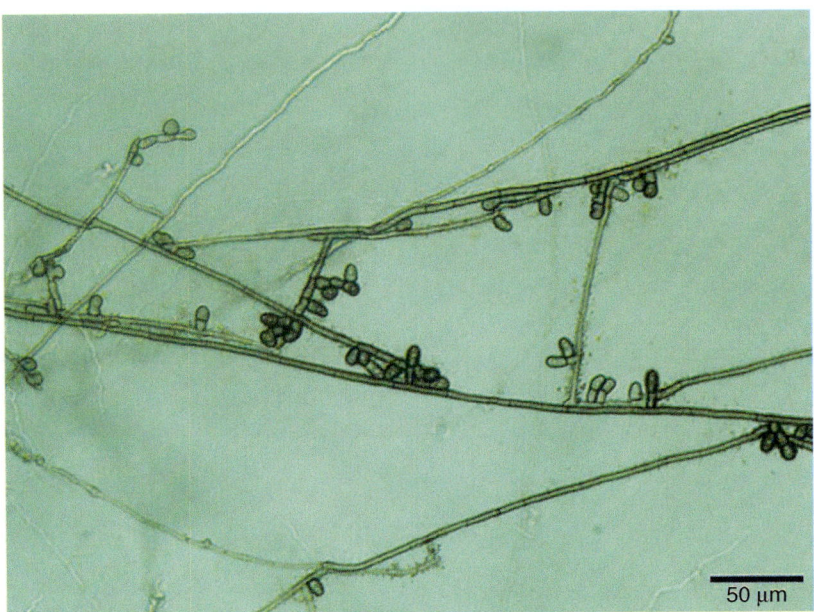

Fig. 6.51. Hyphopobia of *Phialocephala bamura*. (Courtesy of Percy T. Wong.)

Generalized root decline disease profile

Root decline of warm-season grasses can be profiled as a collective of stress-related disease conditions and should be considered incited by one or more genera and species of root-infecting fungi. In certain cases, one ERI fungal species may be the primary pathogen while others secondary pathogens, perhaps even acting synergistically enhancing root decline. This viewpoint particularly applies to highly managed hybrid bermudagrass bowling and golf greens and tennis courts in transitional and temperate regions depending on environmental conditions.

Root decline pathogens can potentially infect and colonize root tissues of warm-season grasses during spring, having an ecological advantage at the expense of their hosts at lower temperatures. The approximate temperature range of growth for root decline pathogens (i.e. 10–35°C) and optimum temperatures for root growth of warm-season grasses (i.e. 24–35°C) partly overlap at higher temperatures. *Gaeumannomyces incrustans* infection and resulting disease, as demonstrated experimentally in growth chambers on Japanese lawngrass (*Zoysia japonica*), ranged between 12 and 25°C with an optimum development at 18°C (Burcher and Wilkinson, 2007).

Members of *Gaeumannomyces*, *Magnaporthiopsis* and other ERI genera are considered facultative saprophytes capable of surviving perhaps as dormant mycelium during unfavourable conditions in the thatch-mat and rootzone regions. Ascospores of *Gaeumannomyces* generally germinate between 20 and 25°C and the role of released ascospores in the infection of turfgrass is unclear. Fungal infection precedes symptom expression by 14–21 days, suggestive that colonization and root deterioration have already occurred. Once inside colonized tissues and under highly favourable conditions of adequate nitrogen, moisture and temperature, ERI fungi are strong competitors against other pathogens and antagonists. ERI fungi are disseminated by infected plants and soil; hence, the necessity of attempting to establish new bowling and golf greens from infection-free nursery plants when possible.

The observed gradual recovery of disfigured bermudagrass on bowling and golf greens over several years in Australia suggests the presence of antagonistic or suppressive microbes. Take-all decline of cereals is well known where

Fig. 6.52. Multiple doughnut-shaped patch symptoms some coalescing on a 'Tifdwarf' bermudagrass bowling green in southern Australia. (Courtesy of Gary W. Beehag.)

the antagonistic phenomenon has been characterized (Wong and Worrad, 1989).

Integrated management of root decline diseases

Effective management of root decline, particularly once established on bermudagrass bowling and golf greens, remains challenging, requiring a concerted and integrated approach. Much of what has been documented about managing root-infecting diseases has been gained through management of bermudagrass decline and spring dead spot in the United States (White, 2004; Martin, 2017) and the latter disease in Australia (Robinson et al., 1995; Martin, 2000).

The overarching objective of minimizing root decline on bermudagrass bowling and golf greens and tennis courts is to establish and maintain a highly functional root system through cultural and possible biological manipulation of the rhizosphere in combination with adoption of proven turfgrass stress-minimizing practices (Box 6.25). The combination of monitoring of weather patterns and overall turfgrass growth and vigour needs to commence with temperature changes in spring and autumn. Future management practices targeting specific root-infecting fungal pathogens will materialize once a greater understanding of the biology and epidemiology of the pathogens is attained.

DISEASE-RESISTANT CULTIVARS Genetic resistance against root decline diseases has not been adequately demonstrated among warm-season grass cultivars. Newer dwarf bermudagrass cultivars (e.g. Champion, TifEagle and MiniVerde), as opposed to Tifdwarf and Tifgreen, produce much shorter leaves; thus tolerate extremely low mowing heights (i.e. less than 3.0 mm) provided there are no limiting factors.

The measured reduction in rhizome weight and root length among established dwarf bermudagrasses (McCarty and Canegallo, 2005) has the potential of reducing environmental stress tolerance and possibly increasing susceptibility of cultivars to root decline diseases. Association between the propensity of certain hybrid ultradwarf bermudagrass cultivars possessing relatively shallow root systems and the potential of bermudagrass decline has been recognized as problematic (Inguagiato and Martin, 2015).

SUNLIGHT REQUIREMENTS Warm-season grasses, as opposed to cool-season grasses, require full sunlight to maximize photosynthetic capacity. 'TifEagle' bermudagrass on an experimental green has been shown to produce acceptable turfgrass quality subject to 8 h of uninterrupted sunlight between 10 a.m. and 6 p.m. (Bunnell and McCarty, 2004). Any cultural practice (e.g. raised mowing height) and procedures (e.g. decrease of shading) will promote the growth of warm-season grasses.

MOWING AND IRRIGATION REGIMES Exceptionally low mowing height less than 2.5 mm or 0.125 inches (bench setting) is attainable for dwarf bermudagrass cultivars, provided there are no limiting factors (e.g. adequate sunlight). Raising the mowing height for nominated periods in conjunction with rolling on bowling and golf greens during high temperatures in summer is an effective stress-minimizing practice.

Maintaining adequate water infiltration on dwarf bermudagrass bowling and golf greens can be challenging. Adequate moisture levels may be maintained by judicious irrigation practices, timely use of narrow-width tines and application of proven soil-wetting agents that may indirectly enhance root growth.

ROOTZONE NUTRITION Timely application of balanced NPK fertilizers at a frequency and rate adequate to maintain consistent but not excessive growth based on seasonal responses, should be followed. Application of certain trace elements (e.g. copper and manganese) may be beneficial against certain root decline pathogens (e.g. *Gaeumannomyces*). A slightly acidic rootzone (pH 6.0–6.5) can help suppress certain disease and caution should be taken if applying lime- and gypsum-containing products.

THATCH-MAT MANAGEMENT The propensity among dwarf bermudagrass cultivars to develop thatch-mat accumulation has been demonstrated experimentally (Guertal and White, 1998; McCarty and Canegallo, 2005) depending on the level of cultural management. Excessive thatch-mat accumulation has been associated with ERI fungi on hybrid couchgrass (bermudagrass) bowling and golf greens in Australia (Beehag and Wong, 2023). Thatch-mat minimization practices (e.g. fraze mowing, hollow tining, scarification and light topdressing) assist directly to maintain moisture–oxygen balance, root

---

Box 6.25. Generalized management factors of root decline of warm-season turfgrasses.

- Monitoring of weather patterns and turfgrass growth commencing in spring and autumn
- Establish new greens by vegetatively propagating from a hygienic nursery
- Modification to the mowing regime (height and frequency)
- Judicious irrigation to encourage a highly functional root system
- Maintain slightly acidic pH, use manganese sulfate and avoid lime
- Minimization of thatch-mat accumulation
- Caution when using pre-emergence herbicides from late summer
- Commencement of registered fungicide application from mid-summer

growth, and indirectly by reducing adsorption of rootzone-targeted fungicides.

BIOLOGICAL MANAGEMENT   Biological suppression of certain ERI fungal species has been demonstrated experimentally under laboratory-controlled conditions (Fig. 6.53). The authors are unaware of any proven or predictable biological management options against root decline of warm-season grasses.

FUNGICIDE SENSITIVITY AND APPLICATION TIMING   Differences in fungicide sensitivity among selected ERI fungal species have been demonstrated experimentally under laboratory-controlled conditions (Butler *et al.*, 2019; Stephens *et al.*, 2019). Application of registered fungicides against root decline diseases should be timed and targeted to uptake by roots during early spring and autumn and should coincide with early fungal activity, depending on geographic location, regional weather patterns and fungicide label statements. Certain fungicides have demonstrated a degree of phytotoxicity on closely mown bermudagrass (Elliott, 1995); hence always consult label directions.

Snow Mould Diseases

The broad term 'snow mould' characterizes low-temperature diseases incited by opportunistic fungal pathogens capable of infecting and colonizing grasses during cold-temperature-induced plant dormancy. Most snow moulds require prolonged snow coverage over many months, with symptoms becoming visible after snow melt (Fig. 6.54). Some snow moulds may have more than one common name and causal agent, depending in which country the diseases occur. Some snow moulds may form a disease complex (e.g. grey snow mould and speckled snow mould) while others form complexes with non-snow mould pathogens in the absence of snow coverage (e.g. *Pythium* spp., *Rhizoctonia* spp. and others).

Snow moulds and their causal agents

Causal agents of snow moulds are fungi classified as either psychrophilic (i.e. 16–20°C) or psychrotrophic (i.e. less than 5°C) based on temperature tolerance. Numerous fungal species, some well-known and others less common, are the causal agents of snow mould diseases on grasses (Table 6.18). The disease pink snow mould, as called in early turfgrass literature, associated with extended snow periods and the disease microdochium patch, which can be common under less snow cover and can be active under cool wet periods, are caused by the same fungus. Microdochium patch has a much wider global distribution.

The optimum temperature of growth among most snow mould pathogens ranges between −5

Fig. 6.53. Experimental suppression of *Phialocephala bamuru* in culture by variable concentrations of a *Trichoderma* sp. (Courtesy of Percy T. Wong.)

Fig. 6.54. Grey snow mould symptoms after snow melt in Canada. (Courtesy of Tom Hsiang.)

Table 6.18. Snow mould diseases and their pathogens and cardinal temperatures.

Snow mould name	Pathogen species	Temperature range (°C)	References
Coprinus snow mould (cottony snow mould)	*Coprinus psychromorbidus*	< –5 to < 25	Hoshino *et al.* (2009); Matsumoto and Hsiang (2016)
Pink snow mould (microdochium patch)	*Monographella nivalis*	> –5 to 30	
Sclerotinia snow mould	*Sclerotinia borealis*	–7 to < 20	
Snow root rot (Pythium snow blight)	*Pythium iwayamai*	< 0 to 30	
	Pythium paddicum	< 0 to 3	
Snow scald	*Myriosclerotinia borealis*	–7 to < 20	
Typhula blight (grey snow mould)	*Typhula incarnata, T. ishikariensis*	< –7 to < 20	
	Tyhula phacorrhiza	< 0 to < 25	

and 25°C while infection and disease development generally occur when temperatures are less than 10°C (Matsumoto and Hsiang, 2016). Temperature differences for fungal activity are often reflective of the geographic origin of the fungal isolates. Emergence of snow mould disease symptoms following snow melt may indicate the presence of one or multiple snow mould pathogens in combination. Verification of snow mould pathogens may require laboratory expertise and genetic (e.g. DNA sequence) techniques.

Geographic distribution and host range

Snow moulds of one type or another widely occur wherever turfgrass is managed in the subarctic, cool temperate regions and at high elevations in warm–temperate locations of the northern hemisphere (Table 6.19). Snow mould diseases occur widely on grasslands and turfgrass throughout the United Kingdom, northern Europe (Denmark, Finland, Iceland, Norway, Sweden and Germany), the northern United States, Canada and northern East Asia.

The distribution of individual snow mould diseases is governed by the interaction between the extent of snow cover and winter temperatures. Pink snow mould (Fig. 6.55), otherwise called microdochium patch in the absence of prolonged snow cover, grey snow mould and speckled snow mould are generally regarded as the most common and disfiguring snow moulds in North America.

Speckled snow mould (Typhula blight) is an economically important snow mould disease throughout the northern hemisphere where snow remains on the ground for several months. Pink snow mould and Typhula blight occur in roughly the same geographical regions of North America and northern Europe. Snow root rot or Pythium snow blight is less common but occurs in northern Japan and the northern United States.

Snow moulds have a wide host range; all cool-season grasses are potential hosts to at least one or more snow moulds but susceptibility varies widely between species. Annual bluegrass, bentgrasses, fine fescues and ryegrasses are highly susceptible species (McBeath, 2003).

Occurrence and distribution of true snow mould diseases in the southern hemisphere is limited. Pink snow mould in the form of microdochium patch occurs in the South Island of New Zealand and southern Australia (Woodcock, 1983; Howard, 1986). Microdochium patch is covered separately in this chapter.

Signs and symptoms

Signs and symptoms of snow moulds generally appear after snow melt and vary depending on host–pathogen combination and environmental

Table 6.19. Documented global distribution of snow mould diseases of turfgrass.

Region	Country	References
East Asia	Japan	Smith (1987); Tani and Beard (1997); Fermanian et al. (2003); Charbonneau and Hsiang (2015); Aamlid et al. (2018); Tredway et al. (2023)
Northern Europe	Denmark, Finland, Iceland, Norway, Sweden and Germany	
North America	Canada, northern United States	
United Kingdom	England, Scotland, Wales and Northern Ireland	

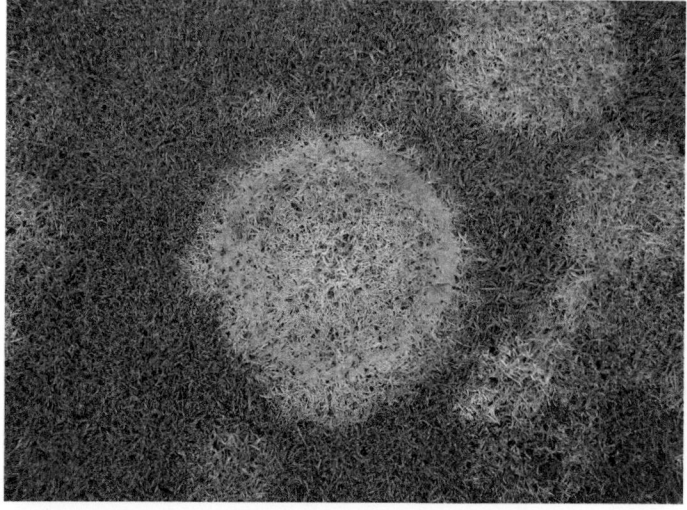

Fig. 6.55. Pink snow mould symptoms in Canada. (Courtesy of Tom Hsiang.)

conditions. Infected foliage may appear in colour from off-white, grey to pink depending on the fungal species present. Coprinus snow mould (cottony snow mould) refers to the presence of white-coloured, cotton-like mycelium among infected foliage. Speckled snow mould (grey snow mould), caused by one of three different *Typhula* species singly or in combination, displays a whitish-grey colour of mycelium.

Snow mould pathogens may produce sclerotia found on dead leaves after snow melt and when present, their shape, size and colour can aid in the identification of the particular snow mould. Snow mould sclerotia vary in shape from roughly spherical, globular to ovate and vary in size and colour depending on the maturity and fungal species causing the damage. Sclerotia colour varies from off-white, amber and orange, brown to black (Fig. 6.56) and size in the range of 1–5 mm depending on the fungal species (Charbonneau and Hsiang, 2015).

Lesions may be present on infected leaves. Infected foliage colour, inside an affected patch and perimeter, ranges from orange-brown, dark reddish-brown to light grey or tan, depending on host–pathogen combination. Snow mould symptoms progress to larger-sized, irregularly shaped, ring-like patches up to 15–30 cm or more in diameter (Fig. 6.57). Advanced symptoms of certain snow mould diseases may vary and may indicate more than one causal agent, depending on the region of occurrence. Certain patches may progress in size coalescing to form very large diseased and blighted areas.

Irrespective of the snow mould disease, extensive areas of infected foliage may become necrotic during conditions highly favourable for the pathogen.

Generalized snow mould disease profile

Snow mould diseases are characterized as appearing following snow melt in spring. Much of what is known about snow moulds in turfgrass systems (e.g. Hsiang *et al.*, 1999; Tredway *et al.*, 2023) has emanated from various grass systems

Fig. 6.56. Mature sclerotia of grey snow mould (*Typhula* sp.). (Courtesy of Tom Hsiang.)

Fig. 6.57. Varying-sized patches of pink snow mould in Canada. (Courtesy of Tom Hsiang.)

(Matsumoto, 2009; Matsumoto and Hsiang, 2016). Snow moulds share many common aspects regarding the disease cycle and epidemiology, providing a generalized disease profile (Table 6.20).

Ecologically speaking, snow mould diseases should be considered the result of an environment abundant in moisture with low temperatures that favour infection and fungal activity in combination with dormant hosts. Snow mould pathogens survive warmer, summer periods as either dormant sclerotia (*Coprinus* spp. and *Typhula* spp.) or spores (*M. nivalis*).

Damage by snow mould diseases can occur any time from late autumn through late spring, depending on the pathogen, temperatures, and depth and duration of snow cover. Strategies minimizing damage caused by snow moulds such as applications of biological agents and fungicides should be implemented before the onset of winter.

Generalized snow mould management

Effective management of most snow mould diseases is highly problematic once established on high-quality turfgrass. Several management strategies in combination can be routinely implemented in an attempt to minimize the intensity and frequency of most snow mould diseases (Box 6.26). Unnecessary application of nitrogenous fertilizers in late autumn should be avoided. For certain snow moulds, timely application of fungicides can be effective in reducing damage caused by the various pathogens.

Disease-resistant cultivars

Improved disease resistance among cool-season grass cultivars remains limited (Kerns and Tredway, 2013; Matsumoto and Hsiang, 2016).

Cultural practices

Excessive application of soluble nitrogen, potassium and certain organic-based fertilizers has been shown to promote snow moulds, most notably microdochium patch (Smiley *et al.*, 2005), particularly if applied within about 6 weeks of a hard frost or first snow fall (Fermanian *et al.*, 2003). Removal of snow cover and organic debris

---

Box 6.26. Generalized management factors of snow mould diseases.

- Monitor weather conditions and turfgrass for signs and symptoms
- Use of less-susceptible cultivars where appropriate
- Mow into late autumn to minimize snow accumulation on foliage
- Minimization of thatch-mat thickness
- Maintain growth using balanced NPK nutrition and check pH
- Judicious irrigation water applications
- Consider protective screens, cloths and remove snow where possible
- Consider application of registered biofungicides
- Use of registered fungicides for the specific snow mould disease

---

Table 6.20. Generalized snow mould profile.

Season of year	Fungal event	References
Late autumn	Germination of sclerotia in response to falling temperatures producing ascocarps (ascomycetes) and basidiocarps (basidiomycetes)	Hsiang *et al.* (1999); Matsumoto and Hsiang (2016)
Early winter	Commencement of infection of turfgrass hosts depending on pathogen species and subspecies and degree of snow cover	
Winter	Infection and colonization by invading mycelium of host coinciding with depleted carbohydrates in the host followed by the production of fungal survival structures	
Spring–autumn	Production of summer survival structures (e.g. oomycetes or sclerotia)	

such as fallen leaves will also aid in reducing the damage caused by snow moulds. Certain animal-product-based composted products have shown some suppression against microdochium patch and Typhula blight but only at relatively high (around 100 kg/100 m^2) application rates (Boulter et al., 2002).

Chemical products and resistance activators (e.g. phosphoric acid, isoparaffins and pigments) have been shown to suppress certain snow moulds (Stricker et al., 2017; Mattox et al., 2021). Plant-growth regulators have been shown to increase winter hardiness or cold acclimatization of turfgrass and can reduce damage caused by certain snow moulds.

Low-temperature-tolerant strains of bacterial and fungal species have been reported as natural antagonists to snow moulds (McBeath, 2003). Several species of soil-borne bacteria and fungi associated with temperate grasses have demonstrated the ability to suppress snow mould pathogens under controlled conditions (Table 6.21). Understanding the pathogen–antagonist relationship remains challenging and short-term results from using microbial antagonists have been mixed.

Lesser-Known, Cool-Weather Fungal Diseases

A suite of other cool-weather fungal pathogens occur widely on wild grasses and grassland systems worldwide. Numerous fungal species primarily infecting the foliage and crown regions can be characterized as lesser important pathogens in turfgrass systems (Table 6.22). Many fungal genera are relatively large and may be closely

Table 6.21. Snow mould diseases and their associated biological agents.

Microbe species	Snow mould	References
Acremonium boreale	Coprinus snow mould Pink snow mould Sclerotinia snow mould Typhula blight	Smith and Davidson (1979)
Pseudomonas fluorescens	Typhula blight	Matsumoto and Tajimi (1987)
Humicola grisea	Pink snow mould	Sato et al. (1999)
Trichoderma atroviride	Sclerotinia snow mould	McBeath (2001)
Typhula phacorrhiza	Typhula blight	Wu and Hsiang (1998)

Table 6.22. Lesser important, cool-weather fungal diseases and their causal agents.

Disease name	Fungal pathogen(s)	References
Ascochyta leaf blight	*Ascochyta* spp., *Neoascochyta* spp.	Smith et al. (1989); Couch (1995); Tredway et al. (2023)
Brown stripe	*Cercosporidium graminis*	
Cephalosporium stripe	*Cephalosporium gramineum*	
Cladosporium eyespot	*Cladosporium phlei*	
Fusarium diseases	*Fusarium* spp.	
Hadrotrichum leaf spot	*Hadrotrichum* spp.	
Mastigosporium diseases	*Mastigosporium* spp.	
Phyllosticta leaf blight	*Phyllosticta* spp.	
Pseudoseptoria leaf spot	*Pseudoseptoria* spp.	
Ramularia leaf spot	*Rhynchosporium* spp.	
Rhynchosporium leaf blotch	*Rhynchosporium orthosporum*	
Septoria leaf spot	*Septoria* spp.	
Spermospora leaf spot	*Spermospora* spp.	
Stagonospora leaf spot	*Stagonospora* spp.	
Yellow ring	*Trechispora alnicola*	

related and the precise taxonomy assignment remains subject to reclassification. Universal disease names may be absent for some, others may have multiple names depending on the country in which the diseases occur. Many of these fungi have an extensive host range on cool-season and warm-season grasses.

Fungal lifestyles can range from weakly pathogenic on senescent tissues to saprophytes on decomposing tissues, depending on host–pathogen combination and changing environmental conditions. Collectively, signs and symptoms and environmental conditions favouring the occurrence of these diseases are generally similar.

Grasses infected by these various fungi may show signs varying from leaf tip dieback and associated dark-coloured spots enlarging to variably shaped lesions causing girdling of leaves displaying chlorosis leading to necrosis, depending on the host–pathogen combination. Overall symptoms may range from unsightly, blighted regions to chlorotic and necrotic areas, depending on favourable temperature and moisture conditions.

Management actions are rarely required for these diseases. The disease status of some of these fungal pathogens may change over time depending on host–pathogen combination and changing environmental conditions.

References

Aamlid, T.S., Thorvaldsson, G., Enger, F. and Pettersen, T. (2012) Turfgrass species and varieties for integrated pest management of Scandinavian putting greens. *Acta Agriculturae Scandinavica, Section B – Plant & Soil Science* 61, 143–152. DOI: 10.1080/09064710.2012.677854

Aamlid, T.S., Espevig, T. and Tronsmo, A. (2016) Microbiological products for the control of *Microdochium nivale* on Nordic golf greens. In: *Proceedings of the 5th European Turfgrass Society Conference, Salgados, Spain, 5–8 June*. European Turfgrass Society, Livorno, Italy, pp. 147–148.

Aamlid, T.S., Niemelaainen, O., Paaske, K., Widmark, D., Ruuttunen, P. and Kedonpera, A. (2018) Evaluation of Civitis One™, alone or in combination with fungicides and potassium phosphite, for control of *Microdochium nivale* on Nordic golf greens. In: *Proceedings of the 6th European Turfgrass Society Conference, Manchester, UK, 2–4 July 2018*. European Turfgrass Society, Livorno, Italy, pp. 68–69.

Abad, Z.G., Shew, H.D. and Lucas, L.T. (1994) Characterisation and pathogen city of *Pythium* species isolated from turfgrass with symptoms of root and crown rot in North Carolina. *Phytopathology* 84, 913–921.

Anon. (2016) Is fairway patch spreading? *New Zealand Turf Management Journal* 33, 4–5.

Anon. (2020) Timeline of turf diseases, plants and species. *New Zealand Turfgrass Management Journal* 37, 20–21.

Baetsen-Young, A.M., Kaminski, J.E., Kasson, M.T. and Davis, D.D. (2016) Thatch collapse in golf course turf. *Golf Course Management* 84, 87–91.

Baird, J.H., Martin, D.L., Taliaferro, C.M., Payton, M.F. and Tisserat, N.A. (1998) Bermudagrass resistance to spring dead spot caused by *Ophiosphaerella herpotricha*. *Plant Disease* 82, 771–774.

Beehag, G.W. and Wong, P.T. (2023) Problem patches. *Australian Turfgrass Management Journal* 25, 32–36.

Bonos, S.A. and Huff, D.R. (2013) Cool-season grasses: biology and breeding. In: Stier, J.C., Horgan, B.P. and Bonos, S.A. (eds) *Turfgrass: Biology, Use, and Management*. Agronomy Monograph No. 56. American Society of Agronomy, Inc. Crop Science Society of America, Inc. and Soil Science Society of America, Inc. Madison, Wisconsin, pp. 591–660.

Bonos, S.A., Clarke, B.B. and Meyer, W.A. (2006) Breeding for diseases resistance in the major cool-season turfgrasses. *Annual Review of Phytopathology* 44, 213–234.

Boulter, J.L., Boland, G.J. and Trevors, J.T. (2002) Assessment of compost for suppression of Fusarium patch (*Microdochium nivale*) and Typhula blight (*Typhula ishikariensis*) snow molds of turfgrass. *Biological Control* 25, 162–172.

Bransgrove, K.L. (2005) Warm-season turfgrass diseases in Australia. *International Turfgrass Society Research Journal* 10, 152–155.

Brien, R.M. (1935) Three fungi causing 'brown patch' of lawns in New Zealand. *New Zealand Journal of Agriculture* 51, 157–158.

Bronzato-Badial, A., King, J. and Tomaso-Peterson, M. (2020) Monitoring ectotrophic root-infecting fungi associated with bermudagrass putting green roots using quantitative multiplex assays. *Plant Health Progress* 21, 144–151.

Bruton, B.D. and Toler, R.W. (1980) Influence of the time and temperature on inoculation and infection of St. Augustinegrass by *Sclerotinia macrospora* (abstract). *Phytopathology* 70, 565.

Bucher, E.S. and Wilkinson, H.T. (2007) The pathogenicity of *Gaeumannomyces incrustans* on turfgrass *Zoysia japonica*. *Canadian Journal of Plant Pathology* 29, 56–62. DOI: 10.1080/07060660709507437

Bunnell, B.T. and McCarty, B. (2004) Sunlight requirements for ultradwarf bermudagrass greens. *Golf Course Management* 72, 92–96.

Burpee, L.L. (1980) *Rhizoctonia cerealis* causes yellow patch of turfgrass. *Plant Disease Reporter* 64, 1114–1116.

Burpee, L.L. and Martin, B.S. (1996) Biology of turfgrass diseases incited by *Rhizoctonia* species. In: Sneh, B., Jabaji-Hare, S., Neate, S. and Dijst, G. (eds) *Rhizoctonia Species: Taxonomy, Molecular Biology, Ecology, Pathology and Disease Control*. Springer, Dordrecht, The Netherlands, pp. 359–368.

Butler, L., Stephens, C., Kerns, J. and Gannon, T. (2019) *Take-All Root Rot: A Complex Disease*. North Carolina State University Extension, Raleigh, North Carolina. Available at: go.ncsu.edu/readext?639666 (accessed 16 October 2023).

Cagas, B., Knot, P. and Zemkova, L. (2010) The effect of different management systems on the incidence or red thread (*Laetisaria fuciforme*) on turf. In: *Proceedings of the 2nd European Turfgrass Society Conference, Angers, France, 11–14 April 2010*. European Turfgrass Society, Livorno, Italy, pp. 62–63.

Cahill, J.V., Murray, J.J. and Dernoeden, P.H. (1983) Interrelationships between fertility and red thread fungal disease of turfgrass. *Plant Disease* 67, 1080–1083.

Canegallo, A.L., Martin, B., Camberato, J.J. and Jeffries, S.N. (2006) Occurrence and control of large patch (*Rhizoctonia solani* AG2-2LP) in seashore paspalum (*Paspalum vaginatum*) in South Carolina. *Phytopathology* 96, S19.

Cassi, O.C., Walker, N.R., Marek, S.M., Enis, J.N. and Mitchell, T.K. (2010) Infection and colonisation of turf-type bermudagrass by *Ophiosphaerella herpotricha* expressing green or red fluorescent proteins. *Phytopathology* 100, 415–423.

Cha, J., Lee, J.H., Kim, J.O., Shim, G.Y., Bae, E. and Kwak, Y.S. (2012) First report of take-all caused by *Gaeumannomyces graminis* var. *graminis* in Korea. *Asian Journal of Turfgrass Science* 26, 72–73.

Chang, T. and Lee, Y.S. (2013) Occurrence of Pythium blight caused by *Pythium aphanidermatum* on chewing fescue. *Weed & Turfgrass Science* 2, 306–311.

Charbonneau, P. (1999) Take-all. *Greenmaster* 34, 30–32.

Charbonneau, P. and Hsiang, T. (eds) (2015) *Integrated Pest Management for Turf*. Publication 845. Ministry of Agriculture, Food and Rural Affairs, Toronto, Canada.

Chastagner, G. and Vassey, W. (1979) Turfgrass disease research report. In: *Proceedings of the 33rd Northwest Annual Turfgrass Conference, Port Ludlow, Washington, 25–27 September 1979*. Northwest Turfgrass Association, Gig Harbor, Washington, pp. 111–115.

Chng, S.F., Cromey, M.G. and Butler, R.C. (2005) Evaluation of the susceptibility of various grass species to *Gaeumannomyces graminis* var. *tritici*. *New Zealand Plant Protection* 58, 261–267.

Christensen, M.J. (1985) Turf disease management – New Zealand. In: *Proceedings of the Third New Zealand Sports Turf Convention, Palmerston North, New Zealand, 26–29 August 1985*. New Zealand Sports Turf Institute, Palmerston North, New Zealand, pp. 106–111.

Christensen, M.J. (1992) *Rhizoctonia* fungi associated with turf diseases in New Zealand. *New Zealand Turf Management Journal* 6(February), 9–11.

Colbach, N., Lucas, P. and Meynard, J.-M. (1997) Influence of crop management on take-all development and disease cycles on winter wheat. *Phytopathology* 87, 26–32.

Cook, R.J. (2003) Take-all of wheat. *Physiological and Molecular Plant Pathology* 62, 73–86.

Couch, H.B. (1995) *Diseases of Turfgrasses*, 3rd. edn. Krieger Publishing Company, Malabar, Florida.

Dernoeden, P.H. (2013) *Creeping Bentgrass Management*. CRC Press, Boca Raton, Florida.

Dernoeden, P.H. and Jackson, N. (1980) Managing yellow tuft disease. *Journal of the Sports Turf Research Institute* 56, 9–17.

Dernoeden, P., Kaminski, J. and Haspel, C. (2011) Thatch collapse: a new disease of fine-leaf fescue. *Golf Course Management* 79, 88–91.

Elliott, M.L. (1991) Determination of an etiological agent of bermudagrass decline. *Phytopathology* 81, 1380–1384.

Elliott, M.L. (1995) Effect of systemic fungicides on a bermudagrass putting green infested with *Gaeumannomyces graminis* var. *graminis*. *Plant Disease* 79, 945–949.

Elmore, W.C., Gooch, M.D. and Stiles, C.M. (2002) First report of *Gaeumannomyces graminis* var. *graminis* on seashore paspalum in the United States. *Plant Disease* 86, 1405.

Endo, R.M. (1961) Turfgrass diseases in Southern California. *Plant Disease Reporter* 45, 869–873.

Endo, R.M. (1963) Influence of temperature on rate of growth of five fungus pathogens of turfgrass and on rate of disease spread. *Phytopathology* 53, 857–861.

Endo, R.M., Ohr, H.D. and Krausman, E.M. (1985) Leptosphaeria korrae, a cause of spring dead spot of bermudagrass. *Plant Disease* 69, 235–237.

Entwistle, C.A. (1999) Turf diseases – what's new. *Turfgrass Bulletin* 294, 26–27.

Evans, B.J., Wong, P.T.W. and Martin, P.M. (2016) The effect of temperature, osmatic potential and pH on the growth of turf pathogens Gaeumannomyces wongoonoo and Magnaporthe griffinii. *Acta Horticulturae* 1122, 49–54. DOI: 10.17660/ActaHortic.2016.1122.7

Fermanian, T.W., Ahring, R.M. and Huffine, W. (1981) The isolation of a toxin from spring dead spot areas in bermudagrass (Cynodon) turf. In: *Proceedings of the 4th International Turfgrass Research Conference, Guelph, Canada, 19–23 July 1981*. International Turfgrass Society, USA, pp. 433–441.

Fermanian, T.W., Shurtleff, M.C., Randell, R., Wilkinson, H.T. and Nixon, P.L. (2003) *Controlling Turfgrass Pests*, 3rd edn. Prentice-Hall, Upper Saddle River, New Jersey.

Fidanza, M.A. (2007) Characterisation of soil properties associated with type-1 fairy ring symptoms in turfgrass. *Biologia* 62, 533–536.

Fidanza, M.A. (2010) Preliminary investigation on the epidemiology of fairy ring in turfgrass. *2010 USGA Turfgrass and Environmental Research Summary*. United States Golf Association, Far Hills, New Jersey, p. 35.

Fidanza, M.A., Colbaugh, P.F., Engelke, M.C., Davis, S.D. and Kenworthy, K.E. (2005) Use of high-pressure injection to alleviate Type-1 fairy ring symptoms in turfgrass. *HortTechnology* 15, 169–172.

Flor, N.C., Harmon, P.F., Kenworthy, K., Raid, R.N., Nagata, R. and Datnoff, L.E. (2017) Screening St. Augustinegrass genotypes for brown patch and large patch disease resistance. *Crop Science* 57, S-89–S-97. DOI: 10.2135/cropsci2016.06.0514

Flores, F.J., Marek, S.M., Anderson, J.A., Mitchell, T.K., Moreno-Zambrano, M. and Walker, N.R. (2017) Reactive oxygen species production and alternative hosts of spring dead spot-causing fungi. *International Turfgrass Society Research Journal* 13, 213–224.

Freeman, J. and Ward, E. (2004) Gaeumannomyces graminis, the take-all fungus and its relatives. *Molecular Plant Pathology* 5, 235–252. DOI: 10.1111/j.1364-3703.2004.00226.x

Geng, J.M., Jiang, S. and Hu, J. (2021) First report of Ophiosphaerella narmari causing spring dead spot of hybrid bermudagrass in China. *Plant Disease*. DOI: 10.1094/PDIS-03-21-0535-PDN

Goetze, P. and Jo, Y.-J. (2022) Reexamining take-all root rot diseases in warm-season turfgrasses. *Golf Course Management* 90, 110–115.

Goss, R.L. and Gould, C.J. (1971) Inter-relationships between fertility levels and Corticium red thread disease of turfgrasses. *Journal of the Sports Turf Research Institute* 47, 48–53.

Gould, C.J. (1973) Ophiobolus patch: bane to bentgrass. *The Golf Superintendent* 41, 44–46

Green, D.E., Fry, J.D., Pair, C. and Tisserat, N.A. (1993) Pathogenicity of Rhizoctonia solani AG-2-2 and Ophiosphaella herpotricha on zoysiagrass. *Plant Disease* 77, 1040–1044.

Gregos, J., Casler, M.D. and Stier, J.C. (2011) Resistance of closely mown fine fescue and bentgrass species to snow mold pathogens. *Plant Disease* 95, 847–852. DOI: 10.1094/PDIS-11-10-0791

Grose, M.J., Parker, C.A. and Sivasithamparam, K. (1984) Growth of Gaeumannomyces graminis and G. tritici in soil: effects of temperature and water potential. *Soil Biology and Biochemistry* 16, 211–216.

Guan, L.J., Xu, B.L., Liang, Q.L. and Xue, Y.Y. (2009) Occurrence of Pythium root rot in turf grass and identifications of the pathogen. *Acta Pratoculturae Sinica* 18, 175–180.

Guertal, B. and White, R. (1998) Dwarf bermudagrasses demand unique care. *Golf Course Management* 66, 58–60.

Gullino, M.L., Mocioni, M. and Titone, P. (2007) First report of Ophiosphaerella korrae causing spring dead spot of bermudagrass in Italy. *Plant Disease* 91, 1200. DOI: 10.1094/PDIS-91-9-1200C

Hallett, P.D., Ritz, K. and Wheatley, R.E. (2001) Microbial derived water repellency in golf green soil. *International Turfgrass Society Research Journal* 9, 518–524.

Han, M., Kim, K.-D., Pyee, J., Choi, S. and Park, D.-S. (2016) Observation of Sclerophthora macrospora causing downy mildew from zoysiagrass with leaf yellowing and excessive tillering (abstract in English). *Weed & Turfgrass Science* 5, 23–28. DOI: 10.5660/WTS.2016.5.1.23

Handoll, C. (1966) Turf disease notes. *Journal of the Sports Turf Research Institute* 42, 65–68.

Hawkes, N. (1987) Spring dead spot of Tifdwarf turf in South Australia. *Journal of the Sports Turf Research Institute* 63, 136–140.

Hayakawa, T., Kobayshi, M. and Yaguchi, S. (2007a) Necrotic ring spot of zoysiagrass in Japan. *Journal of Japanese Society of Turfgrass Science* 35, 86–94. DOI: 10.11275/turfgrass1972.35.86

Hayakawa, T., Kobayashi, M., Sasaki, N. and Yaguchi, S. (2007b) Bentgrass dead spot caused by *Ophiosphaerella agrostis* in Japan. *Journal of Japanese Society of Turfgrass Science* 36, 12–19. DOI: 10.11275/turfgass1972.36.12

Heckman, J.R., Clarke, B.B. and Murphy, J.A. (2003) Optimising manganese fertilisation for the suppression of take-all patch disease on creeping bentgrass. *Crop Science* 43, 1395–1398.

Hernández-Restrepo, M., Groenewald, J.Z., Elliott, M.L., Canning, G., McMillan, V.E. and Crous, P.W. (2016) Take-all or nothing. *Studies in Mycology* 83, 19–48.

Hodges, C.F. and Campbell, D.A. (1994) Infection and adventious roots of *Agrostis palustris* by *Pythium* species at different temperature regimes. *Canadian Journal of Botany* 72, 378–383.

Hofgaard, I.S., Molteberg, B. and Tronsmo, A.M. (2008) *Report from the project 'Improved Strategy for Control of* Microdochium nivale *on Golf Courses' (2006-2008), funded by STERF*. Scandinavian Turfgrass Environmental Research Foundation, Stockholm.

Hood, M. (2006) Pythium disease. *New Zealand Turf Management Journal* 21, 17–20.

Hoshino, T., Xiao, N. and Tkachenko, O.B. (2009) Cold adaptation in the pathogenic fungi causing snow molds. *Mycoscience* 50, 26–38. DOI: 10.1007/s10267-008-0452-2

Howard, D.R. (1986) Fusarium patch. *New Zealand Turf Management Journal* 1, 23.

Hsiang, T., Wu, C., Yung, L. and Liu, L. (1995) Pythium root rot associated with cool-season dieback of turfgrass in Ontario and Quebec. *Canadian Plant Disease Survey* 72, 191–195.

Hsiang, T., Matsumoto, N. and Millet, S.M. (1999) Biology and management of Typhula blight snow mold of turfgrass. *Plant Disease* 83, 789–798.

Hyakumachi, M. and Hayakawa, T. (2008) New isolates of *Rhizoctonia* diseases of turfgrasses. *Floriculture and Ornamental Biotechnology* 2, 14–24.

Ichitani, T., Tani, T. and Umakoshi, T. (1986) Identification of *Pythium* spp. pathogenic on manilla grass (*Zoysia matrella*). *Transactions of the Mycological Society of Japan* 27, 41–50.

Inguagiato, J.C. and Martin, S.B. (2015) Diseases of cool- and warm-season putting greens. *USGA Green Section Record* 53, 1–19.

Iriarte, F.B., Wetzel, H.C., Fry, J.D., Martin, D.L. and Tisserat, N.A. (2004) Genetic diversity and aggressiveness of *Ophiosphaerella korrae*, a cause of spring dead spot of bermudagrass. *Plant Disease* 88, 1341–1346.

Jackson, N. (1981) Take-all patch (Ophiobolus patch) of turf grasses in the northeastern United States. In: *Proceedings of the Fourth International Turfgrass Research Society Conference, Guelph, Canada, 19–23 July 1981*. International Turfgrass Society, USA, pp. 421–424.

Jackson, N. (1993) Geographic distribution, host range, and symptomology of patch diseases caused by soilborne ectotrophic fungi. In: Clarke, B.B. and Gould, A.B. (eds) *Turfgrass Patch Diseases Caused by Ectotrophic Root-Infecting Fungi*. The American Phytopathological Society, St. Paul, Minnesota, pp. 17–40.

Kaminski, J.E. (2008) A review of dead spot on creeping bentgrass and a road map for future research. In: Pessarakli, M. (ed.) *Handbook of Turfgrass Management and Physiology*. CRC Press, Boca Raton, Florida, pp. 245–256.

Kaminski, J.E. and Dernoeden, P.H. (2002) Geographic distribution, cultivar susceptibility, and field observations of bentgrass dead spot. *Plant Disease* 86, 1253–1259.

Kaminski, J.E. and Dernoeden, P.H. (2005) Dead spot of bentgrass and hybrid bermudagrass. *Applied Turfgrass Science* 2, 1–8. DOI: 10.1094/ATS-2005-0419-01-DG

Kaminski, J.E. and Hsiang, T. (2006) first report of *Ophiosphaerella agrostis* infecting creeping bentgrass in Canada. *Plant Disease* 90, 1114. DOI: 1094/PD-90-1114B

Kaminski, J.E., Dernoeden, P.H. and Fidanza, M.A. (2007) Environmental monitoring and exploratory development of a predictive model for dead spot of creeping bentgrass. *Plant Disease* 91, 565–573.

Kammerer, S.J. and Harmon, P.F. (2008) The importance of early and accurate diagnosis of *Rhizoctonia* diseases. *Golf Course Management* 76, 92–98.

Kammerer, S.J., Burpee, L.L. and Harmon, P.F. (2011) Identification of a new *Waitea circinata* variety causing basal leaf blight of seashore paspalum. *Plant Disease* 95, 515–522.

Keighley, J.M. (2017) The epidemiology and integrated control of fairy rings on golf courses. PhD thesis, Harper Adams University, Newport, UK.

Kerns, J.P. and Butler, E.L. (2018) Cream leaf blight ultradwarf bermudagrass. *Golf Course Management* 86, 66–69.

Kerns, J.P. and Tredway, L.P. (2008) Pathogenicity of *Pythium* species associated with Pythium root dysfunction of creeping bentgrass and their impact on root growth and survival. *Plant Disease* 92, 862–869. DOI: 10. 1094/PDIS-92-6-0862

Kerns, J.P. and Tredway, L.P. (2013) Advances in turfgrass pathology since 1990. In: Stier, J.C., Horgan, B.P. and Bonos, S.A. (eds) *Turfgrass: Biology, Use and Management*. Agronomy Monograph No. 56. American Society of Agronomy, Inc., Crop Science Society of America, Inc. and Soil Science Society of America, Inc., Madison, Wisconsin, pp. 733–776.

Khemmuk, W., Geering, A.D.W. and Shivas, R.G. (2016) *Wongia* gen. nov. (Papulosaceae, Sordariomycetes), a new generic name for two root-infecting fungi from Australia. *IMA Fungus* 7, 247–252.

Kim, J.-W. and Park, E.W. (1999) Occurrence and pathogenicity of *Pythium* species isolated from leaf blight symptoms of turfgrasses of golf courses in Korea. *Plant Pathology Journal* 15, 112–118.

Kozelnicky, G.M. (1974) Updating 20 years research: spring dead spot. *USGA Green Section Record* 12, 12–14.

Krausz, J.P., White, R.H., Tisserat, N.A. and Dernoeden, P.H. (2001) Bermudagrass dead spot: a new disease of bermudagrass caused by *Ophiosphaerella agrostis*. *Plant Disease* 85, 1286.

Kvalbein, A., Espevig, T., Waalen, W. and Aamlid, T.S. (2017) *Turf Grass Winter Stress Management*. Scandinavian Turfgrass and Environment Research Foundation, Stockholm.

Landschoot, P.J., Gould, A.B. and Clarke, B.B. (1993) Ecology and epidemiology of ectotrophic root-infecting fungi associated with patch diseases of turfgrasses. In: Clarke, B.B. and Gould, A.B. (eds) *Turfgrass Patch Diseases Caused by Ectotrophic Root-Infecting Fungi*. The American Phytopathological Society, St. Paul, Minnesota, pp. 73–105.

Latin, R. (2005) Managing take-all patch of creeping bentgrass on sand-based greens. *Golf Management* 73, 101–105.

Liang, J., Li, G., Zao, M. and Lei, C. (2019) A new leaf blight of turfgrasses caused by *Microdochium poae*, sp. nov. *Mycologia* 111, 265–273. DOI: 10.1080/00275514.2019.1569417

Luc, J.E., Canegallo, A. and Martin, B. (2005) Spring dead spot in bermudagrass greens in Argentina and South Carolina. *Golf Course Management* 73, 92–95.

Lucas, L.T. (1980) Spring dead spot of bermudagrass. In: Joyner, B.G. and Larsen, P.O. (eds) *Advances in Turfgrass Pathology*. Harcourt Brace Jovanovich, Inc., Duluth, Minnesota, pp. 183–187.

McAlpine, D. (1898) *Fairy Rings and the Fairy Ring Puff-Ball*. Victoria Department of Agriculture Report, Melbourne, Australia, 14 May.

McAlpine, D. (1906) A new hymenomycete, the so-called *Isaria fuciformis* Berk. *Annales Mycologici* 4, 541–551.

McBeath, J.H. (2001) Effects of *Trichoderma atroviride* on snow mold diseases of turfgrasses in interior Alaska. In: Huber, D.M. (ed.) *Biocontrol in a New Millennium. Proceedings of the Third Joint National Biocontrol Conference, Estes Park Center, Colorado*, pp. 98–101.

McBeath, J.H. (2003) Snow mold winter diseases nemesis. *Golf Course Management* 71, 121–124.

Maccaroni, M., Corazza, L., Buonaurio, R. and Cappelli, C. (2002) Occurrence of pink patch of perennial ryegrass caused by *Limonomyces roseipellis* in Italy. *Plant Disease* 8, 6. DOI: 10.1094/PDIS.2002.86.1.74A

McCarty, L.B. and Canegallo, A. (2005) Tips for managing ultradwarf bermudagrass greens. *Golf Course Management* 73, 90–95.

McCarty, L.B. and Lucas, L.T. (1989) *Gaeumannomyces graminis* associated with spring dead spot of bermudagrass in the southeastern United States. *Plant Disease* 73, 659–661.

McCarty, L.B. and Miller, G. (2002) *Managing Bermudagrass Turf: Selection, Construction, Cultural Practices and Pest Management Strategies*. Ann Arbor Press, Ann Arbor, Michigan.

Machielse, P.L. (1968) Greenkeeping and its problems in South Africa. *Journal of the Sports Turf Research Institute* 44, 61–65.

Mahendra, R., Ahmed, K. and Abd-Elsalam, K.A. (2020) *Pythium: Diagnosis, Diseases and Management*. CRC Press, Boca Raton, Florida.

Mann, R. (2004) A review of the main turfgrass diseases in Europe and their best management practices at present. *Journal of Turfgrass and Sports Science* 80, 2–18.

Mann, R. (2011) Top of the queries. *Bulletin for Sports Surface Management* 252, 24–25.

Mann, R.L. and Newell, A.J. (2005) A survey to determine the incidence and severity of pests and diseases on golf course putting greens in England, Ireland, Scotland and Wales. *International Turfgrass Society Research Journal* 10, 224–229.

Martin, B. (2017) Take-all root rot on the increase in the Carolinas. *Golf Course Management* 85, 72–77.

Martin, D.L., Taliaferro, C.M., Tisserat, N.A., Bell, G.E., Baird, J.H., et al. (2001) Hardy bermudagrass sought with resistance to spring dead spot. *Golf Course Management* 69, 75–79.

Martin, P. (2000) Spring dead spot in couch. Presented at *TGAA Seminar on Pests and Diseases and Nematodes, Canberra, 26 July 2000*. Turfgrass Association of Australia, Canberra.

Marvasti, F.B. and Banihashemi, Z. (2011) Identification and pathogenicity of turfgrass-infecting fungi in Shiraz landscape. *Iranian Journal of Plant Pathology* 47, 127–129.

Matsumoto, N. (2009) Snow molds: a group of fungi that prevail under snow. *Microbes and Environments* 24, 14–20. DOI: 10.1264/jsme2.ME09101

Matsumoto, N. and Hsiang, T. (2016) *Snow Mold: The Battle Under Snow Between Fungal Pathogens and Their Plant Hosts*. Springer, Singapore.

Matsumoto, N. and Tajimi, A. (1987) Bacterial flora associated with the snow mold fungi *Typhula incarnata* and *T. ishikariensis*. *Annals of the Phytopathological Society of Japan* 53, 250–253.

Mattox, C.M., Kowalewski, A.R., McDonald, B.W., Lambrinos, J.G., Davidson, B.L. and Pscheidt, J.W. (2017) Nitrogen and iron sulphate affect Microdochium patch severity and turf quality on annual bluegrass putting greens. *Crop Science* 57, 1–8.

Mattox, C., McDonald, B., Braithwaite, E. and Kowalewski, A. (2021) Controlling Microdochium patch with nontraditional fungicides. *USGA Green Section Record* 59(October), 01.

Melvin, B.P. and Vargas, J.M. (1994) Irrigation frequency and fertiliser type influence necrotic ring spot of Kentucky bluegrass. *HortScience* 29, 1028–1030.

Miller, G.L., Grand, L.F. and Tredway, L.P. (2011) Identification and distribution of fungi associated with fairy rings on golf putting greens. *Plant Disease* 95, 1131–1138.

Monteith, J. and Dahl, A.S. (1932) Turf diseases and their control. *Bulletin of the United States Golf Association* 12, 87–187.

Nelson, E.R. and Craft, C.M. (1991) Identification and comparative pathogenicity of *Pythium* spp. from roots and crowns of turfgrasses exhibiting symptoms of root rot. *Phytopathology* 81, 1529–1536.

Niemczyk, H.D. and Krueger, H.R. (1980) Binding of insecticides on turfgrass thatch. In: Joyner, B.G. and Larsen, P.O. (eds) *Advances in Turfgrass Pathology*. Harcourt Brace Jovanovich, Inc., Duluth, Minnesota, pp. 61–64.

Nilsson, H.E. and Smith, J.D. (1981) Take-all of grasses. In: Asher, M.J.C. and Shipton, P.J. (eds) *Biology and Control of Take-All*. Academic Press, New York, pp. 433–451.

Nus, J.L. and Shashikumar, K. (1993) Fungi associated with spring dead spot reduces freezing resistance in bermudagrass. *HortScience* 28, 306–307.

Ormsby, D. (2021) Take-all patch: an old disease making a comeback. *New Zealand Turf Management Journal* 40, 7–9.

Patton, A.J., Schwartz, B.M. and Kenworthy, K.E. (2017) Zoysiagrass (*Zoysia* spp.) history, utilisation and improvement in the United States: a review. *Crop Science* 57, S-37–S-72. DOI: 10.2135/cropsci2017.02.0074

Peixoto, C.N., Ottoni, G., Philippi, M.C.C., Silva-Lobo, V.L. and Prabhu, A.S. (2013) Biology of *Gaeumannomyces graminis* var. *graminis* isolates from rice and grasses and epidemiological aspects of crown sheath rot of rice. *Tropical Plant Pathology* 38, 495–504.

Plant Health Australia (2001) Australian Plant Pest Database [online]. Assessed with assistance of the New South Wales Department of Primary Industry Biosecurity Collections, Australia (accessed 12 June 2020).

Plumley, K.A., Gould, A.B. and Clarke, B.B. (1997) Impact of temperature, osmotic potential and osmoregulant on the growth of three ectotrophic root-infecting fungi on Kentucky bluegrass. *Plant Disease* 81, 873–879.

Rillig, M.C. (2005) A connection between fungal hydrophobins and soil water repellency. *Pedobiologica* 49, 395–399.

Robinson, M., Neylan, J., Hinch, J. and Parish, J. (1995) *Pests of Bowling Greens in Western Australia*. Final Report TU301. Horticultural Research & Development Corporation, Sydney, Australia.

Ryan, J. (1986) Spring dead patch – *Leptosphaeria narmari* its prevention and cure. *The Bowling Greenkeeper* 19(July), 14–15.

Sarniguet, A. and Lucas, P. (1992) Evaluation of populations of fluorescent pseudomonads related to decline of take-all patch on turfgrass. *Plant and Soil* 145, 11–15.

Sato, A., Kageyama, K. and Hyakumachi, M. (1999) Biological control of snow mold diseases in turfgrass by *Humicola grisea* var. *grisea*. *Annals of the Phytopathological Society of Japan* 65, 354.

Schoevers, T.A.C. (1937) Some observations on turf diseases in Holland. *Journal of the Board of Greenkeeping Research* 5, 23–26.

Schroeder, K.L., Martin, F.N., de Cock, A.W.A.N., Levesque, C.A., Spies, C.F.J., et al. (2013) Molecular detection and quantification of *Pythium* species: evolving taxonomy, new tolls and challenges. *Plant Disease* 97, 4–20. DOI: 10.1094/PDIS-03-12-0243-FE

Scott, D.B. (1989) *Gaeumannomyces graminis* var. *graminis* on Gramineae in South Africa. *Phytophylactica* 21, 251–254.

Shantz, H.L. and Piemeisel, R.L. (1917) Fungus fairy rings in eastern Colorado and their effect on vegetation. *Journal of Agricultural Science* 11, 191–246.

Shim, G.-Y. and Kim, H.-K. (1995) Identification and pathogenicity of *Rhizoctonia* spp. isolated from turfgrasses in golf courses in Korea (abstract in English). *Korean Journal of Turfgrass Science* 9, 235–252.

Sifers, S.I., Beard, J.B. and Kim, K.S. (1985) Spring root decline (SRD): a research summary. In: *Texas Turfgrass Research – 1985*. Texas Agricultural Experiment Station, Texas A&M University, College Station, Texas, pp. 19–30.

Siviour, T. (1972) Ophiobolus patch. *Bowls in New South Wales* 35, 28.

Smiley, R.W. (1993) Historical perspective of research on ectotrophic root-infecting pathogens of turfgrasses. In: Clarke, B.B. and Gould, A.B. (eds) *Turfgrass Patch Diseases Caused by Ectotrophic Root-Infecting Fungi*. The American Phytopathological Society, St. Paul, Minnesota, pp. 1–15.

Smiley, R.W., Dernoeden, P.H. and Clarke, B.B. (2005) *Compendium of Turfgrass Diseases*, 3rd edn. The American Phytopathological Society, St. Paul, Minnesota.

Smith, A.M. (1965) *Ophiobolus herpotricha*; a cause of spring dead spot in couch. *The Agricultural Gazette* 76, 753–758.

Smith, A.M. (1969) An oat-attacking strain of take-all in New South Wales. *Journal of the Australian Institute of Agricultural Science* 35, 270.

Smith, A.M. (1971) Control of spring dead spot on couch grass turf in New South Wales. *Journal of the Sports Turf Research Institute* 47, 60–65.

Smith, J.D. (1980) Fairy rings: biology, antagonism and possible new control methods. In: Joyner, B.G. and Larson, P.O. (eds) *Advances in Turfgrass Pathology*. Harcourt Brace Jovanovich, Inc., Duluth, Minnesota, pp. 81–85.

Smith, J.D. (1987) *Winter-Hardiness and Overwintering Diseases of Amenity Turfgrasses with Special Reference to the Canadian Prairies*. Technical Bulletin 1987-12E. Research Branch, Agriculture Canada, Saskatoon, Canada.

Smith, J.D. and Davidson, J.C.N. (1979) *Acremonium boreale* n. sp., a sclerotial, low-temperature-tolerant, snow mold antagonist. *Canadian Journal of Botany* 57, 2122–2139.

Smith, J.D. and Rupps, R. (1978) Antagonism in *Marasmius oreades* fairy rings. *Journal of the Sports Turf Research Institute* 54, 97–105.

Smith, J.D., Jackson, N. and Woolhouse, A.R. (1989) *Fungal Diseases of Amenity Turf Grasses*, 3rd edn. E. & F.N. Spon, London.

Stalpers, J.A. and Loerakker, W.M. (1982) *Laetisaria* and *Limonomyces* species (Corticiaceae) causing pink diseases in turf grasses. *Canadian Journal of Botany* 60, 529–537.

Stalpers, J.A., Redhead, S.A., May, T.W., Rossman, A.Y., Crouch, J.A., et al. (2021) Competing sexual–asexual generic names in *Agaricomycotina* (*Basidiomycota*) with recommendations for use. *IMA Fungus* 12, 22. DOI: 10.1186/s43008-021-00061-3

Stephens, C.M. and Kerns, J.P. (2020) First report of *Gaeumannomyces graminicola* causing bermudagrass decline of ultradwarf bermudagrass putting greens in North Carolina. *Plant Disease*. DOI: 10.1094/PDIS-10-19-2147-PDN

Stephens, C.M., Kerns, J.P. and Gannon, T.W. (2019) *In vitro* sensitivity and pathogenicity of organism causing take-all root rot on ultradwarf bermudagrass putting greens. Presented at *ASA, CSSA and SSA International Annual Meeting, San Antonio, Texas, 1–13 November 2019*.

Stephens, C.M., Gannon, T.W., Cubeta, M.A., Sit, T.L. and Kerns, J.P. (2022) Characterization and aggressiveness of take-all root rot pathogens isolated from symptomatic bermudagrass putting greens. *Phytopathology* 112, 811–819. DOI: 10.1094/PHYTO-05-21-0215-R

Stirling, M. (2001) *Ectotrophic Root-Infecting Fungi on Golf Turf in Queensland*. Project No. TU00005. Horticulture Australia Ltd, Sydney, Australia.

Stricker, S., Hsiang, T. and Bertrand, A. (2017) Reaction of bentgrass cultivars to a resistance activator and elevated CO_2 levels when challenged with *Microdochium nivale*, the cause of Microdochium patch. *International Turfgrass Society Research Journal* 13, 229–232.

Stuttard, R.D., Mann, R.L. and Newell, A. (2005) Does *Gaeumannomyces graminis* exhibit specialisation to *Agrostis* spp. and *Festuca* spp.? *International Turfgrass Research Society Annexe – Technical Paper* 10, 28–29.

Tani, T. and Beard, J.B. (1997) *Color Atlas of Turfgrass Diseases*. Wiley, Hoboken, New Jersey.

Tisserat, N., Fry, J. and Green, D. (1994) Managing Rhizoctonia large patch. *Golf Course Management* 62, 58–61.

Titone, P., Mocioni, M., Landschoot, P.J. and Gullino, M.L. (2004) Survey of ectotrophic root infecting fungi associated with turfgrass patch diseases in Italy. *Acta Horticulturae* 661, 491–498. DOI: 10.17660/ActaHortic.2004.661.65

Tkaczyk, M. (2020) Phytopythium; origin, differences and meaning in modern plant pathology. *Folia Forestalia Polonica, Series A – Forestry* 62, 227–232. DOI: 10.2478/ffp-2020-0022

Tran, N.T., Taggart, A.R., Drenth, A., Shivas, R.G., Loch, D.S., *et al.* (2020) Couch smut, an economically important diseases of *Cynodon dactylon* in Australia. *Australasian Plant Pathology* 49, 87–94.

Tredway, L.P., Soika, M.D. and Clarke, B.B. (2001) Red thread development in perennial ryegrass in response to nitrogen, phosphorus, and potassium fertiliser applications. *International Turfgrass Society Research Journal* 9, 715–718.

Tredway, L.P., Tomaso-Peterson, M., Perry, H. and Walker, N.R. (2009) Spring dead spot of bermudagrass: a challenge for researchers and turfgrass managers. *Plant Health Progress*. DOI: 10.1094/PHP-2009-0710-01-RV

Tredway, L.P., Tomaso-Peterson, M., Kerns, J.P. and Clarke, B.B. (2023) *Compendium of Turfgrass Diseases*, 4th edn. The American Phytopathological Society, St. Paul, Minnesota.

Tronsmo, A.M., Hsiang, T., Zhao, G. and Griffith, M. (2001) Low temperature diseases caused by *Microdochium nivale*. In: Iriki, N., Gaudet, D.A., Tronsmo, A.M., Matsumoto, N., Yoshida, M. and Nishimune, A. (eds) *Low Temperature Plant Microbe Interactions Under Snow*. Hokkaido National Agricultural Experiment Station, Sapporo, Japan, pp. 75–86.

Tucker, M.A., Bronzato-Badial, A., King, J., McCurdy, J.D., Vines, P.L. and Tomaso-Peterson, M. (2022) Identification, frequency of occurrence and, inoculum density of select ectotrophic root-infecting fungi within ultradwarf bermudagrass greens in Mississippi. *International Turfgrass Society Research Journal* 14, 902–910. DOI: 10.1002/its2.104

Turgeon, A.J. and Kaminski, J.E. (2019) *Turfgrass Management, Edition 1.0*. Turfpath LLC, State College, Pennsylvania.

van der Plaats-Niterink, A.J. (1981) Monograph of the genus *Pythium*. *Studies in Mycology* 21, 1–242.

Vargas, J.M. (2005) *Management of Turfgrass Diseases*, 3rd edn. Wiley, Hoboken, New Jersey.

Venkatasubbaiah, P., Tisserat, N.A. and Chilton, W.S. (1994) Metabolites of *Ophiosphaerella herpotricha*, a cause of spring dead spot of bermudagrass. *Mycopathologia* 128, 155–159.

Vincelli, P. and Williams, D. (1998) *Managing Spring Dead Spot of Bermudagrass*. Publication No. ID-130. Cooperative Extension Service, Agriculture and Natural Resources, University of Kentucky, Lexington, Kentucky.

Vines, P.L., Hoffman, F.G., Meyer, F., Allen, T.W. and Tomaso-Peterson, M. (2021) *Gaeumannomyces nanograminis*, sp. nov., a hyphopodiate fungus identified from diseased roots of ultradwarf bermudagrass in the United States. *Mycologia* 113, 938–948. DOI: 10.1080/00227514.2021.1911192

Wadsworth, D.F. and Young, H.C. (1960) Spring dead spot of bermudagrass. *Plant Disease Reporter* 44, 516–518.

Walker, J. (1975) Take-all diseases of Gramineae: a review of recent work. *Review of Plant Pathology* 54, 113–144.

Walker, J. and Smith, A.M. (1972) *Leptosphaeria narmari* and *L. korrae* spp. nov., two long-spored pathogens of grasses in Australia. *Transactions of the British Mycological Society* 58, 459–469.

Walker, N.R., Mitchell, T.K., Morton, A.N. and Marek, S.M. (2006) Influence of temperature and time of year on colonisation of bermudagrass roots by *Ophiosphaerella herpotricha*. *Plant Disease* 90, 1326–1330.

Wang, S.J. (2015) First report of take-all root rot caused by *Gaeumannomyces graminis* var. *graminis* on *Paspalum vaginatum* in China. *Plant Disease* 99, 1858. DOI: 10.1094/PDIS-01-15-0083-PDN.

Watschke, T.L., Dernoeden, P.H. and Shetlar, D.J. (2013) *Managing Turfgrass Pests*, 2nd edn. CRC Press, Boca Raton, Florida.

Weibel, E.N., Tredway, L.P. and Clarke, B.B. (2002) Impact of temperature and fungal isolate on the susceptibility of bentgrass cultivars to take-all patch. *Phytopathology* 92, S146–S147.

Weste, G. (1970) Factors affecting vegetative growth and production of perithecia in culture by *Ophiobolus graminis*. *Australian Journal of Botany* 18, 11–28.

White, R.H. (2004) Environment and culture affect bermudagrass growth and decline. *USGA Green Section Record* 42, 21–24.

Whiteley, G. and Baldwin, N.A. (1990) Variations in soil strength in the zones of fairy rings formed by *Marasmius oreades*. *Journal of the Sports Turf Research Institute* 66, 109–114.

Wollaston, W. (1807) On fairy rings. *Philosophical Transactions of the Royal Society of London* 97, 133–138.

Wong, F.P. and Kaminski, J.E. (2007) A new Rhizoctonia disease of bluegrass putting greens. *Golf Course Management* 75, 98–103.

Wong, F.P., Gelernter, W., Stowell, L. and Tisserat, N.A. (2003) First report of *Gaeumannomyces graminis* var. *graminis* on kikuyugrass (*Pennisetum clandestinum*) in the United States. *Plant Disease* 87, 600. DOI: 10.1094/PDIS.2003.87.5.600A

Wong, P.T.W. (2002) *Gaeumannomyces wongoonoo* sp. nov., the cause of a patch disease of buffalo grass (St. Augustine grass). *Mycological Research* 106, 857–862.

Wong, P. (2019) ERI diseases: new and problematic patch diseases. *New South Wales Golf Course Superintendents' Association Newsletter* (Summer), 12–17.

Wong, P.T.W. and Baker, R. (1986) Biological suppression of Ophiobolus patch of turfgrasses. *Journal of the Sports Turf Research Institute* 62, 141–146.

Wong, P.T.W. and Siviour, T. (1979) Control of Ophiobolus patch in Agrostis turf using avirulent fungi and take-all suppressive soils in pot experiments. *Annals of Applied Biology* 92, 191–197.

Wong, P.T.W. and Worrad, D. (1989) Preventive control of take-all patch of bentgrass using fungicides and an antagonistic fungus. In: Martin, P.M. (ed.) *Proceedings of 1st Australasian Turf Researchers Seminar, Bermagui, NSW, Australia, 4–6 April 1989*. Australian Turfgrass Research Institute, Sydney, pp. 35–37.

Wong, P.T.W., Worrad, D.J. and Beehag, G.W. (1988) Recent advances in the control of take-all (Ophiobolus) patch of bentgrass using triazole fungicides and biological antagonists. In: *Proceedings of the 10th National Turf Conference, Perth, Western Australia, 12–17 June 1988*. Australian Golf Course Superintendents Association, Melbourne. Australia.

Wong, P.T.W., Tan, M.K. and Beehag, G.W. (2000) Confirmation of take-all patch of turfgrass in Tifdwarf hybrid couch grass (bermudagrass) by morphological and DNA methods. *Australasian Plant Pathology* 29, 19–23.

Wong, P.T.W., Dong, C., Stirling, A.M. and Dickinson, M.L. (2012) Two new *Magnaporthe* species pathogenic to warm-season turf grasses in Australia. *Australasian Plant Pathology* 41, 321–329.

Wong, P.T.W., Tan, Y.P. and Shivas, R.G. (2014) *Penicillifer martinii*. Fungal Planet 280 (10 June 2014).

Wong, P.T.W., Dong, C., Martin, P.M. and Sharp, P.J. (2015a) Fairway patch – a serious disease of couch (syn. Bermudagrass) [*Cynodon dactylon*] and kikuyu (*Pennisetum clandestinum*) turf in Australia caused by *Phialocephala bamuru* Wong, P.T.W. & C. Dong sp. nov. *Australasian Plant Pathology* 44, 545–555. DOI: 10.1007/s13313-015-0369-0

Wong, P.T.W., Khemmuk, W., Geering, A.D.W. and Shivas, R.G. (2015b) *Budhanggurabania cyndonticola*. Fungal Plant 359 (10 June 2015).

Wong, P.T.W., Tan, Y.P., Weese, T.L. and Shivas, R.G. (2022) *Magnaporthiopsis* species associated with patch diseases in turfgrasses in Australia. *Mycosphere* 13, 602–611. DOI: 10.5943/mycosphere/13/1/5

Woodcock, T. (1983) Patch diseases of turfgrass. In: *Proceedings of the Turfgrass Disease Seminar, 12–13 July 1983*. Victorian College of Agriculture and Horticulture (Burnley) and Turf Research Institute (Frankston), Victoria, Australia, pp. 13–29.

Worf, G.L., Stewart, J.S. and Avenius, R.C. (1986) Necrotic ring spot disease of turfgrass in Wisconsin. *Plant Disease* 70, 453–458.

Wu, C. and Hsiang, T. (1998) Pathogenicity and formulation of *Typhula phacorrhiza*, a biocontrol agent of gray snow mold. *Plant Disease* 82, 1003–1006.

York, C.A. (1998) *Turfgrass Disease and Associated Disorders*. The Sports Turf Research Institute, Bingley. UK.

York, C.A. and Canaway, P.M. (2000) Water repellent soils as they occur on UK golf greens. *Journal of Hydrology* 231–232, 126–133.

Zhang, W., Hu, M., Liu, G., Gao, Z., Li, M. and Nan, Z. (2015a) Investigation and characterisation of red thread and pink patch on warm-season turfgrasses in Hainan Province, tropical China. *European Journal of Plant Pathology* 141, 311–325.

Zhang, W., Nan, Z., Tian, P., Hu, M., Gao, Z., Li, M. and Liu, G. (2015b) *Microdochium paspali*, a new species causing seashore paspalum disease in southern China. *Mycologia* 107, 80–90. DOI: 10.3852/14-119

Zidek, M.J., Yu, L., Jochum, M. and Jo, Y.-K. (2021) Complexity of *Gaeumannomyces* species causing take-all root rot of St Austinegrass in Texas. *Mycologia* 113, 599–611. doi: 10.1080/00275514.2021.1881735

Zong, J., Chen, J., Geng, J., Dong, Y., Liu, J., *et al.* (2022) First report of take-all root rot caused by *Gaeumannomyces graminis* on hybrid bermudagrass in China. *Plant Disease* 106, 2267. DOI: 10.1094/PDIS-12-21-2841-PDN

7

Warm-Weather Fungal Diseases of Mown Grasses

Abstract
This chapter covers the multitude of fungal pathogens occurring primarily during the warm and hotter months of the year when conditions are conducive to their growth and development to cause disease. However, certain diseases caused by fungal pathogens may persist at any time of year, depending on environmental conditions. Some fungal diseases are associated with the foliage, others with the stems and roots of their hosts.

Many warm-weather diseases are common and readily recognizable while other diseases share common symptoms often leading to a misdiagnosis. Many fungal diseases result in intractable and disfiguring symptoms, once established, and can be problematic and challenging to manage. Effective management of many fungal diseases requires an integrated approach encompassing genetic resistance and cultural, possible biological and chemical technologies.

Warm-weather fungal diseases of mown grasses are generally incited by mesophilic fungi able to survive and infect at moderate to relatively high temperatures causing disease during warm to hot periods. These fungal pathogens can infect and colonize host tissues from early spring through summer even to early autumn under environmental conditions that favour fungal growth. Warm-weather fungal diseases may infect the seedheads, foliage, stem and/or roots, depending on the specific host–pathogen interaction.

Many warm-weather fungal diseases share many common environmental and management-induced factors as outlined in Chapter 5 (this volume). Certain fungicides and biofungicides have registration in many countries for turfgrass application against the most common fungal diseases. Details of the fungicide groups and their characteristics are provided in Chapter 11 (this volume). Readers are strongly encouraged to consult fungicide and biofungicide labels for current information about specific product registrations.

Foliage-borne fungal diseases

Numerous fungal pathogens, some widely distributed and others more restricted, cause turfgrass diseases. Late-stage symptoms range from small-sized, randomly scattered spots some coalescing to form mosaic regions of chlorotic and necrotic patches with no definitive shape or boundary. Fungal infection typically begins on individual leaves then spreads to nearby leaves of other plants. Certain foliage- and soil-borne diseases and their associated symptoms may be the result of more than one active pathogen species, often complicating disease symptoms and a precise diagnosis.

Colletotrichum-Incited Diseases

Colletotrichum is a large genus comprising hundreds of described fungal species and is an important genera of plant-pathogenic fungi worldwide. The genus contains species sharing similar characteristics and has been the subject of continued taxonomic reclassification. *Colletotrichum* species gain their nutrition as saprophytes, endophytes or pathogens with transitory lifestyles, depending on host–fungus combination and environmental conditions (Jayawardena *et al.*, 2021; Liu *et al.*, 2022).

Numerous graminicolous *Colletotrichum* species are associated with cool-season and warm-season turfgrasses (Table 7.1), causing anthracnose diseases on turfgrass systems worldwide. Anthracnose describes foliage diseases caused by fungi that produce acervuli containing sterile hair-like structures (setae), aiding identification of the diseases.

Research has shown genetic variation exists within certain *Colletotrichum* populations isolated from different turfgrass species, resulting in a degree of host specificity. New *Colletotrichum* species continue to be isolated from non-symptomatic and symptomatic grass hosts and differentiated using morphological and molecular-based methodologies (e.g. DNA sequencing and others). Many *Colletotrichum* species have been isolated from warm-season grass species managed under low cultural intensity in tropical and subtropical regions.

Much of what is known about the identity, basic biology and ecology of *Colletotrichum* species associated with turfgrass systems emanates from a greater understanding of the biology and epidemiological factors of pathogenic species investigated in crop systems (Prusky *et al.*, 2000; Crouch and Beirn, 2009). The global distribution and virulence of many *Colletotrichum* species remain unknown. The widespread cultivation of numerous forage and pasture grasses, some known to be hosts of *Colletotrichum* species, poses a future potential risk of anthracnose diseases emerging on specific turfgrass species in many countries.

Anthracnose on cool-season and warm-season grasses

Anthracnose has long been known to occur, becoming problematic once established, on certain cool-season turfgrasses. Anthracnose on cool-season turfgrasses is caused by *Colletotrichum cereale* (previously named *Colletotrichum graminicola*) and may be expressed in two phases, foliar blight and basal rot (crown and stems) disease (Tredway *et al.*, 2023), depending on the host–fungi combination and season of the year.

Rapid disease progression under highly favourable conditions and potential fungicide resistance on mixed annual bluegrass–creeping bentgrass golf greens accounts for anthracnose causing severe disfiguration and being highly challenging to effectively manage (Fig. 7.1). *C. cereale* may be isolated from the same infected plants in combination with other foliar pathogens (e.g. *Alternaria* spp., *Curvularia* sp. and

Table 7.1. *Colletotrichum* species and their principal host grasses.

Colletotrichum species	Host grasses	References
C. axonopodi	Carpetgrasses	Baxter *et al.* (1983); Tanaka *et al.* (1999); Fuke *et al.* (2006); Crouch and Beirn (2009); Crouch *et al.* (2009); Crouch (2014); Shivas *et al.* (2016); Zhang *et al.* (2020); Alizadeh *et al.* (2021)
C. capsici	Bermudagrass	
C. caudatum	Centipede grass, zoysiagrass	
C. cereale	Bentgrass, bluegrasses, fescues, crested dog's tail, ryegrass	
C. eremochloae	Centipede grass	
C. hanaui	*Digitaria* spp.	
C. nicholsonii	*Paspalum* spp.	
C. paspali	Bahiagrass	
C. sublineola	Centipede grass	
C. zoysiae	Zoysiagrass	

Microdochium bolleyi), suggestive of a disease complex of primary and secondary pathogens.

Anthracnose on warm-season grasses has only been known in recent decades. Early reports of anthracnose caused by *Colletotrichum caudatum* on manilla grass appeared during the 1990s and on centipede grass in the early 2000s (Tanaka *et al.*, 1999; Fuke *et al.*, 2006). Anthracnose is known to occur on less-intensively managed centipede grass and carpetgrasses (Crouch *et al.*, 2009; Crouch and Tomaso-Peterson, 2012).

Geographic distribution and host range

Anthracnose on cool-season turfgrasses has a worldwide distribution in subtropical, temperate and transitional regions where they are managed. The disease occurs throughout the northern central regions of the United States, southern Canada, Western Europe, the United Kingdom, East Asia and Australasia (Table 7.2). Anthracnose undoubtedly occurs in other countries under cool and humid conditions and is misdiagnosed. Annual bluegrass is the primary host and creeping bentgrass to a lesser degree (Dernoeden, 2013), depending on fungal strain and environmental conditions. The global distribution of anthracnose on warm-season turfgrasses remains unclear but no doubt is more widespread in subtropical and tropical regions than currently documented.

Signs and symptoms

Anthracnose signs and symptoms are well known on closely mown cool-season grasses (i.e. annual bluegrass, creeping bentgrass and fine fescues) but to a much lesser extent on warm-season grasses (e.g. carpetgrasses and centipede grass). Symptomatic, cool-season turfgrass leaves may

Fig. 7.1. Severe damage caused by anthracnose to plants on a putting green. (Courtesy of Nathan R. Walker.)

Table 7.2. Reported global distribution of anthracnose on turfgrass.

Region/country	References
Canada	Charbonneau and Hsiang (2015)
United States	Kerns and Tredway (2013); Tredway *et al.* (2023)
Europe	Mann (2004)
United Kingdom	Smith *et al.* (1989); Mann and Newell (2005)
Australasia	Falloon (1975); Shakesby (1989); Bransgrove (2005)
East Asia	Tani and Beard (1997); Fuke *et al.* (2006); Zhang *et al.* (2020)

have numerous, dark-coloured setae observed microscopically protruding from infected leaves, which is a characteristic sign of anthracnose. Large numbers of acervuli may also be observed microscopically on dead leaves in the upper thatch-mat region and is indicative of the saprophytic stage of the fungus.

Anthracnose symptoms on closely mown, cool-season turfgrasses vary depending on the stage and phase of the disease. Overall, early-stage symptoms of anthracnose on closely mown golf greens appear as a random mosaic of light yellow to reddish-brown areas and general thinning (Fig. 7.2) depending on the distribution and density of annual bluegrass and bentgrass plants.

Progression of the disease under conducive conditions often results in individual leaves displaying elongated, chlorotic spots enlarging to orange- or yellow-coloured and ranging 6–12 mm in size depending on mowing height and environmental conditions (Fig. 7.3). Gradual coalescing of adjacent spots results in larger-sized, random-scattered areas displaying a yellow or reddish-brown coloration and thinning coverage resembling other foliar diseases (e.g. leaf spot or bacterial wilt). The presence of acervuli may be observed on infected leaves and sheaths.

Late-stage symptoms of anthracnose result in significant patchiness and surface disfiguration and large, non-uniform regions of discoloured and dead turfgrass (Fig. 7.4). Bentgrass stolons may also be affected in severe cases. Symptoms are observed either on annual bluegrass or creeping bentgrass but rarely on both grass species on the same infected green.

Symptom expression of anthracnose on warm-season turfgrasses is not well documented. Individual infected leaves may develop greyish-white to light brown, oblong-shaped lesions with distinctive borders, depending on host–pathogen combination. Enlarging legions may coalesce forming larger, irregularly sized and shaped sections on the leaves. Symptom development causes leaf blighting progressing downwards on the infected leaves resulting in a mosaic of yellow- to reddish-brown-coloured and necrotic leaves, depending on grass species and environmental conditions. Stem and root tissues are generally not affected.

Disease profile

Much of what is known about the biology and lifecycles of anthracnose pathogens on grasses

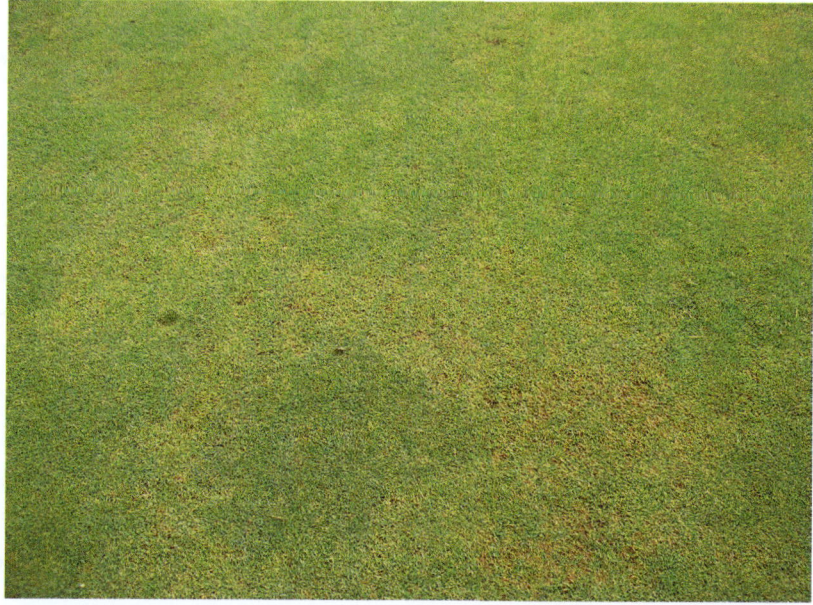

Fig. 7.2. Early-stage symptoms of anthracnose on a mixed annual bluegrass–creeping bentgrass golf green. (Courtesy of Ben Evans.)

Fig. 7.3. Close-up view of damage caused by anthracnose. (Courtesy of Ben Evans.)

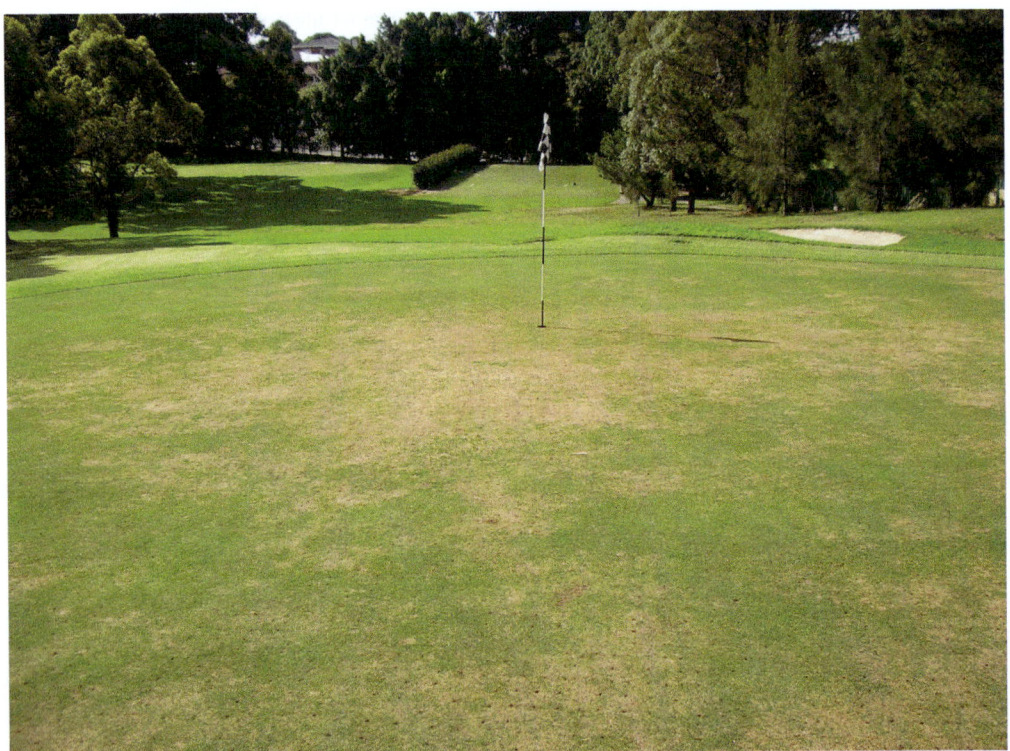

Fig. 7.4. Late-stage symptoms of anthracnose on a mixed annual bluegrass–creeping bentgrass golf green. (Courtesy of Ben Evans.)

emanates from the disease caused by *C. cereale* and *Colletotrichum sublineolum* on cereal crops (Prusky *et al.*, 2000) and *C. cereale* on golf greens (Tredway *et al.*, 2023) primarily in the United States. Anthracnose on golf greens comprising cool-season turfgrasses is generally considered a polycyclic, stress-induced (senectopathic) foliar and crown disease facilitated by intense mowing and reduced nutrition and irrigation regimes (Kerns and Tredway, 2013). Key aspects of the biology and ecology of *Colletotrichum* species associated with warm-season turfgrasses remain unclear.

C. cereale is generally considered a facultative parasite usually gaining its nutrition by a hemibiotrophic lifestyle encompassing biotrophic and necrotrophic stages. Furthermore, on cool-season turfgrasses the fungus can undergo two distinct phases: foliar blight or basal (crown) rot, depending on environmental conditions. The fungus is considered to survive unfavourable conditions saprophytically in decomposing tissues as mycelium, acervuli or dormant structures on host tissues.

C. cereale and *C. sublineolum* can penetrate leaf tissue of susceptible turfgrass hosts within 12–24 h following contact, either directly via intact surfaces and stomates or indirectly by open wounds (Smith *et al.*, 1989; Khan and Hsiang, 2003) depending on host–pathogen combination and environmental conditions. Initial infection following penetration results in a relatively short biotrophic infection phase (12–24 h) during which the invading fungus grows systemically inside the plant gaining nutrition without causing discernible injury nor eliciting host defence responses. The necrotrophic stage of gaining nutrition begins after several days, causing cell death and becoming evident with the outward development of symptoms.

The foliar and basal rot phases of anthracnose on cool-season turfgrasses occur at different times of the year depending on host–fungi combination and environmental conditions. Key aspects of the relationship between the seasonal occurrence of foliar blight and basal rot phases remain unclear. The foliar phase on mixed annual bluegrass–creeping bentgrass golf greens is associated with high-temperature stress during the warmer seasons. Progression from the foliar to the basal rot or crown phase can occur during different seasons, depending on temperature.

The basal rot stage commences when fungal mycelium colonizes the leaf sheaths at the base of the tillers. Anthracnose basal rot on annual bluegrass generally begins to develop during winter–spring and progressing into summer and is generally much more destructive on closely mown annual bluegrass compared to creeping bentgrass. The basal rot phase of anthracnose does not generally occur before high temperatures in summer. Anthracnose basal rot on golf greens is associated with anaerobic conditions in the thatch-mat region influenced by rainfall events and excessive irrigation (Landschoot and Hoyland, 1995). Anthracnose basal rot is the more destructive phase often causing turfgrass death. The fungus is disseminated by water or wind to adjacent plants.

Environmental influences

Several key environmental and management-induced factors have been shown to influence development of anthracnose (Box 7.1). Management-induced stress is the underlying factor on closely mown cool-season grasses.

TEMPERATURE AND MOISTURE Generally, fungal infection of cool-season turfgrasses during the foliar blight phase develops between 29 and 35°C and the basal rot phase between 15 and 24°C (Couch, 1995) depending on the *C. cereale* isolate. Free moisture in the form of high humidity or prolonged periods of leaf wetness and an anaerobic rhizosphere contribute to the frequency and severity of anthracnose.

CULTURAL PRACTICES Extremely close mowing, frequent lightweight rolling, reduced nutrition and irrigation, plant growth regulators, verticutting and sand topdressing, singly and in combination, have increased the importance of anthracnose on golf greens in the United States and probably elsewhere where the disease occurs.

Extremely low mowing heights less than 2.00 mm (bench setting) and reduced irrigation

---

Box 7.1. Key environmental factors of anthracnose.

- Elevated temperatures (> 18°C)
- Long periods of leaf wetness (12–18 h)
- Relatively low leaf nitrogen content
- Relatively low soil moisture level

have generally increased anthracnose severity (Crouch and Clarke, 2012; Hempfling et al., 2017, 2020). Mowing frequency in combination with lightweight rolling generally has minimal or no measurable, detrimental impacts on anthracnose (Roberts et al., 2012).

Reduced plant-available nitrogen and potassium are generally associated with increased anthracnose severity and use of plant-growth regulators singly or in combination with nitrogen has either intensified or marginally reduced anthracnose severity on annual bluegrass–creeping bentgrass (Murphy et al., 2018).

FOLIAGE WOUNDS The impact of foliage and stem wounds (i.e. mowing, aerification or sand topdressing and foot traffic) on golf greens remains unclear. Limited laboratory-controlled studies have generally shown that puncture wounding of the crown region of annual bluegrass followed by artificial inoculation of C. graminicola isolates predisposed the plants to greater disease as opposed to non-wounded plants (Landschoot and Hoyland, 1995).

Sand topdressing impacts from sand texture, rate and frequency have generally shown measurable reductions of anthracnose severity using weekly or fortnightly, light applications of medium to medium-fine sands (Inguagiato et al., 2013). Subtle wounding from mowing and sand topdressing generally is not a significant factor affecting anthracnose severity and any perceived increase in anthracnose associated with topdressing is likely due to indirect effects rather than wounding as anthracnose is not a wound-related disease (Murphy et al., 2018).

AERIFICATION PRACTICES Investigations into the impact of verticutting have generally shown relatively shallow operations conducted using triplex attachments as opposed to more deeper operations with walk-behind units to result in minimal measurable anthracnose (Dernoeden, 2013; Murphy et al., 2018).

Integrated management

Environmental conditions favouring anthracnose infection and virulence on mixed cool-season grass golf greens are relatively broad. Ultimately the prevalence of anthracnose will be determined partly by the host–pathogen combination and partly by the effectiveness of cultural and chemical management.

Effective management of anthracnose caused by *C. cereale* is key to maintaining high-quality, mixed cool-season grass golf greens (Box 7.2). Minimizing annual bluegrass populations in temperate regions is a key strategy of effectively managing anthracnose in creeping bentgrass–fine fescue swards. Proven cultural, biological and fungicidal options combined with frequent monitoring of disease signs and changing weather patterns are required. Appropriate and careful programming of fungicides is necessary to minimize potential of fungicide resistance.

Copper Spot

Copper spot, also called zonate leaf spot, is a relatively uncommon foliar disease but which may occur simultaneously with dollar spot during warm conditions. The disease is not widespread but can prove problematic once established on susceptible turfgrasses.

Causal agent

Microdochium sorghi (formerly *Gloecercospora sorghi*) is the causal agent of copper spot (Tredway et al., 2023) and an economically important pathogen of sorghum (Hernández-Restreopo et al., 2016); hence the specific name.

Geographic distribution and host range

Copper spot occurs in the eastern and southwestern regions of the United States. Bentgrass

---

Box 7.2. Integrated management factors of anthracnose.

- Monitoring turfgrass health for early signs and symptoms
- Maintain consistent growth with use of a balanced NPK nutrition
- Minimization of thatch-mat thickness
- Judicious irrigation water applications
- Targeted application of registered fungicides and biofungicides
- Reduced frequency of growth regulators
- Increase mowing height

and bermudagrass are host grasses (Smith *et al.*, 1989; Tredway *et al.*, 2023). The global distribution of copper spot remains unknown.

Signs and symptoms

The signs and late-stage symptoms of copper spot on golf greens may closely resemble those of dollar spot. Foliar lesions of copper spot are red to brown in colour. Late-stage symptoms appear as randomly scattered, small-sized, orange- to copper-coloured spots or patches 3.0–8.0 cm in diameter but with ill-defined margins (Fig. 7.5). Mycelium may be observed during warm, wet conditions.

Disease profile

Copper spot is generally not considered an economically important turfgrass disease. Many aspects of the biology and epidemiology have not been investigated. The fungus survives unfavourable

Fig. 7.5. Late-stage copper spot symptoms on a creeping bentgrass golf green. (Courtesy of Nathan R. Walker.)

conditions as black-coloured sclerotia or thick-walled mycelium in infected tissues. Fungal infection occurs between 20 and 30°C reportedly by direct penetration or through stomates.

Fungal inoculum and disease development can occur rapidly during favourable conditions of temperature, prolonged leaf wetness and release of plant exudates following mowing. Dissemination of conidia can occur via infected leaves, moving water and machinery. Infected turfgrass recovers with the onset of cooler conditions in autumn.

Integrated management

There are few specific management recommendations for copper spot. Management practices for copper spot can be adopted from dollar spot, i.e. optimizing soluble nitrogen application, maintaining slightly acidic rootzones and application of registered fungicides when required. Succinate dehydrogenase inhibitor (SDHI) fungicides have been shown to increase the incidence of copper spot, thus should be avoided (Turgeon and Kaminski, 2019).

Dollar Spot

Dollar spot ranks among one of the most widespread and destructive foliar diseases worldwide on closely mown turfgrass. Rapid appearance of symptoms coupled with multiple fungal species and strains, an extensive grass host range and potential fungicide resistance accounts for the continual development of improved turfgrass cultivars and cultural and fungicide management tactics against this disease.

The disease name was originally given to individual symptoms on bentgrass golf greens (Fig. 7.6) in the United States approximately the size of a dollar coin. Dollar spot represents one of the most intensely studied of all turfgrass diseases and was subject to recent taxonomic revision. Key aspects of the biology and genetics and the relative susceptibility of the dollar spot pathogens to fungicides continue to be investigated.

Causal agents

The dollar spot pathogen (formerly *Sclerotinia homoeocarpa*) has been reclassified under a new fungal genus, *Clarireedia* (Salgado-Salazar et al., 2018). Reclassification of the pathogen to *Clarireedia* based on multiple fungal isolates collected from North America, Europe and East Asia accounted for morphological differences between numerous fungal isolates. Currently six *Clarireedia* species are associated with cool-season and warm-season grasses worldwide (Table 7.3). Mycelium of the isolates of dollar spot pathogens

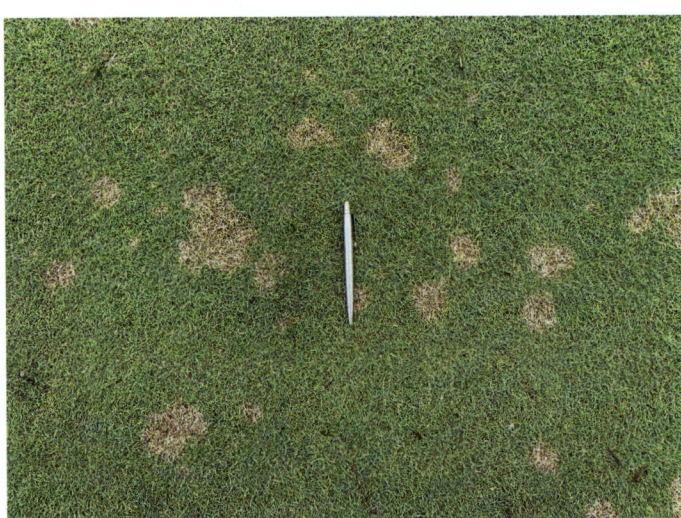

Fig. 7.6. Dollar spot symptoms on a creeping bentgrass golf green. (Courtesy of Jyri Kaapro.)

Table 7.3. *Clarireedia* species by host grass group and region.

Clarireedia species	Host group	Region/country	References
C. bennettii	C3	North America, UK, Europe, East Asia	Salgado-Salazar *et al*. (2018); Ayanardi *et al*. (2019); Hu *et al*. (2021); Zhang *et al*. (2022)
C. homoeocarpa	C3	UK, East Asia	
C. jacksonii	C3/C4	USA, UK, Netherlands, Europe, East Asia	
C. monteithiana	C3/C4	The Americas, East Asia	
C. paspali	C4	East Asia	
C. hainanense	C4	East Asia	

varies in colour from off-white to shades of grey, brown or yellow as the culture ages, depending on their geographic origin.

The specific identity of *Clarireedia* species in most countries where the disease occurs remains unclear. Limited studies have tentatively shown certain *Clarireedia* species (e.g. *Clarireedia jacksonii*) are associated with certain cool-season grasses while others (e.g. *Clarireedia monteithiana*) appear to be limited to certain warm-season grasses (Ayanardi *et al*., 2019; Zhang *et al*., 2022). Further work will clarify host–fungus combinations in many countries where the disease occurs.

Co-colonization of more than one *Clarireedia* species on the same infected plant and reclassification to *Clarireedia* spp. will have practical implications in developing new turfgrasses for dollar spot resistance, correct identification and proving virulence of fungi, as well as fungicide efficacy and resistance among fungal species.

Geographic distribution

Dollar spot has a global distribution (Dernoeden, 2013; Tredway *et al*., 2023) and probably occurs on most turfgrass species wherever managed in tropical, subtropical and transition zones of the world.

THE AMERICAS Dollar spot is among one of the earliest known and now most important turfgrass diseases on creeping bentgrass golf greens in the United States (Monteith and Dahl, 1932; Dernoeden, 2013). Dollar spot has a wide distribution throughout the continental United States and occurs in Hawaii, southern Canada, Central America and South America (Smith *et al*., 1989; Charbonneau and Hsiang, 2015).

AUSTRALASIA Dollar spot is widespread throughout the warmer regions of Australia (Beehag, 1999; Bransgrove, 2005) having been officially recognized in Queensland in the early 1930s (Bennett, 1937; Simmonds, 1966) on Queensland blue couch. In New Zealand, dollar spot was officially recorded for the first time in the 1960s on bentgrass and chewing's fescue (Boesewinkel, 1977). The distribution and current importance of dollar spot in New Zealand remain unclear.

EUROPE AND GREAT BRITAIN Dollar spot occurs widely in Europe and is known as 'coin spot' in Scandinavia (Espevig *et al*., 2022), becoming widespread in Nordic countries due to global warming and other factors. Dollar spot is relatively widespread in southern England (Mann and Perris, 2005).

SOUTH AFRICA AND EAST ASIA Dollar spot has been verified in recent years in South Africa (Schoeman *et al*., 2001) and in East Asia (Hu *et al*., 2021).

Host range and susceptibility

Collectively, the dollar spot pathogens have an extensive host range of primary and secondary cool-season and warm-season grasses worldwide. Cool-season grasses (e.g. bentgrass, bluegrasses, fescues and ryegrasses) are widely known and highly susceptible hosts. Warm-season grass species of bermudagrass (couchgrass) (Fig. 7.7) and centipede grass, kikuyugrass, Queensland blue couch, St. Augustinegrass, seashore paspalum and zoysiagrass are known hosts to dollar spot (Couch, 1995; Bransgrove, 2005; Tredway *et al*., 2023).

Fig. 7.7. Severe outbreak of dollar spot on a couchgrass (bermudagrass) apron. (Courtesy of Gary W. Beehag.)

Signs and symptoms

Early-morning presence of cobweb-like, whitish-coloured, fungal mycelium on closely mown turfgrass is characteristic of actively growing dollar spot (Fig. 7.8). Fungal mycelium may be present on uncut foliage during highly favourable conditions but may also characterize other foliar pathogens. Individual infected, fine-textured leaves may have hourglass-shaped lesions and girdling. Oval-shaped lesions with tan- to reddish-brown-coloured outer edges may occur along the margins of coarse-textured leaves; however, these same symptoms can be caused by other diseases.

Fungal mycelium enveloping several infected leaves may be observed during cool nights and on humid and windless early mornings in the presence of dew. Mycelium may persist during the day under cool and humid conditions and cloudy days, disappearing during the day as leaves become drier for whatever reason.

Dollar spot symptoms vary widely depending on turfgrass leaf maturity and mowing regime. Under relatively low height/high frequency mowing regimes (e.g. bowling and golf greens) immature symptoms are characterized by randomly scattered clusters of roughly circular bleached spots. Initially about 2.0 cm diameter, individual spots can enlarge up to 5.0 cm diameter or more in close proximity with each other.

Left unmanaged, individual spots typically coalesce forming larger-sized, mature straw-coloured spots as the disease progresses (Fig. 7.9). Mature spots form slight depressions which impacts on the aesthetics and playing quality of the surface. Reasons for the restricted size of individual dollar spots is unknown.

In contrast, under relatively high height/low frequency mowing regimes (e.g. residential lawns and playing fields) dollar spot symptoms may appear as large-sized, straw-coloured spots or small-sized patches up to 15 cm diameter, or even non-discrete areas of diseased turfgrass, depending on circumstances (Fig. 7.10).

Fig. 7.8. Fungal mycelium of a dollar spot pathogen on infected leaves. (Courtesy of Jyri Kaapro.)

Fig. 7.9. Late-stage, coalescing dollar spot symptoms on a creeping bentgrass golf green. (Courtesy of Jyri Kaapro.)

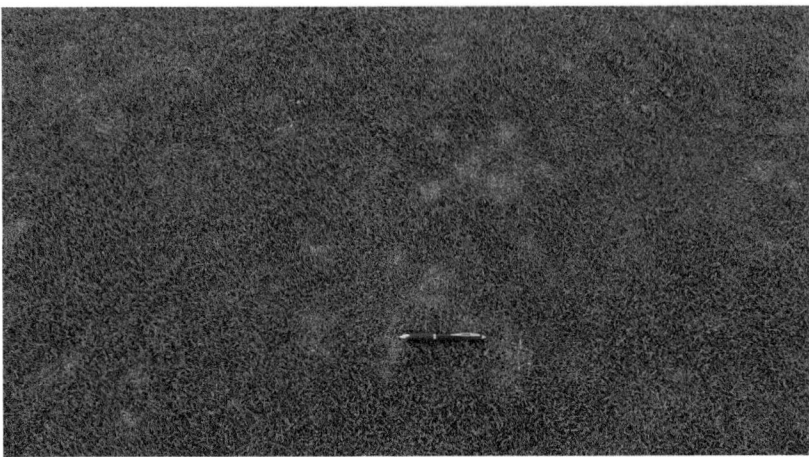

Fig. 7.10. Late-stage symptoms of dollar spot on a Queensland blue couch lawn in Australia. (Courtesy of Gary W. Beehag.)

Characteristic colour and lesions on infected foliage vary between turfgrass species.

Disease profile

Dollar spot is a polycyclic disease and knowledge of ecology and epidemiology of the disease has been extensively documented in North America (Fermanian *et al.*, 2003; Tredway *et al.*, 2023). During unfavourable conditions or in late winter, dormant mycelium in leaf lesions from previous disease epidemics or dark-coloured, sclerotia-like stroma surviving saprophytically in decaying host tissues or thatch-mat region, aid survival of the fungus.

Occurrence of dollar spot demonstrates seasonal fluctuations or epidemics from spring through to autumn, even early winter, depending on environmental conditions. Commencing in spring during highly favourable conditions, fungal mycelium grows radially from infected leaves and possibly uncollected leaf clippings bridging adjacent foliage. Fungal mycelium on warm-season turfgrass at higher mowing height may be more difficult to distinguish, depending on environmental conditions.

On contact with foliage, fungal mycelium has been commonly reported able to penetrate intact leaf surfaces, natural leaf openings (e.g. stomates), and cut leaf ends and mechanical wounds (Endo, 1966). Unequivocal evidence confirming penetration of cut leaf ends by the dollar spot fungus remains limited. Penetrating fungal hyphae colonize host cells and infection is expressed as oval- and hourglass-shaped leaf lesions.

Dissemination of fungal mycelium or infected leaf tissue is purported to result by movement of natural (e.g. wind and water) or anthropogenic (e.g. equipment) means. Fungal-contaminated seed of creeping bentgrass may provide a potential source of dollar spot (Rioux *et al.*, 2014).

Environmental influences

Several key environmental and management-induced factors have been shown to specifically influence development of dollar spot (Box 7.3). Atmospheric temperature and moisture are the two key determinant factors of disease occurrence.

TEMPERATURE AND MOISTURE The temperature range for growth and infection of the dollar spot pathogens is generally considered a reflection of the geographic origin of different species and isolates. This point explains partly why the seasonal frequency, duration and severity of dollar spot vary widely between and among countries in tropical, subtropical and temperate regions.

American fungal isolates have been generally shown to cause infection anywhere between 14 and 35°C (Endo, 1963; Skorulski, 2014) depending on the fungal isolate. Dollar spot epidemics have occurred after two consecutive wet

> **Box 7.3.** Key environmental factors of dollar spot.
>
> - Moderate to high temperatures (20–30°C)
> - High atmospheric humidity (> 85%)
> - Prolonged periods of leaf wetness (12–14 h)
> - Relatively low leaf nitrogen content
> - Relatively low soil moisture levels

> **Box 7.4.** Integrated management factors of dollar spot.
>
> - Disease-resistant cultivars
> - Monitoring turfgrass health and early disease signs and symptoms
> - Maintain consistent growth with use of a balanced NPK nutrition
> - Judicious irrigation water applications
> - Minimize leaf wetness duration
> - Foliage application of registered fungicides and biofungicides

days when the average temperature for the period was above 22°C or after three consecutive wet days when the average temperature for the period was 15°C or greater (Hall, 1984). Maximum virulence of the dollar spot pathogens generally occurs in a temperature range 21–27°C and a relative humidity above 85% (Couch, 1995).

HOST NUTRITION Low nitrogen fertility and relatively dry rootzone conditions have long been associated with occurrence of dollar spot. However, observations and experience show dollar spot can occur on susceptible turfgrasses under both deficient and excessive nitrogen. Light applications of iron sulfate suppress the disease (Shelton et al., 2021). Rootzone pH level appears to have no direct effect on dollar spot.

Application of processed organic-based and synthetic slow-release and soluble nitrogen sources on occurrence of dollar spot, primarily on creeping bentgrass, has resulted in conflicting results (Boulter et al., 2002; Davis and Dernoeden, 2002). Most studies support the view that soluble nitrogen sources applied at low rates can suppress dollar spot to manageable levels when applied continually throughout the growing season.

Integrated management

Environmental conditions favouring dollar spot infection and virulence are relatively broad and their combined interaction will determine duration, severity and frequency of the disease. Effective management of dollar spot on high-value turfgrass necessitates an integrated approach (Box 7.4). Regular monitoring of turfgrass health and weather variations as well as use of disease-resistant varieties in conjunction with proven cultural, biological and fungicidal options are required. Appropriate and careful programming of registered fungicides is a necessity given the high risk of fungicide resistance to the disease.

Disease-resistant cultivars

Genetic resistance against dollar spot has been widely documented among cultivars and interspecific hybrids of colonial and creeping bentgrass (Ryan et al., 2012; Bonos and Huff, 2013; Thompson et al., 2019), fine fescue (Braun et al., 2020; Espevig et al., 2022) and seashore paspalum (Flor et al., 2013; Steketee et al., 2017). Interspecific hybrids of colonial and creeping bentgrass (Belanger et al., 2004) and blends of resistant and susceptible cultivars (Abernathy et al., 2001) offer alternative approaches to minimize dollar spot infection provided other performance requirements are met.

Grey Leaf Spot

Grey leaf spot was a relatively minor disease on St. Augustinegrass in the United States until a change in the pathogen occurred, allowing it to broaden its host range. The name grey leaf spot was ascribed to the leaf lesions and sporulation on St. Augustinegrass. Grey leaf spot has long been associated with warm-season grasses but emergence of disease epidemics and severe blighting on perennial ryegrass and tall fescue during the 1990s in the United States represented a shift in host species.

Existence of fungicide-resistant strains of the pathogen makes fungicide selection and application a critical part of the integrated management of the disease on perennial ryegrass.

Causal agent

Pyricularia oryzae (*Magnaporthe oryzae*) is the currently accepted name of the causal agent of grey leaf spot (Tredway *et al.*, 2023). The taxonomy of *Pyricularia* has been the subject of revision (Uddin *et al.*, 2003) and a related species *Pyricularia grisea* also occurs on grasses.

Geographic distribution

Grey leaf spot has a global distribution and a wide turfgrass host range (Table 7.4). In the United States, the pathogen is probably now well established in areas where susceptible hosts are grown throughout the north-eastern, central, and south-eastern and western regions (Harmon and Latin, 2005; Kerns and Tredway, 2013). In Europe, grey leaf spot is widely distributed from Spain, Portugal and France through Germany, Austria and Switzerland and in the United Kingdom (K. Entwistle, England, 2021, personal communication).

Host range and susceptibility

Bahiagrass, bermudagrass, centipede grass, kikuyugrass, St. Augustinegrass, perennial ryegrass, hard fescue and tall fescue are reported hosts of grey leaf spot (Couch, 1995; Tredway *et al.*, 2023). Perennial ryegrass cultivars are highly susceptible to grey leaf spot and the disease is generally more damaging to relatively immature as opposed to more established swards. The disease is more severe on stands mown at relatively higher heights and where clippings are not collected (Uddin *et al.*, 2003).

Signs and symptoms

The presence of numerous small-sized, coloured spots or oblong-shaped lesions (1–3 mm long) on infected leaves is characteristic of grey leaf spot, depending on turfgrass species and stage of infection (Fig. 7.11). The fungus is capable of producing large numbers of asexual spores or conidia and their development on infected leaves presents a grey-velvet coloration.

On warm-season grasses individual spots are generally located along the mid-vein and are tan to grey with purple- to brown-coloured borders. Lesions may appear blue grey during more humid periods due to the presence of massed conidia. Numerous lesions may be present on an individual leaf and adjacent lesions may coalesce creating larger-sized lesions extending across the leaf.

On cool-season species, initial symptoms are small, water-soaked lesions which rapidly become necrotic. Lesions vary in size, shape and colour depending on infection stage. Lesions tend to be oblong-shaped and grey to light brown in colour with purple to dark brown borders or surrounded by a yellow-coloured halo, depending on the stage of infection. Grey leaf spot lesions on infected tall fescue are generally less than 0.5 mm long. Numerous lesions may exist on a single leaf and individual leaves may suffer complete blighting often with misshapen and twisted leaf ends. Cool-season turfgrasses suffer extensive damage resulting in plant death due to the disease.

Overall symptoms of grey leaf spot on closely mown swards appear as irregular and random necrotic spots to coalesced, blighted patches. Advanced symptoms on cool-season turfgrasses may resemble those of severe drought or heat stress. On golf courses, late-stage symptoms may be observed on perennial ryegrass in the higher-mown roughs followed by fairways mowed at lower heights.

Disease profile

Grey leaf spot is a polycyclic, foliar disease capable of devastating perennial ryegrass swards if

Table 7.4. Documented global distribution of grey leaf spot and host grass groups.

Region	Host group	References
North America	C3/C4	Yamanaka (1982); Smith *et al*. (1989); Bransgrove (2005); Kerns and Tredway (2013); Milazzo *et al*. (2018); Tredway *et al*. (2023)
Central/South America	C4	
Africa	C4	
East Asia	C3/C4	
Australasia	C4	
Western Europe	C3/C4	
Great Britain	C3	

Fig. 7.11. Grey leaf spot lesions on common St. Augustinegrass (buffalo grass) leaves. (Courtesy of Gary W. Beehag.)

left unmanaged. The biology, ecology and epidemiology of grey leaf spot on turfgrass in the United States are relatively well understood (Kerns and Tredway, 2013; Watschke et al., 2013).

In the United States, grey leaf spot on perennial ryegrass typically develops in summer (July/August) and continues through to autumn (September/October) depending on regional conditions. The grey leaf spot pathogen survives unfavourable conditions saprophytically as dormant mycelium or conidia on decomposing or infected plant tissues. Conidia are the primary source of inoculum under favourable temperature conditions of 25–28°C and humidity levels around 90%. Wind-blown dispersal of spores during stressful hot and dry periods has been proposed as an important underlying cause of disease epidemics on perennial ryegrass (Wong, 2006).

Sporulation and infection occur during warm to hot weather on environmentally stressed turfgrass at ambient temperatures greater than 25°C. Infection occurs rapidly and penetration of the leaf surface may take place directly on intact leaves, through stomates, or indirectly through wounds such as cut leaf ends. Infection and colonization are restricted to the immediate area of fungal penetration. Fungal lesions may appear 5–7 days following infection. Under favourable conditions, infected perennial ryegrass plants are devastated within several days.

Environmental influences

Several key environmental and management-induced factors have been shown to influence development of grey leaf spot (Box 7.5). Temperature and leaf wetness are key factors.

TEMPERATURE AND MOISTURE A close relationship between temperature, humidity and leaf wetness duration influences the occurrence of grey leaf spot (Uddin et al., 2003). Generally, temperatures favourable for sporulation, fungal infection and disease development have similar ranges. Optimum temperatures reported for fungal infection and disease development on warm-season grasses range around 21–30°C, depending on host species, humidity and leaf wetness period. Temperature variation in warmer regions is relatively small; thus high humidity and prolonged dew formation contribute to the development of grey leaf spot. Night-time temperatures must be considered given the fact that dew periods primarily occur at night.

CULTURAL PRACTICES Excessive application of soluble forms of nitrogen, relatively high mowing heights and certain pre-emergent herbicides

> **Box 7.5.** Key environmental factors of grey leaf spot.
>
> - Moderate temperatures (20–28°C)
> - High humidity (> 90%)
> - Prolonged leaf wetness (21–26 h)
> - Light and dark cycles
> - Excessive nitrogen
> - Certain herbicide formulations

> **Box 7.6.** Integrated management factors of grey leaf spot.
>
> - Disease-resistant varieties
> - Monitoring of rainfall and temperature conditions
> - Maintain consistent growth with use of a balanced NPK nutrition
> - Minimize leaf wetness periods
> - Judicious irrigation water applications
> - Caution when using certain pre-emergent herbicides
> - Use of registered fungicides and avoiding fungicide resistance

have been shown to influence severity of grey leaf spot, particularly on perennial ryegrass and tall fescue (Williams et al., 2001; Uddin et al., 2003).

Silica accumulation in epidermal cells of cereal crops has long been reported to be associated with resistance to rice blast possibly due partly to effects of a physical barrier or reduced absorption of applied nitrogen (Ou, 1985). Silicon (silicate) is transformed and absorbed as monosilicic acid and has been shown to reduce severity of grey leaf spot on St. Augustinegrass (Datnoff et al., 2005) and perennial ryegrass turfgrass (Rahman et al., 2015).

Pre-emergent herbicides dithiopyr and ethofumesate use has been associated with increased severity of grey leaf spot on perennial ryegrass (Uddin and Soika, 2000; Clarke and Vaiciunas, 2001). Application of the plant growth regulator trinexapac-ethyl may aid the management of grey leaf spot, depending on fungicide inputs (Uddin and Soika, 2000).

BIOLOGICAL ANTAGONISTS Antagonistic strains of the bacteria *Bacillus lentimorbus* and *Pseudomonas aeruginosa* and non-pathogenic isolates of *Rhizoctonia* sp. have shown promising results under controlled conditions in growth chambers and field plots in suppressing grey leaf spot (Uddin et al., 2003). Available formulations of microbial antagonists for application against grey leaf spot have yet to be commercialized.

Integrated management

Effective management of grey leaf spot on cool-season turfgrasses in particular necessitates a concerted and integrated approach (Box 7.6). Monitoring of disease signs and changing weather patterns in conjunction with proven cultural and fungicidal options are required to minimize the disease.

Disease-resistant varieties

Genetic resistance against grey leaf spot has been documented among perennial ryegrass, St. Augustinegrass and tall fescue cultivars (Bonos et al., 2005; Bonos and Huff, 2013; Kerns and Tredway, 2013). Physiological strains or pathotypes restrict the host range of the pathogen to specific grass species and high genetic variability of the grey leaf spot fungus presents challenges in attempting to develop highly resistant turfgrass varieties.

Helminthosporium Group Diseases

The word *helminthosporium* is of Greek origin, meaning 'worm spore', acknowledging the overall shape of fungal spores. Helminthosporium is no longer a valid fungal genus whereas now graminicolous species of fungi producing sexual or asexual spores formerly grouped under the genus are segregated into four fungal genera (Fig. 7.12). Previous adoption of a dual nomenclature system (i.e. anamorph and teleomorph) for the reproductive states (asexual and sexual) of certain fungi is no longer accepted among fungal taxonomists.

The anamorphic or asexual state of *Bipolaris* and *Curvularia* (teleomorph *Cochliobolus*), *Exsorohilum* (teleomorph *Setosphaeria*) and the teleomorph or sexual stage of *Pyrenophora* (anamorph *Drechslera*) are now assigned based on small differences in spore morphology and

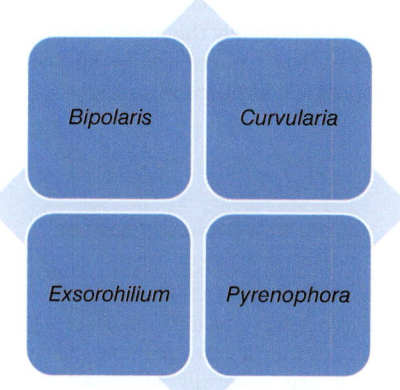

Fig. 7.12. Fungal genera previously grouped under Helminthosporium.

recently advanced molecular methodologies and systematics (Goh *et al.*, 1998; Manamgoda *et al.*, 2012, 2014). Further reclassifications may occur because of continual isolation of undescribed isolates and improvements in genetic-based methodologies.

Differentiation between each of these fungal genera and delineation into species for diagnosticians remain challenging. Conventional microscopy techniques to identify fungal isolates, some with overlapping spore size and shape, using published fungal taxonomic keys and adoption of genetic-based techniques (e.g. DNA sequencing) often prove inconsistent because of intermediatory fungal species and strains.

As a group, *Bipolaris, Curvularia, Exsorohilum* and *Pyrenophora* cause leaf spot diseases (Fig. 7.13). One or more *Bipolaris, Curvularia, Exsorohilum* or *Pyrenophora* species together with other foliage-infecting fungi (e.g. *Alternaria, Colletotrichum* or *Leptosphaerulina*) may be simultaneously isolated from the same infected turfgrass. Co-colonization of multiple fungi raises the possibility of a disease complex of primary and secondary pathogens making a precise diagnose of the causal agent based on field symptoms problematic.

Bipolaris, Curvularia, Exsorohilum and *Pyrenophora* cause leaf spotting diseases on a wide range of cool-season and warm-season grasses. Symptomology on susceptible host grasses often displays many similarities with some slight

Fig. 7.13. Multiple leaf spot lesions on kikuyugrass. (Courtesy of Nadeem Zreikat.)

differences, depending on the host–fungi combination and environmental factors.

Reclassification of fungal species previously under Helminthosporium is of more than academic interest, having several practical implications for turfgrass managers. Some fungicide labels may simply state the word 'Helminthosporium' often without any reference to the fungal species. Other product labels may state the disease name and fungal genera (species) depending on product registration in specific states or countries. The potential for various host–fungus combinations partly explains the difficulties in predictable management by fungicides.

Geographic distribution and host range

Bipolaris, Curvularia, Exsorohilum and *Pyrenophora* are cosmopolitan and among the most widely distributed fungal pathogens associated with crop and turfgrass systems worldwide (Sivanesan, 1987; Smith *et al.*, 1989). With few exceptions, most described species of *Bipolaris* and *Exsorohilum* occur on warm-season grasses while most *Pyrenophora* (*Drechslera*) species occur on cool-season grasses. *Curvularia* species are associated with cool-season and warm-season grasses, depending on fungal species. (e.g. *Pyrenophora catenaria* and *Pyrenophora poae*). *P. poae* is probably the most important disease of Kentucky bluegrasses despite advances made in disease resistance.

Bipolaris and Exsorohilum

Several well-known *Bipolaris* and one *Exsorohilum* species cause leaf spotting diseases on turfgrass (Table 7.5). *Bipolaris* and *Exsorohilum* possess conidia tapered at both ends and often darkly pigmented. *Bipolaris* are relatively common and certain species (e.g. *Bipolaris sorokiniana*) have proven virulence on numerous turfgrass species while the virulence among others (e.g. *Bipolaris micropus*, *Bipolaris nobleae*, *Bipolaris stenospila* and *Bipolaris triseptata*) remains unclear (Fermanian *et al.*, 2003). *B. sorokiniana* ranks among the most important and destructive species among the group.

An increasing number of *Bipolaris* species (e.g. *Bipolaris axonopicola*, *Bipolaris microlaena*, *Bipolaris peregianensis* and *Bipolaris simmondsii*) continue to be isolated from symptomatic grasses (Tan *et al.*, 2016; Wang *et al.*, 2016). Their virulence remains imprecise.

Pyrenophora (Drechslera)

Numerous *Pyrenophora* (formerly *Drechslera*) species cause leaf spotting on many principal turfgrass species worldwide (Table 7.6). *Pyrenophora* possess dark brown mycelia and conidia and certain species are well-known pathogens

Curvularia-incited diseases

A multitude of *Curvularia* species are associated with leaf spotting diseases on many grasses (Table 7.7). *Curvularia* produce masses of spores and are generally identified by their curved conidia which have two or three cells of varying colour (light brown to brown), one of which is enlarged. *Curvularia* are common on cool-season and warm-season grasses and the degree of virulence among many species remains uncertain. *Curvularia* are generally considered to be relatively poor competitors on healthy plants and characterized as secondary pathogens or facultative saprophytes causing senectopathic diseases on temperature-stressed turfgrass.

Several *Curvularia* species are well known (e.g. *Curvularia geniculata*, *Curvularia intermedia*, *Curvularia lunata*, *Curvularia trifolii* and *Curvularia spicifera*) while others (e.g. *Curvularia malina* and *Curvularia stenotaphri*) are newly described on turfgrass. *C. lunata* is among the most important species of the group.

The full extent of *Curvularia*-incited diseases and virulence of their fungal species remain unclear (Brecht *et al.*, 2007). *Curvularia* species probably occur worldwide wherever turfgrasses are managed in warm and humid regions. Two specific *Curvularia*-incited diseases on turfgrasses are recognized: Curvularia leaf blight and ink spot.

Table 7.5. Principal *Bipolaris* and *Exsorohilum* species and host grass groups.

Bipolaris and *Exsorohilum* species	Host group	References
B. buchloes	C4	Fermanian
B. cynodontis	C4	et al. (2003);
B. gigantea	C3/C4	Vargas
B. heveae	C4	(2005);
B. micropus	C4	Tredway et al.
B. sorokiniana	C3/C4	(2023)
B. spicifera	C4	
B. stenospila	C4	
E. rostratum	C3/C4	

Table 7.6. Principal *Pyrenophora* species and host grass group.

Pyrenophora species	Host group	References
P. catenaria	C3	Vargas (2005);
P. dictyoides		Fermanian
P. erythrospila		et al. (2003);
P. lolii		Tredway
P. grahamii		et al.
P. poae		(2023)

Table 7.7. Principal *Curvularia* species and host grass group.

Curvularia species	Host group	References
C. affinis	C4	Smith *et al.* (1989); Couch (1995); Huang *et al.* (2005); Coelho *et al.* (2009); Crouch and Tomaso-Peterson (2012); Tran *et al.* (2012); Tomaso-Peterson *et al.* (2016); Zhang *et al.* (2018); Gilardi *et al.* (2022); Tredway *et al.* (2023)
C. americana	C4	
C. cymbopogonis	C4	
C. geniculata	C3/C4	
C. inaequalis	C3/C4	
C. intermedia	C3/C4	
C. lunata	C3/C4	
C. malina	C4	
C. platzii	C4	
C. pallescens	C3/C4	
C. protuberata	C3	
C. spicifera	C3/C4	
C. tropicalis	C4	
C. stenotaphri	C4	
C. trifolii	C3, C4	

Curvularia leaf blight

Curvularia leaf blight is attributed to several *Curvularia* species (Tredway *et al.*, 2023). Dog footprint is the alternative name in Japan (Tani and Beard, 1997). Symptoms of Curvularia leaf blight are similar and readily misdiagnosed to other leaf spot-type diseases. The presence of dark-coloured lesions, chlorotic and blackened leaves and stems, and profuse sporulation of conidia can be characteristic of Curvularia leaf blight. Microscopic examination of infected foliage may reveal fungal mycelium extending out from severed leaf ends.

Ink spot

Ink spot caused by *C. malina* is a relatively new turfgrass disease on bermudagrass and zoysiagrasses in the southern United States (Tomaso-Peterson *et al.*, 2016; Tomaso-Peterson, 2018) and southern China (Zhang *et al.*, 2018). Pathogenicity has been shown experimentally on centipede grass, seashore paspalum and zoysiagrass. Symptoms consistent with ink spot have been reported in other countries but the confirmed global distribution of the disease remains unclear.

The unusual descriptive name acknowledges individual, charcoal- to black-coloured spots 2.0–15.0 cm in diameter observed on closely mown turfgrasses. Microscopic examination of infected leaves reveals purplish-black-coloured lesions and mycelium. Significantly, *C. malina*, as opposed to other *Curvularia* species which produce conidia or asexual spores, does not produce spores and is said to be sterile, making morphological identification problematic.

Signs and symptoms of leaf spot diseases

Signs and symptoms of the principal leaf spotting, crown and root rot diseases caused by *Bipolaris*, *Curvularia*, *Exserohilum* or *Pyrenophora* species have many similarities with some differences, depending on host–pathogen interaction. Presence of irregularly rounded to oval-shaped lesions can be characteristic signs of leaf spot diseases (Fig. 7.14). Lesions may occur in close proximity to each other, and others enlarge around or along the leaf blade down to the crown region, depending on pathogen and host. Lesion size varies widely between 2 and 3 mm in diameter, depending on fungal species. Stem lesions are purple to black in colour.

Colour of fungal lesions and their surrounding border on host foliage varies depending on pathogen and turfgrass host (Table 7.8). The overall, late-stage symptoms of most leaf spot diseases are a general decline or diffuse pattern of yellow-, reddish- to blackish-brown-coloured turfgrass becoming straw-coloured, thin and wilted.

Overall symptom expression of melting-out on Kentucky bluegrass may appear yellow under low nitrogen and blackish-brown under high nitrogen. Net blotch on tall fescue has dark-coloured

Fig. 7.14. *Bipolaris* sp. lesions on infected foliage. (Courtesy of Nathan R. Walker.)

Table 7.8. Characteristic lesion colour of principal leaf spot diseases.

Disease/causal agent	Colour of foliar lesions/spots	Principal hosts	References
Brown blight (*Pyrenophora siccans*)	Brown spots and streaks	Bluegrasses, fescues, ryegrasses, buffalo grass	Fermanian *et al.* (2003); Vargas (2005); Watschke *et al.* (2013); Tredway *et al.* (2023)
Brown stripe (*Bipolaris heveae*)	Reddish-brown	Bermudagrass, zoysiagrass	
Curvularia leaf blight (*Curvularia* spp.)	Tan to light brown	Bermudagrass, zoysiagrass	
Fading-out (*Curvularia* spp.)	Tan centres, red-brown margins	Bentgrasses, bluegrasses, fescues	
Ink spot (*Curvularia malina*)	Chocolate brown to black	Bermudagrass	
Leaf blotch (*Pyrenophora cynodontis*)	Olive green spots	Zoysiagrass	
Leaf blight (*Pyrenophora catenaria*)	Dark-coloured spots	Bermudagrass, carpetgrass, kikuyugrass, St. Augustinegrass, zoysiagrass	
Leaf spot (*Bipolaris sorokiniana*)	Dark purple to black spots	Bentgrass, bluegrasses, fescues, ryegrass	
Melting-out (*Pyrenophora poae*)	Black to purple spots	Bluegrasses, bentgrass, fescues, ryegrass, bermudagrass, buffalo grass	
Net blotch (*Pyrenophora dictyoides*)	Black-coloured spots	Bluegrasses, ryegrass, fescues, buffalo grass	
Red leaf spot (*Pyrenophora erythrospila*)	Dark red-brown spots and borders	Bentgrasses, fescues, ryegrasses	
Zonate leaf spot (*Pyrenophora gigantea*)	Black-coloured spots	Bentgrasses, bluegrasses, bermudagrass, buffalo grass, zoysiagrass	

threads across the infected leaves. Infection of the crown and stems results in dark brown- to black-coloured dry rotting, leading to death of affected tissues.

Bipolaris, Curvularia, Exsorohilum and Pyrenophora diseases profile

Many aspects of the disease cycle and epidemiology of diseases caused by *Bipolaris*, *Curvularia*, *Exsorohilum* or *Pyrenophora* are generally considered similar (Vargas, 2005; Watschke *et al.*, 2013; Tredway *et al.*, 2023). Plant-pathogenic *Bipolaris*, *Curvularia*, *Exsorohilum* or *Pyrenophora* species can be profiled collectively as agents of fungal diseases of all plant parts (i.e. foliage, crown and stems and their seeds) on a wide range of cool-season and warm-season grasses (Fig. 7.15). The principal difference between many leaf spot diseases is the plant parts infected and the principal seasons of occurrence.

Bipolaris, *Curvularia*, *Exsorohilum* and *Pyrenophora* species are considered to survive unfavourable conditions saprophytically as dormant mycelia or conidia on decomposing or infected plant tissues. The disease cycle of leaf spot, crown and root rot diseases occurs with alternating periods of cool to warm and hot temperature, moist conditions and leaf wetness.

Infection generally occurs during warm to hot weather on turfgrass when ambient temperatures exceed around 25°C, depending on host–fungi combination and other environmental factors. Sporulation and fungal infection occur rapidly, and host penetration generally takes place through cut leaf ends or wounds. Infection and colonization are restricted to the immediate area of fungal penetration.

Combination of high temperatures, humidity and extended periods of leaf wetness are key factors for most leaf spot diseases. Infection and virulence of *C. lunata* and *C. trifolii* occurs at daytime temperatures of 20–35°C (Roberts and Tredway, 2008; Coelho *et al.*, 2009). Infection of adventitious roots of creeping bentgrass by inoculated *C. lunata* isolates has been demonstrated experimentally between 23 and 48°C (Hodges and Campbell, 1995). The temperature for infection by *B. sorokiniana* is 26°C while that for *Pyrenophora dictyoides* is lower at 20°C.

Pyrenophora-incited diseases primarily occur during cool and wet periods. The temperature response has been well documented for 'melting-out' (Box 7.7) caused by *P. poae* on Kentucky bluegrass (Couch, 1995). Generally speaking, infection and disease development of most leaf spotting diseases are more severe when nitrogen regimes are in excess of sustainable levels.

Disease development of most leaf spotting diseases peaks during favourable conditions any time from late spring to late summer, depending on geographic region and host–pathogen relationship. Fungal spores are disseminated by wind, water and machinery.

Integrated management

The foregoing information highlights many leaf spot diseases are complexes of one or more fungal

Fig. 7.15. Severe leaf spotting on kikuyugrass leaves. (Courtesy of Ben Evans.)

Box 7.7. Key temperature values of *Pyrenophora poae*.

- Fungal growth optimum: 18°C
- Fungal growth range: 3–30°C
- Sporulation optimum: 13–19°C
- Sporulation range: 6–20°C
- Spore germination and penetration: 6–30°C
- Penetration optimum: 18–24°C

genera and species subject to different combinations of environmental and management-induced variables which ultimately determine severity and frequency of disease. Prevalence of each leaf spot disease will be determined partly by the host–pathogen combination and partly by the effectiveness of cultural and chemical management (Box 7.8).

Disease-resistant varieties

Genetic resistance against *Bipolaris*- and *Pyrenophora*-incited diseases has been documented among Kentucky bluegrass (Bonos *et al.*, 2006) and against *Bipolaris*- and *Exserohilum*-incited diseases among bermudagrass cultivars (Tomaso-Peterson and Young, 2010).

Leptosphaerulina Leaf Blight

Leptosphaerulina species have long been associated with cultivated grasses and are generally considered relatively weak pathogens whose diseases are senectophytic because of environmental stresses. Leptosphaerulina leaf blight singly or in combination with other foliar pathogens (e.g. *Alternaria* spp., *Bipolaris* spp. and others) can cause inexplicable damage on cool-season grass bowling and golf greens during conducive conditions.

Causal agent

Leptosphaerulina australis (synonymous with *Leptosphaerulina trifolii*) is recognized as the principal causal agent of Leptosphaerulina leaf blight (Tredway *et al.*, 2023). The name of the type species, *L. australis*, denotes Australia from where the fungus was initially described and *L. trifolii* denotes being associated with clover (*Trifolium* spp.).

Geographic distribution

Leptosphaerulina leaf blight occurs worldwide (Table 7.9). In the United States, the disease occurs in the northern eastern, mid-Atlantic, southern and western regions of the country. Leptosphaerulina leaf blight is undoubtedly more widespread in subtropical and temperate regions than currently reported.

Host range and susceptibility

Leptosphaerulina leaf blight has a widely reported host range. Bermudagrass, Queensland blue couch and zoysiagrass are warm-season hosts and bentgrasses, bluegrasses, fescues and ryegrasses are cool-season hosts (Bransgrove, 2005; Watschke *et al.*, 2013). Proven pathogenicity of *Leptosphaerulina* sp. on turfgrasses based on Koch's postulates has not been widely reported.

Signs and symptoms

Presence of flask-shaped, dark-coloured fruiting bodies (pseudothecia) containing ascospores or sexual spores partly embedded in infected leaves is characteristic of the Leptosphaerulina leaf blight pathogen. Pseudothecia may be observed on infected leaves using a 10× hand lens. Late-stage symptoms of Leptosphaerulina leaf blight resemble those of other foliage diseases and may be commonly misdiagnosed (Fig. 7.16).

Initial symptoms on individual, infected leaves are seen as dieback and necrosis from the tip associated with yellow-, brown- to reddish-brown-coloured lesions extending down the infected leaf. Late-stage symptoms are random and non-descript blighted regions or ill-defined spots similar in appearance to other leaf blight and

---

Box 7.8. Generalized management factors of leaf spot diseases.

- Selection of disease-resistant varieties
- Maintain consistent growth with use of a balanced NPK nutrition
- Possible use of soluble silicon against certain leaf spot diseases
- Minimize leaf wetness periods and improve air circulation
- Judicious irrigation water applications
- Monitor plants for key fungal lesions
- Use of registered fungicides and biofungicides

Table 7.9. Reported global distribution of Leptosphaerulina leaf blight.

Region/country	References
United States	Watschke *et al.* (2013)
Southern Canada	Mitkowski and Browning (2004)
Western Europe	Smith *et al.* (1989)
Australia	Bransgrove (2005)
East Asia (Korea)	Kim *et al.* (2010)

Fig. 7.16. Damage caused by a *Leptosphaerulina* sp. and a *Curvularia* sp. on a mixed annual bluegrass–creeping bentgrass golf green. (Courtesy Gary W. Beehag.)

spot diseases. Symptoms are limited largely to mature or senescent leaves.

Disease profile

Key aspects of the ecology and epidemiology of Leptosphaerulina leaf blight remain unclear (Abler and Jung, 2004; Tredway *et al.*, 2023). *Leptosphaerulina* species are facultative parasites, and the disease is best described as senectophytic generally capable only of infecting and colonizing mature or plant tissue during periods of heat and/or drought stress.

In the United States, Leptosphaerulina leaf blight occurs any time from spring (March–May) all the way through to autumn (September–November) under favourable warm, humid weather and stressful turfgrass conditions. Viable ascospores may be wind- or water-dispersed. The fungus presumably survives overwintering as dormant pseudothecia on infected tissues or mycelium on decomposing tissues.

Environmental influences

Certain environmental factors have been shown to influence development of Leptosphaerulina leaf blight (Box 7.9). Moderate temperature and

> **Box 7.9.** Key environmental factors of Leptosphaerulina leaf blight.
>
> - Moderate temperatures
> - High humidity
> - Senescent hosts

high humidity are two key determinant factors of Leptosphaerulina leaf blight.

The moderate infection temperature of the fungus partly explains its limited physiological advantage during higher temperatures unless the infected cool-season hosts are subject to external environmental stress, thus reducing their physiological competitiveness.

Integrated management

Proven options for the effective management of Leptosphaerulina leaf blight remain limited but the disease can be treated similarly as for dollar spot. Monitoring of weather conditions, early disease signs, balanced NPK nutrition, judicious morning irrigation and minimizing turfgrass stress are paramount in managing the disease.

Nigrospora Leaf Blight

Nigrospora is a widespread fungal genus associated with crop and turfgrass systems worldwide. Nigrospora blight has been generally described as a senectophyte or at best a weak plant pathogen and the disease is relatively minor on turfgrass.

Causal agent

Nigrospora sphaerica is the causal agent of Nigrospora blight (Tredway *et al.*, 2023). The fungus is a facultative parasite accredited as the most common *Nigrospora* species on infected turfgrass. Additional *Nigrospora* species (*Nigrospora oryzae* and *Nigrospora osmanthi*) have demonstrated virulence on certain turfgrasses (Zheng *et al.*, 2012; Mei *et al.*, 2019).

Geographic distribution

Nigrospora blight occurs in North America (Thompson *et al.*, 1982; Zheng *et al.*, 2012) and East Asia (Mei *et al.*, 2019). Global distribution of Nigrospora blight on turfgrass remains uncertain.

Fig. 7.17. Massed, whitish-coloured mycelium of Nigrospora leaf blight. (Courtesy of Percy T. Wong.)

Host range and susceptibility

Creeping red fescue, Kentucky bluegrass, perennial ryegrass and St. Augustinegrass and perhaps zoysiagrass are hosts in the United States (Fermanian *et al.*, 2003; Tredway *et al.*, 2023). Bluegrass, fescue and ryegrasses vary in their susceptibility to the disease. Bermudagrass and possibly zoysiagrass are hosts of Nigrospora blight in East Asia (Choi *et al.*, 2000).

Signs and symptoms

Girdling of individual infected leaves of cool-season grasses can be a characteristic sign of Nigrospora blight having some similarities to that of dollar spot. Whitish-coloured mycelium may be present during warm, humid conditions during early morning (Fig. 7.17) depending on the mowing height of the host grass.

Symptoms of Nigrospora blight may be easily confused with those of several other foliage diseases. Infected leaves die back from the tip and may show tan-coloured lesions with purple to reddish-brown margins. Early-stage symptoms may appear as irregularly shaped patches 1–20 cm diameter.

Late-stage symptoms of the disease may display a purple colour of dying leaves below girdling. Overall symptoms of Nigrospora blight are generalized blighting and thinning of entire areas of necrotic leaves. On St. Augustinegrass, dark-coloured lesions typically girdle infected stolons ultimately causing overall thinning and patchiness. Symptoms of infected zoysiagrass show a yellow colour (Choi *et al.*, 2000).

Disease profile

Many key aspects of the ecology and epidemiology of Nigrospora blight remain limited. The pathogen appears to survive unfavourable conditions as dormant conidia and mycelium in decomposing host tissue. Conidia germinate with the onset of warm and wet conditions and disseminate by wind, water and possibly machinery to infect adjacent plants. Abundant mycelium is produced during highly favourable conditions. Severe outbreaks coincide with warm, humid nights. The pathogen has been shown to produce a biochemical compound demonstrated experimentally to cause toxicity to bentgrass and bermudagrass leaves (Choi *et al.*, 2000). The importance of the toxin in disease infection is unclear.

Host infection occurs between 23 and 28°C (Zheng *et al.*, 2012; Mei *et al.*, 2019). Reported seasonal occurrence and severity of Nigrospora blight on turfgrass vary depending on geographic location and host–pathogen combination. In the United States, Nigrospora blight has been reported on cool-season turfgrasses from spring (March–May) to summer (June–August) (Fermanian *et al.*, 2003; Tredway *et al.*, 2023). In East Asia (Korea) *N. sphaerica* was isolated from infected zoysiagrass in June (summer) coinciding with high temperature and rainfall events (Choi *et al.*, 2000).

Environmental influences

Moderate temperature and high humidity are two key determinant factors of Nigrospora blight. The relatively high infection temperature of the Nigrospora blight fungus explains partly its physiological advantage to cause blighting during stressful summer periods.

Integrated management

There are few proven or predicable management strategies for Nigrospora blight on high-value turfgrass but an integrated approach similar to other foliar-blighting diseases can be adopted (Box 7.10).

Lesser-Known, Warm-Weather, Foliage Fungal Diseases

Numerous warm-weather fungal diseases, primarily infecting the foliage and crown regions, can be characterized as of lesser importance on managed grasses (Table 7.10). Many fungal genera are relatively large and closely related to the extent their precise taxonomy remains subject to reclassification. Universal disease names may be absent for some and others may have multiple names depending on the country in which they occur. Many of these fungi have an extensive host range of cool-season and warm-season grasses.

Fungal lifestyles range from weakly pathogenic to senescent or saprophytes on decomposing tissues, depending on host–pathogen combination and changing environmental conditions. Collectively, signs and symptoms and environmental conditions favouring their occurrence generally are similar among these diseases.

Infected grasses show dieback from leaf tips extending downwards and associated dark-coloured spots enlarging into variably shaped lesions causing girdling of leaves displaying chlorosis leading to necrosis, depending on host–pathogen combination. Overall symptoms range from unsightly, blighted regions to chlorotic and necrotic areas, depending on favourable temperature and moisture conditions. Management actions are rarely required for these diseases. The status of some of these lesser-known fungal diseases may change over time depending on the specific host–pathogen combination and changing environmental conditions.

Box 7.10. Integrated management factors of Nigrospora leaf blight.

- Monitoring turfgrass health and early disease signs
- Maintain consistent growth with use of a balanced NPK nutrition
- Judicious irrigation water applications
- Minimize leaf wetness duration
- Avoid or minimize herbicide application during stressful periods
- Foliage application of registered fungicides and biofungicides

---

Table 7.10. Less-important warm-season, foliage fungal diseases.

Disease name	Fungal pathogen(s)	References
Cercospora leaf spot	*Cercospora* spp.	Smith *et al.* (1989); Couch (1995); Bransgrove (2005); Tredway *et al.* (2023)
Physoderma leaf spot	*Physoderma* spp.	
Setosphaeria leaf spot	*Exsorohilum* spp.	
Tar spot	*Phyllachora* spp.	
White blight	*Melanotus phillipsii*	

Stem- and root-infecting fungal diseases

Stem- and root-infecting or soil-borne fungal pathogens infecting stems and roots of their plant hosts cause symptoms which can range from roughly circular spots and patches, partial or 'doughnut'-shaped rings some which may coalesce forming larger-sized, roughly circular-shaped patches with unaffected centres, depending on the host–fungi combination and environmental conditions. Late-stage symptoms can result in severe disfiguration and be problematic to effectively manage once established.

Fungal infection on stems or roots begins well prior to symptom expression. The frequency and duration of soil-borne diseases is governed by rhizosphere factors, notably fertility, temperature, moisture and microbial antagonists. Many warm-season fungal diseases of stem and root tissues share many common environmental and management-induced factors; hence, many cultural practices are common to warm-weather, stem- and root-infecting, fungal diseases.

Kikuyu Yellows

Kikuyu yellows is the most important fungal disease of kikuyu pasture and turfgrass systems in the warmer regions of Australia. The disease name was first ascribed to chlorotic symptoms on infected foliage of kikuyugrass pastures (Fig. 7.18). Early incidence of the disease in northern New South Wales was one of the driving forces behind many commercial sod farms changing to cultivating couchgrass (bermudagrass).

Causal agent

Verrucalvus flavofaciens, an oomycete, is the causal agent of kikuyu yellows in Australia. The undescribed oomycete fungus was initially isolated from infected kikuyugrass collected from two coastal locations in New South Wales (Wong, 1975) and later identified as *V. flavofaciens* (Dick *et al.*, 1984).

Geographic distribution

Kikuyu yellows in Australia is distributed in south-eastern Queensland and throughout the northern coastal regions of New South Wales. Undescribed chlorotic symptoms were initially reported in the late 1950s in kikuyu pastures on the mid-north coast of New South Wales (Anon., 1959) and the early 1960s in Queensland (Wong, 1982). The fungal-like microbe is considered indigenous to Australia. Kikuyu yellows in Australia should not be confused with the more recent *Pythium*-incited disease of the same name in South America (Grijalba *et al.*, 2017).

Fig. 7.18. An early photograph of kikuyu yellows symptoms on kikuyugrass pasture in Australia. (Courtesy of Percy T. Wong.)

The disease was named kikuyu yellows because it was thought without verification that it was probably a viral disease. Widespread cultivation of vegetatively propagated, common kikuyugrass on sod farms throughout southern mainland Australia poses a significant risk in the spread of the disease.

Host range and susceptibility

The natural host range of kikuyu yellows appears to be restricted to common and seeded kikuyu varieties. Research demonstrates that the majority of common ecotypes and hybrids of kikuyugrass are susceptible to kikuyu yellows (Wong, 2011; Fulkerson *et al.*, 2021).

Signs and symptoms

Early-stage symptoms of kikuyu yellows appear as individual, infected leaves displaying chlorosis from the tip gradually progressing downwards (Fig. 7.19). Over time small-sized spots (5–10 cm diameter) appear and under favourable conditions, the disease progresses radially over several years forming chlorotic and necrotic patches. Close examination of infected leaves may reveal brown flecking on yellow leaves.

Late-stage symptoms on mown turfgrass are seen as random, roughly circular patches covering up to several metres in diameter (Fig. 7.20). Centre portions consist of chlorotic and necrotic plants and can be colonized by broad-leaved weeds or disease-resistant grasses such as paspalum (*Paspalum dilatatum*). The disease patches during conducive conditions are usually surrounded by a ring of bright yellow leaves.

Disease profile

The biology of the causal agent and epidemiology of kikuyu yellows in Australia have not been extensively investigated. Kikuyu yellows is a warm-weather disease that persists under favourable conditions of rainfall during late spring to early autumn. The zoospores are highly motile in water; hence the disease is readily spread over mown turfgrass by surface-moving water.

Environmental influences

Several key environmental factors appear to strongly influence the incidence and severity of kikuyu yellows on turfgrass (Box 7.11).

Integrated management

Effective management of kikuyu yellows on highly managed kikuyu is problematic in the absence of predicable cultural management practices and a lack of efficacious fungicides. The current range of commercial fungicides have failed to provide effective suppression of the pathogen and management of the disease (Wong and Tesoriero, 1990). Disease symptoms may be masked temporarily by a programme of balanced NPK nutrition. The kikuyugrass pasture variety Noonan possesses a high level of tolerance (Wong and Wilson, 1983).

Fig. 7.19. Chlorotic leaves on kikuyugrass caused by kikuyu yellows. (Courtesy of Percy T. Wong.)

Magnaporthiopsis-Incited Diseases

Magnaporthiopsis is a recently designated genus of necrotrophic fungal parasites infecting the roots of cool-season and warm-season grasses. The genus *Magnaporthiopsis* (formerly *Magnaporthe*) was originally established to accommodate *Magnaporthiopsis poae* (formerly *Magnaporthe poae*) and two other fungal species, *Magnaporthiopsis incrustans* (formerly *Gaeumannomyces incrustans*) and *Magnaporthiopsis rhizophila* (formerly *Gaeumannomyces rhizophila*) (Luo and Zhang, 2013). *Magnaporthiopsis* now encompasses additional newly described species associated with grasses in several countries (Table 7.11).

Magnaporthiopsis are relatively slow-growing fungi and weak competitors but capable of surviving

Fig. 7.20. Late-stage symptoms of kikuyu yellows on kikuyugrass in eastern Australia. (Courtesy of Percy T. Wong.)

Box 7.11. Key environmental factors of kikuyu yellows.

- Warm temperatures (20–30°C)
- Relatively high moisture levels
- Relatively high humidity
- Low nitrogen nutrition

Table 7.11. *Magnaporthiopsis* species associated with turfgrasses by host group and region.

Magnaporthiopsis species	Host group	Region/country	References
M. agrostidis	C3	Australia	Landschoot (1993); Luo and Zhang (2013); Wong *et al.* (2015, 2022); Vines *et al.* (2020)
M. cynodontis	C4	USA	
M. dharug	C3/C4	Australia	
M. gadigal	C4	Australia	
M. gumbaynggirr	C4	Australia	
M. incrustans	C3/C4	USA	
M. meyer-festucae	C3	USA	
M. poae	C3	USA	
M. rhizophila	C3	USA	
M. yugambeh	C3	Australia	

adverse conditions. *Magnaporthiopsis* have an ectotrophic growth habit and each species varies from weakly to strongly virulent depending on host–pathogen combination and environmental conditions.

In the United States, certain *Magnaporthiopsis* species (e.g. *Magnaporthiopsis meyer-festucae*, *M. poae* and *M. rhizophila*) are associated primarily on cool-season grasses (e.g. bentgrass, bluegrass and fescues) and one species (e.g. *Magnaporthiopsis cynodontis*) is associated with warm-season grasses. Other *Magnaporthiopsis* species (i.e. *M. incrustans*) occur on bermudagrass.

In Australia, several newly described *Magnaporthiopsis* are associated with cool-season grasses and/or warm-season grasses (Wong et al., 2022). An unknown *Magnaporthiopsis* sp. causing patch symptoms (Fig. 7.21) has been identified on mixed cool-season bentgrass golf greens in eastern Australia (P.T. Wong, New South Wales, 2023, personal communication). Undescribed *Magnaporthiopsis* may form a disease complex with other root-infecting fungi as a collective of root rot diseases of warm-season grasses, as covered in Chapter 6 (this volume).

Summer patch

Magnaporthiopsis poae (formerly *Magnaporthe poae*) is the causal agent of summer patch (Turgeon and Kaminski, 2019; Tredway et al., 2023). Summer patch is a problematic and disfiguring disease of cool-season grasses, especially Kentucky bluegrass in North America. Summer patch is one of the most extensively studied turfgrass diseases.

The disease name was given based on symptoms occurring during warm to hot periods during summer conditions (Fig. 7.22). *M. poae* is capable of surviving without a host for periods of time. Symptomatology of summer patch can be similar to that of several other patch diseases caused by other ectotrophic root infecting fungi.

Fig. 7.21. Patch symptoms caused by *Magnaporthiopsis* sp. on a mixed cool-season grass golf green in eastern Australia. (Courtesy of Gary W. Beehag.)

Fig. 7.22. Summer patch on a creeping bentgrass golf green in the United States. (Courtesy of Nathan R. Walker.)

Geographic distribution

Summer patch is widely distributed throughout the northern regions of North America and in East Asia (Table 7.12). In North America, the disease occurs from the north-eastern seaboard through the mid-western and central Rocky Mountain states and in Washington and southern California in the United States and in southern Canada. The distribution of summer patch in North America is undoubtedly much wider wherever Kentucky bluegrass is managed. The global distribution of the disease is unknown.

Host range and susceptibility

Annual bluegrass, Kentucky bluegrass and fine fescues are the primary hosts, followed by creeping bentgrass and perennial ryegrass (Dernoeden, 2013; Tredway *et al.*, 2023). The causal agent of summer patch has been shown experimentally to infect and cause disease on bermudagrass (Elliott, 1991) but not in landscape sites.

Table 7.12. Reported global distribution of summer patch.

Region/country	References
United States	Jackson (1993); Tredway *et al.* (2023)
Canada	Charbonneau and Hsiang (2015)
East Asia (China, Korea)	Han *et al.* (2016); Lee *et al.* (2017)

Signs and symptoms

The presence of mycelium growing along infected stems, roots and crowns may be observed but is not a reliable indicator of the pathogen. Fungal lesions are not associated with summer patch. Initial symptoms commence as random, small-sized chlorotic spots 3.0–8.0 cm in diameter and wilted and thinned grass. Expression of late-stage symptoms depends on mowing height and turfgrass species. Disease progression results in individual spots enlarging and coalescing resulting in large, diffused areas of diseased and dead turfgrass; however, distinct patches of dead grass occur. On higher-mown swards, entire areas may display severe chlorosis intermixed with live and dead plants.

Late-stage symptoms of lower-mown swards range from randomly scattered individual patches and coalesced shapes, crescent-shaped rings to distinct, roughly circular doughnut-type rings, depending on circumstances. On Kentucky bluegrass fairways in the United States, doughnut-type rings range 30–100 cm in diameter with unaffected centres ('frog eye' symptom). Crescent rings and circular patches may eventually coalesce forming larger areas of diseased turfgrass (Fig. 7.23). Infected perimeter leaves range in colour from light green, bronze or copper to reddish-brown depending on stage of disease. Symptoms continue to expand in size as long as temperatures are elevated.

Late-stage symptoms on mixed cool-season grass golf greens appear as randomly scattered assortments of individual, blighted patches of dead and live plants together with coalesced, roughly circular patches. Individual patches range up to 30 cm diameter ranging in pattern and shape. The overall disease pattern depends on the botanical uniformity of annual bluegrass and creeping bentgrass; annual bluegrass being more susceptible to damage.

Disease profile

Summer patch is a warm-weather, root-, crown- and stolon-infecting disease. The ecology and epidemiology of summer patch have been extensively documented in the United States (Kerns and Tredway, 2013; Tredway *et al.*, 2023).

The summer patch pathogen is considered to survive unfavourable winter conditions as mycelium saprophytically inside infected host tissues. Resumption of warmer conditions in late spring or early summer stimulates mycelium to grow ectotrophically and to colonize turfgrass plants; thus resulting in disease. Symptoms

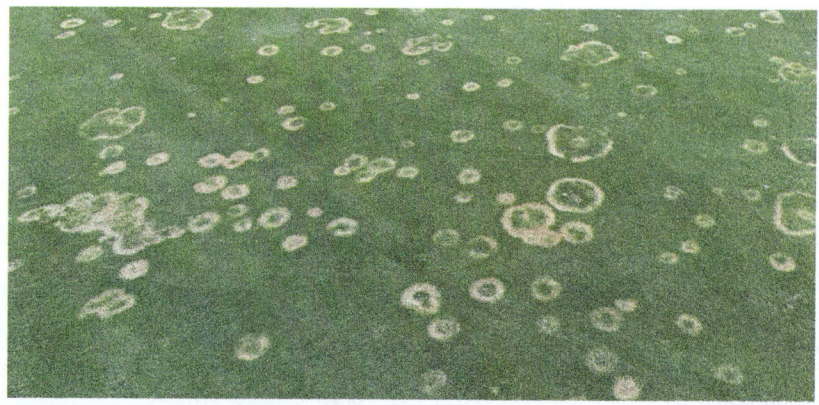

Fig. 7.23. Close-up view of symptoms of summer patch on a mixed annual bluegrass–creeping bentgrass golf green in the United States. (Courtesy of Nathan R. Walker.)

become severe during warm to hot conditions often following rainfall events. Drought stress can enhance the severity of disease once established. The disease continues through the summer and early autumn even through to the following spring, depending on seasonal conditions.

Environmental influences

Several environmental and management-induced factors have been shown to specifically influence development of summer patch (Box 7.12). Soil temperature and moisture are the two most important factors of the disease.

TEMPERATURE AND MOISTURE High temperature in combination with moisture stress, the rate of pathogen growth and host root development are key factors governing disease development. Host infection under field conditions has been shown to occur between 18 and 21°C.

HOST NUTRITION AND PH Interactions of soluble and coated forms of nitrogen fertilizer and soil pH influence the development of summer patch. Generally, acidifying nitrogen forms (e.g. ammonium sulfate and sulfur-coated urea) suppress summer patch (Davis and Dernoeden, 1991; Thompson and Clarke, 1992). The mechanisms behind the influence of pH on summer patch remain unclear and indirect mechanisms (e.g. shifts in microbial population) may be involved (Kerns and Tredway, 2013).

MOWING AND IRRIGATION The combination of higher mowing height (7.6 cm) and infrequent irrigation, as opposed to lower mowing height (3.8 cm) and more frequent mowing, generally results in less summer patch in Kentucky bluegrass (Davis and Dernoeden, 1991).

ARSENICAL HERBICIDES Applications of an early arsenical-based herbicide (calcium arsenate) to Kentucky bluegrass infected by summer patch resulted in increased severity of the disease (Smiley *et al.*, 1985). The impacts of more recent arsenical-based herbicides (e.g. monosodium methanearsonate (MSMA) and disodium methanearsonate (DSMA)) on the disease are unknown.

MICROBIAL ANTAGONISM AND SUPPRESSION Suppression of summer patch on Kentucky bluegrass has been demonstrated experimentally using several strains of bacterial and fungal antagonists from take-all-suppressive soil (Thompson and Clarke, 1992). Future research may determine the potential efficacy of antagonistic microbes against summer patch.

Integrated management

Effective management of summer patch on high-value turfgrass necessitates an integrated approach (Box 7.13). Hence, regular monitoring of turfgrass health and weather variations, use of disease-resistant varieties in conjunction with proven cultural, biological and fungicidal options is required for effective management of summer patch.

Disease-resistant varieties

Limited genetic resistance against summer patch has been documented among fine fescue cultivars (Braun *et al.*, 2020).

Warm-Weather *Pythium* Diseases

Pythium blight, Pythium crown and root rot, and Pythium root dysfunction are named

---

Box 7.12. Key environmental factors of summer patch.

- High soil temperature
- Low soil moisture
- Nitrogen forms and pH
- Mowing regime and irrigation
- Excessive rootzone compaction

Box 7.13. Integrated management factors of summer patch.

- Disease-resistant cultivars
- Monitoring turfgrass health and early disease symptoms
- Maintain consistent growth with use of a balanced NPK nutrition
- Judicious irrigation water applications
- Minimization of thatch-mat thickness and organic matter content
- Application of registered fungicides and biofungicides

warm-weather diseases. Knowledge of many aspects of Pythium crown and root rot remains unclear and the symptoms can be difficult to precisely diagnose from Pythium root dysfunction.

Pythium blight

Pythium blight is a destructive disease, primarily of the foliage, once established on cool-season grasses (Fig. 7.24) because of the rapidity of disease progression. Alternative names are spot blight, grease spot and cottony blight. Pythium blight can destroy significant portions of a turfgrass sward in a single 24-h period during highly favourable, hot and humid conditions.

Once diagnosed, Pythium blight requires rapid modification of cultural practices combined with targeted application of appropriate fungicides to minimize further damage. Fungicide rotation and tank mixing must form part of an overall disease management strategy to minimize the potential of fungicide resistance in the pathogen.

Signs and symptoms

Early-stage symptoms of Pythium blight on closely mown turfgrass under favourable conditions appear rapidly as randomly scattered, circular-shaped infection centres 2.0–5.0 cm or more in diameter (Fig. 7.25). Infected foliage displays a reddish-brown to orange-bronze colour. A 'smoke ring' may be present during early morning under highly favourable conditions for fungal infection.

Presence of dark-coloured, water-soaked leaves with an 'oily-like' feel on infected foliage when rubbed between the fingers in the early morning can be characteristic of severe infection by Pythium blight (Fig. 7.26). Initial disease outbreaks occur in locations where moisture levels are high or where dew formations remain throughout the morning.

Disease progression sees a rapid increase and enlargement of the number and size of infected spots throughout a green. Late-stage symptoms appear as widespread loss of verdure forming irregular-shaped patches and mosaics (Fig. 7.27). Blighted symptoms may follow water movement in accordance with surface contouring or machinery movement. On higher-

Fig. 7.24. Damage by Pythium blight on a fine fescue lawn in the United States. (Courtesy of Nathan R. Walker.)

Fig. 7.25. Pythium blight on a creeping bentgrass golf green. (Courtesy of Ben Evans.)

Fig. 7.26. Damage caused by *Pythium* spp. on creeping bentgrass in the United States. (Courtesy of Nathan R. Walker.)

mown turfgrass, initial spots are larger in diameter and irregularly shaped. Infected foliage may collapse and form matted foliage covered in off-white or grey-coloured massed mycelium, acknowledging the other name of cottony blight.

Environmental influences

Fungal infection commences when the ambient temperature reaches around 18°C (Latin, 2021). Foliar blighting on cool-season turfgrasses proceeds most rapidly at temperatures around

Fig. 7.27. Severe Pythium blight symptoms on a sloped creeping bentgrass golf green. (Courtesy of Ben Evans.)

29–35°C, humidity levels exceeding 90% and leaf wetness periods in excess of 14 h (Couch, 1995; Tredway *et al.*, 2023) depending on *Pythium* species. Temperatures above 10°C during humid and rainy periods are conducive to infection on warm-season turfgrasses.

Pythium root dysfunction

Pythium root dysfunction is a disease recognized primarily of bentgrass golf greens in the south-eastern United States during periods of high temperatures. The disease is manifested by a decline of the root system and the apparent inability of dysfunctional roots to perform water conduction and solute absorption normally. Pythium root dysfunction, as opposed to the physiologically stress-driven summer bentgrass decline, is considered a distinct disease (Dernoeden, 2013). The distribution of this form of *Pythium*-incited disease outside the United States is unclear.

Near pure-sand, alkaline rootzones, lack of microbial antagonists, presence of root abrasion caused by certain sand types and subsequent nutrient leakage, and exceptionally low mowing heights have all been implicated in the occurrence of Pythium root dysfunction (Hodges, 1990).

Signs and symptoms

Signs and symptoms of Pythium root dysfunction can be highly variable and problematic to accurately diagnose. Foliar symptoms may commence any time during spring and autumn during warm and dry conditions. Early-stage symptoms may appear as roughly circular or irregularly shaped areas, up to 15 cm in diameter, of chlorotic and wilted leaves. Late-stage symptoms during summer progress to irregularly shaped, larger-sized areas, 20–50 cm in diameter, of yellow- to orange- or reddish-brown-coloured foliage. Significant loss of surface results if the disease is left untreated.

Late-stage symptoms may resemble other infectious diseases (e.g. take-all patch), even non-infectious conditions (e.g. drought stress or hydrophobicity). Close examination of washed roots in affected areas may reveal light-tan-coloured roots, presence of bulbous root tips but lack of normal root depth and volume and lack of root hairs. Laboratory staining techniques of infected roots during summer does not always confirm the presence of *Pythium* oospores.

Key environmental influences of Pythium-incited diseases

The combination of temperature and moisture are the key environmental factors responsible for most *Pythium*-incited diseases. Temperature adaptation among *Pythium* species varies widely depending on the geographic origin of the isolate. Most *Pythium* species associated with grasses are mesophilic causing infection between 30 and 35°C, while some others are psychrophilic infecting between 13 and 18°C (Hsiang *et al.*, 1995; Kim and Park, 1999), depending on the *Pythium* species and strain.

Generalized Pythium-incited disease profile

Pythium-incited diseases on turfgrass may be best described as a syndrome of continual infection and symptom expression of the foliage, crown, stem and root tissues by multiple pathogenic species during conducive environmental conditions of temperature and moisture. *Pythium*-incited diseases can cause a rapid decline to both seedling and mature swards, becoming highly problematic once established on susceptible turfgrasses.

Ecologically, *Pythium* species are generally considered necrotrophic pioneer colonists but are relatively poor competitors in the presence of diverse microbial communities. *Pythium* species are highly tolerant of anaerobic conditions. High soil moisture favours saprophytic growth of *Pythium* species on decaying organic matter. *Pythium* species can readily gain nutrition from actively growing seedlings or mature feeder roots of their hosts. Pathogenic *Pythium* species cause rapid seedling, foliar and root decline diseases alone as primary pathogens or in combination with secondary pathogens (e.g. *Alternaria* spp., *Bipolaris* spp. and *Pyrenophora* spp.).

> **Box 7.14.** Generalized management factors of *Pythium*-incited diseases.
>
> - Monitor for early disease symptoms
> - Reduce water application and improve drainage
> - Minimization of thatch-mat thickness and organic matter content
> - Monitoring of recycled irrigation water for *Pythium* species
> - Avoid mowing during periods when surfaces are wet
> - Maintain consistent growth with use of a balanced NPK nutrition
> - Judicious irrigation water applications

Integrated management of Pythium-incited diseases

Effective management of warm-weather *Pythium* diseases necessitates an integrated approach to minimize infection of newly planted and established swards (Box 7.14). Frequent monitoring of disease signs and changing weather patterns combined with cultural, biological and fungicidal options are required for effective management of each disease while minimizing fungicide resistance. Existence of *Pythium* disease complexes associated with turfgrass roots presents significant challenges in the choice and implementation of predictable management options.

Warm-Weather *Rhizoctonia*- and *Waitea*-Incited Diseases

Brown patch, brown ring patch, basal leaf blight, and leaf and sheath spot are named warm-weather diseases caused by species either of *Rhizoctonia* or *Waitea*. Symptoms of leaf and sheath rot and brown ring patch are similar, depending on environmental conditions, which can pose difficulties for precise causal agent diagnosis.

Brown patch

Brown patch is the most widely distributed and most intensively studied *Rhizoctonia*-incited turfgrass disease. The disease name describes late-stage symptoms of roughly circular-shaped, brown-coloured patches (Fig. 7.28) and the

Fig. 7.28. Multiple, advanced brown patch symptoms on a creeping bentgrass golf green. (Courtesy of Jyri Kaapro.)

associated smoke ring during conducive conditions, originally observed on bentgrass golf greens in the United States. It is among the earliest recognized turfgrass patch diseases, otherwise known as Rhizoctonia blight in Western countries or Rhizoctonia brown patch in Eastern countries. Brown patch occurs on seedling and established turfgrasses and has one of the most extensive host ranges of all turfgrass diseases.

Causal agent

Rhizoctonia solani (AG1-1A, 1-1B and AG2-2IIIB) are the causal agents of brown patch (Burpee and Martin, 1996). The fungus is highly variable, best described as a group comprising many closely related subgroups. Mature mycelium grown in culture varies in colour from off-white, grey to dark brown and sclerotia are coloured off-white to dark brownish-white (Tredway *et al.*, 2023).

Geographic distribution

Brown patch has a global distribution particularly in warm temperate, subtropical and tropical regions (Table 7.13). The name brown patch

Table 7.13. Reported global distribution of brown patch (*Rhizoctonia solani*).

Region/country	References
United States	Tredway *et al.* (2023)
Canada	Charbonneau and Hsiang (2015)
United Kingdom	Mann and Newell (2005)
Western Europe	Yildiz *et al.* (1990); Massimo *et al.* (2005)
Australasia	Christensen (1992); Bransgrove (2005)
East Asia	Huang *et al.* (2005); Shim and Lee (2018)
Western Asia	Yildiz *et al.* (1990)
Middle East	Marvasti and Banihashemi (1990)

was ascribed in the United States to distinguish the disease from other patch disease symptoms then known as small brown patch (Monteith and Dahl, 1932). The incidence of brown patch is likely under-reported given that *R. solani* is a ubiquitous, soil-borne pathogen.

Brown patch is widely distributed throughout the United States and southern Canada. In Europe, brown patch is more common in the warmer central and southern regions (e.g.

France, Italy and Germany) but less common in maritime and northern countries. Brown patch is widely distributed throughout the warmer regions of mainland Australia. The adoption of brown patch in Australia was loosely ascribed from the 1930s to brown-coloured symptoms observed on bowling and golf greens in New South Wales and Queensland (Scott, 1937). In New Zealand, brown patch was first recorded in the 1930s (Brien, 1935). The disease status of brown patch on golf greens in New Zealand is unclear.

Brown patch has been reported in East Asia in the warmer regions of China, Japan and Korea. The disease Rhizoctonia brown patch caused by *R. solani* on temperate turfgrasses has been adopted by the Phytopathology Society of Japan (Tani and Beard, 1997).

Host range and susceptibility

The brown patch pathogen has an extensive host range of primary and secondary cool-season and warm-season grasses worldwide. Creeping bentgrass, perennial ryegrass, fescue, Queensland blue couch, kikuyugrass and seashore paspalum are among the most susceptible turfgrasses to the disease (Couch, 1995; Loch *et al.*, 2013).

Signs and symptoms

Signs and symptoms of brown patch on specific hosts can be distinctive during highly conducive conditions for growth of the pathogen. Examined microscopically, the presence of dark-coloured, spear-shaped ends of infected stems (Fig. 7.29) and tan- to brown-coloured lesions (Fig. 7.30) can be characteristic of *R. solani*.

Field symptoms of brown patch vary widely depending on the mowing height of the host and environmental conditions. On closely mown surfaces (e.g. bentgrass bowling and golf greens), early-stage symptoms appear as single or multiple roughly circular patches ranging from 5 to 30 cm or more in diameter (Fig. 7.31). Onset of mild weather conditions or a cessation of rainfall and reduced temperatures can result in a decline of symptom expression and grass regrowth, depending on management inputs.

Individual patches grow radially at different rates depending on the fungal strain and environmental conditions. Infected leaves may initially have water-soaked areas with dark purplish to black margins that change rapidly to light brown lesions during favourable conditions. The outer perimeter (1.0–10.0 cm margin) of infected leaves can rapidly change in colour to greyish-brown to dark purple or reddish-purple characteristic of the classic advancing smoke ring under highly favourable warm and humid conditions.

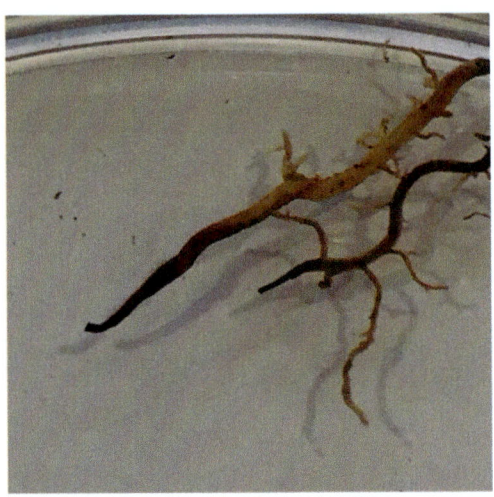

Fig. 7.29. Stem and root infection by *Rhizoctonia solani* on infected kikuyugrass. (Courtesy of Percy T. Wong.)

Fig. 7.30. Dark-coloured lesions of *Rhizoctonia solani* on kikuyugrass. (Courtesy of Percy T. Wong.)

Fig. 7.31. Early-stage symptom of brown patch on a creeping bentgrass golf green. (Courtesy of Gary W. Beehag.)

The smoke ring may not always be present but can appear during calm and warm early mornings in the presence of dew, when the cobweb-like mycelium is actively growing outwardly from infected foliage. The smoke ring gradually disappears when the foliage begins to dry with increasing temperatures and during the morning. Late-stage symptoms appear more distinct, varying in size, and certain patches may under conducive conditions eventually coalesce forming irregularly, larger-sized damaged areas.

On higher-mown turfgrass (e.g. golf fairways and sportsgrounds) late-stage symptoms of brown patch may not display well-defined perimeters (Fig. 7.32) and are often larger in diameter (Fig. 7.33) depending on host species and environmental conditions.

Disease profile

Brown patch is generally considered a foliage-borne, polycyclic disease capable of causing extensive damage and highly problematic to effectively manage on high-quality turfgrass. The pathogen survives unfavourable conditions on decomposing plant tissue by producing survival structures called bulbils and resumes growth once favourable temperature and moisture conditions occur. Germinating sclerotia (bulbils) and emerging mycelium rapidly grow radially on to adjacent foliage.

Infection occurs either directly through epidermal cells or indirectly through natural leaf openings (e.g. stomates) and mower wounds depending on environmental conditions. Penetration is achieved by a combination of hydrostatic pressure of fungal structures (e.g. appressoria) and biochemical degradation through the secretion of numerous enzymes and toxins. Once inside the infected leaf, invading hyphae grow both within plant cells (i.e. intracellularly) and between host cells (i.e. intercellularly), causing cell collapse.

Environmental influences

Several key environmental and management-induced factors have been shown to specifically influence development of brown patch (Box 7.15). The pathogen has relatively wide temperature ranges for hyphal growth, host infection and disease development, depending on environmental conditions.

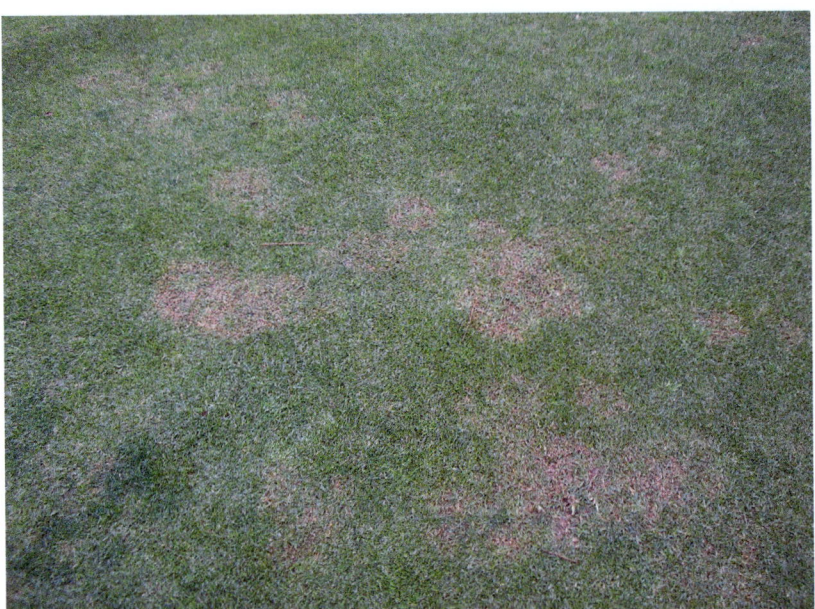

Fig. 7.32. Late-stage brown patch symptoms of a kikuyugrass fairway in Australia. (Courtesy of Ben Evans.)

Fig. 7.33. Brown patch symptoms on a kikuyugrass sportsground with couchgrass recolonization in the centre. (Courtesy of Percy T. Wong.)

TEMPERATURE AND MOISTURE Optimal temperature for host infection by *R. solani* ranges from 20 to 32°C. Development of brown patch disease can coincide with favourable conditions of temperature and moisture between spring even into early autumn depending on epidemiological conditions.

Box 7.15. Key environmental factors of brown patch.

- Moderate temperatures (20–25°C)
- High humidity (> 95%)
- Prolonged periods of leaf wetness (minimum 10 h)

HOST NUTRITION   Excessive or imbalanced nitrogen fertilization is generally associated with the frequency and severity of brown patch, provided temperature and moisture are favourable (Fidanza and Dernoeden, 1996; Teuton *et al.*, 2007). The combination of high nitrogen fertility and relatively moist soils increases plant growth that is succulent and more prone to fungal infection.

Disease-resistant varieties

Genetic resistance against brown patch has been documented among cultivars of creeping bentgrass (Bonos and Huff, 2013), Kentucky bluegrass and tall fescue (Bonos *et al.*, 2006), perennial ryegrass (Watkins *et al.*, 2001) and St. Augustinegrass (Flor *et al.*, 2013). The newer, high-density creeping bentgrass cultivars, as opposed to the older cultivars, appear to be more susceptible to brown patch (Dernoeden, 2013).

Brown ring patch

Brown ring patch (as the disease is known in the United States) or Waitea patch (in some other countries) is a relatively new disease associated on creeping bentgrass putting greens. The name was proposed in Japan describing symptoms of yellow- to brownish-coloured rings on creeping bentgrass.

Causal agent

Waitea circinata (formerly *W. circinata* var. *circinata*) is the causal agent of brown ring patch (Stalpers *et al.*, 2021). Mature mycelium grown in culture varies in colour from off-white, grey to dark brown and sclerotia are coloured off-white to dark brownish-white (Tredway *et al.*, 2023).

Geographic distribution

Brown ring patch has been documented in North America, East Asia, New Zealand and southern Europe (Table 7.14). Distribution of the disease has increased throughout the United States in recent decades attributed to changing environmental conditions or management-induced practices. Global occurrence of brown ring patch is probably under-reported.

Brown ring patch was first reported in East Asia. In the United States, brown ring patch occurs throughout the country from the cool, humid north-eastern to mid-western and southern regions (Kerns and Tredway, 2013). Symptoms consistent with the disease were probably first described in the 1980s in California and later in 2003 from Washington (Wong and Kaminski, 2007). Brown ring patch occurs in southern Canada (de la Cerda *et al.*, 2007) and in Western Europe in Spain (Gomez de Barreda *et al.*, 2019). Waitea patch has been identified on creeping bentgrass golf greens in Australia (P.T. Wong, New South Wales, 2023, personal communication).

Host range

Annual bluegrass, Kentucky bluegrass, rough bluegrass, creeping bentgrass and perennial ryegrass are known hosts of brown ring patch (Tredway *et al.*, 2023). Annual bluegrass is particularly susceptible to brown ring patch.

Symptoms

Symptoms of brown ring patch vary depending on host species and environmental conditions. Initial symptoms of brown ring patch commence as relatively small-sized, yellow-coloured regions (1.0–5.0 cm diameter) gradually expanding to partial or fully formed, irregularly shaped rings with unaffected green centres. Disease progression results in formation of larger-diameter rings with brown- to orange-coloured outer perimeters 2.0–5.0 cm in width. Late-stage symptoms are seen as coalescing crescents up to 30 cm in diameter that can form a mosaic-like pattern (Fig. 7.34). Symptoms on mixed greens are often irregularly shaped when the actively growing disease encounters areas of creeping bentgrass. Sunken areas may be associated with the disease.

Table 7.14. Reported global distribution of brown ring patch.

Region/country	References
North America	de la Cerda *et al.* (2007); Wong and Kaminski (2007)
Western Europe	Gomez de Barreda *et al.* (2019)
East Asia	Toda *et al.* (2005); Ni *et al.* (2012)
New Zealand	Hannan and Cushnahan (2010)

Waitea patch symptoms may not display the typical foliage coloration and pattern under unfavourable conditions for pathogen growth (Fig. 7.35).

Disease profile

Brown ring patch is a disease of the foliage and crown of hosts and is most common during cool and humid conditions. Brown ring patch has been reported occurring across a wide range of daytime temperatures around 16–35°C, with most infection and damage occurring between 18 and 29°C coinciding with high humidity and extended periods of leaf wetness (Wong and Kaminski, 2007; Kammerer and Harmon, 2008).

Basal leaf blight

Basal leaf blight is a relatively new *Waitea*-incited disease of warm-season turfgrasses in the United States. The descriptive disease name acknowledges severe blighting of mature leaves.

Causal agent

Waitea prodiga (formerly *W. circinata* var. *prodiga*) is the causal agent of basal leaf blight (Kammerer *et al.*, 2011; Stalpers *et al.*, 2021). Mature mycelium grown in culture varies in colour from yellow to pink off-white and sclerotia are coloured salmon, yellow to pink (Tredway *et al.*, 2023). The pathogen is closely related to a *Waitea* species damaging to cool-season grasses in Japan (Kammerer *et al.*, 2011).

Host range and pathogenicity

Bermudagrass, kikuyugrass, seashore paspalum, creeping bentgrass and roughstalk bluegrass are known hosts. The fungus is highly pathogenic on Sonesta bermudagrass, Sea Dwarf and SeaIsle Supreme seashore paspalum, but less pathogenic on creeping bentgrass and roughstalk bluegrass (Kammerer *et al.*, 2011).

Geographic distribution

Basal leaf blight is distributed in California and Florida in the United States. The pathogen and disease were initially reported in 2007 in Florida

Fig. 7.34. Waitea patch symptoms on a creeping bentgrass golf green. (Courtesy of Jyri Kaapro.)

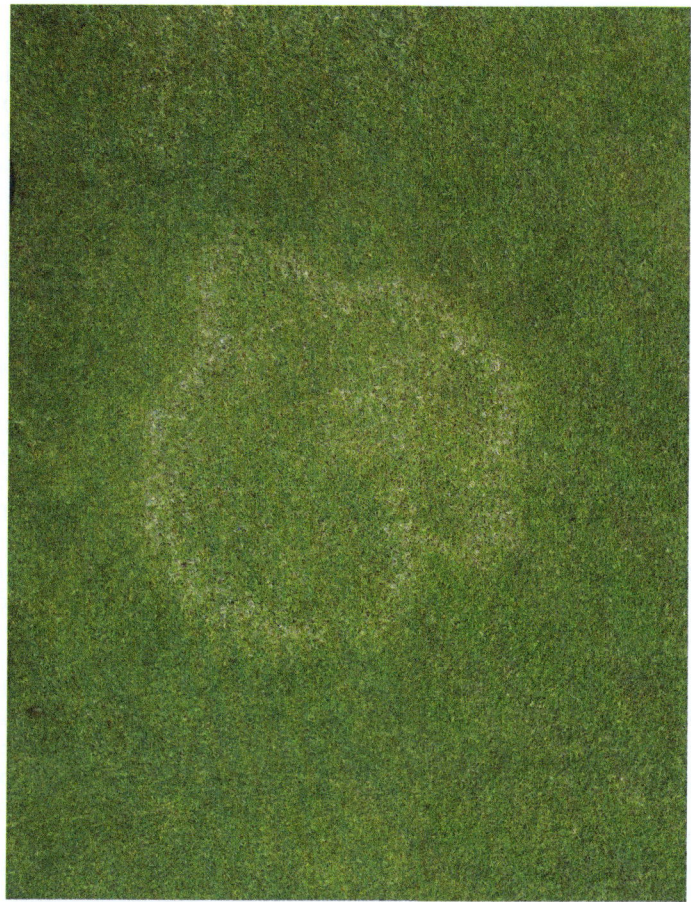

Fig. 7.35. Symptom of Waitea patch under less favourable conditions on a creeping bentgrass golf green in Australia. (Courtesy of Ben Evans.)

and in 2008 in southern California (Kammerer et al., 2011). Global distribution of brown ring patch is unknown.

Symptoms

Late-stage symptoms of basal leaf blight appear as irregularly shaped, blighted areas of chlorotic and necrotic patches up to 100 cm or more in size. Symptoms appear rapidly during highly favourable conditions resulting in brown-coloured sheaths covered by fungal mycelium.

Disease profile

Knowledge of the ecology and epidemiology of basal leaf blight continues to evolve. Basal leaf blight is a foliar disease of the lower canopy during warm and humid conditions. Colonization and infection of plants occur rapidly at and above 30°C (Kammerer et al., 2011). The high infection temperature of basal leaf blight overlaps with that with other diseases such as brown patch and leaf and sheath blight; and identification of the pathogen for proper management of the disease may be required.

Leaf and sheath spot

Leaf and sheath rot, also called 'mini-ring' in some countries, represents the third major *Rhizoctonia/Waitea*-incited turfgrass disease in the United States. The disease has become problematic since the 1990s in the central United States associated with increased adoption of hybrid bermudagrasses. The disease name was proposed to specifically describe roughly circular-shaped, scalloped or fully formed rings on bermudagrass golf greens.

Causal agent

Leaf and sheath spot is caused by two species of *Waitea*: *Waitea oryzae* (formerly *W. circinata* var. *oryzae*) and *Waitea zeae* (formerly *W. circinata* var. *zeae*) (Stalpers et al., 2021). Mature mycelium of *W. oryzae* grown in culture varies in colour from off-white to salmon pink and sclerotia colour varies from off-white to salmon pink. Mature mycelium of *W. oryzae* is white to cream to salmon pink, the colour of the sclerotia is salmon pink to orange (Tredway et al., 2023). *W. zeae* is generally regarded to be the more common species on bermudagrass and *W. oryzae* a less common species.

Geographic distribution and host range

Leaf and sheath spot is documented in the United States and East Asia (Zhang et al., 2014; Turgeon and Kaminski, 2019). In the United States hosts are hybrid bermudagrasses (Kammerer and Harmon, 2008). In China, hosts are seashore paspalum and zoysiagrass (Zhang et al., 2014). Global distribution of the disease is unclear.

Signs and symptoms

Symptoms on cool-season and warm-season grasses may vary depending on environmental conditions and perhaps the pathogenicity of the dominant *Rhizoctonia*/*Waitea* species. Symptoms on closely mown bermudagrass putting greens emerge in late summer to form randomly scattered spots 2–10 cm in diameter that gradually expand in size forming crescents or doughnut-shaped rings up to 0.5 m in diameter with a bronze-coloured perimeter.

Late-stage symptoms are seen as multiple rings some enlarging and coalescing forming a mosaic of irregularly shaped patches over a green. In severe infections the perimeters of crescents or doughnut-shaped rings may be sunken causing severe disfiguration to the putting surface. Presence of smoke rings is uncommon.

Symptoms on cool-season turfgrasses may appear as irregularly shaped patches of chlorotic and necrotic plants depending on height of cut and turfgrass species. The presence of orange-coloured sclerotia embedded in unmown infected foliage can be characteristic during advanced stages of the disease.

Generalized profile of Rhizoctonia/Waitea diseases

Certain *Rhizoctonia* and *Waitea* pathogens on turfgrass share many commonalities in their overall disease cycles, which can be reasonably assumed to be similar. Other *Rhizoctonia* and *Waitea* diseases display certain differences. Different *Rhizoctonia*- and *Waitea*-incited diseases occur at different times of the year, depending on the host–pathogen combination.

The biology and ecology of *Rhizoctonia* have been extensively documented and are relatively well understood (e.g. Parmeter, 1970; Sneh et al., 1996) but specific aspects of the disease cycles on turfgrass continue to be investigated (e.g. Hyakumachi and Hayakawa, 2008; Kammerer and Harmon, 2008). *Rhizoctonia* species are characterized as being capable of rapid growth, are associated with seedling and mature turfgrass, and possess saprophytic, non-pathogenic or pathogenic lifestyles, depending on the species and strain.

Rhizoctonia species are generally considered necrotrophs of plant roots, crowns and foliage. During unfavourable conditions, either cool or warm depending on species, the fungus survives as dormant, thick-walled mycelium and sclerotia-like structures in the rootzone and thatch-mat regions. Resumption of highly favourable environmental conditions and the presence of a highly susceptible host allow the fungi to rapidly infect hosts. Many *Rhizoctonia* pathogens are generally considered weak ecological competitors in the absence of highly susceptible hosts and against other facultative parasites (e.g. *Bipolaris* spp. and *Curvularia* spp.) and microbial antagonists (e.g. non-pathogenic *Rhizoctonia* sp.).

Brown patch caused by *R. solani* is the most investigated disease of the group. Brown patch is foliar-borne and generally considered a polycyclic disease. The pathogen survives unfavourable conditions on decomposing plant tissue, produces survival structures called bulbils and resumes growth once favourable temperature and moisture conditions occur. Germinating sclerotia and emerging mycelium rapidly grow radially on to adjacent foliage.

Infection occurs either directly through epidermal cells or indirectly through natural leaf openings (e.g. stomates) and mower wounds depending on environmental conditions. Penetration is achieved by a combination of hydrostatic pressure of fungal structures (e.g. appressoria) and biochemical degradation through the secretion of numerous enzymes and toxins. Once inside the infected leaf, invading hyphae grow both within plant cells (i.e. intracellularly) and between host cells (i.e. intercellularly), causing cell collapse.

Large patch (*R. solani* AG2-2LP) is a disease linked to the growth of warm-season grasses in regions subject to winter dormancy. Large patch disease occurs during cooler seasons when the host begins to enter cool-temperature-induced dormancy in the autumn or when growth resumes with increasing temperatures in spring. Development of large patch of zoysiagrass is supressed at soil temperatures above 30°C. Some disease patches are seasonal, occurring in either spring or autumn. Other disease patches are perennial, developing in the same location year after year for several years (Green *et al.*, 1993; Obasa *et al.*, 2017).

Yellow patch (*R. cerealis*) is a disease of the foliage and crown occurring primarily during prolonged cool to cold and moist conditions but may occur any time between early autumn to late spring. A detailed understanding of the ecology, epidemiology and management of yellow patch remains limited (Fermanian *et al.*, 2003; Kammerer and Harmon, 2008).

The biology and ecology of *Waitea*-incited diseases continue to be investigated. Brown ring patch is a disease of the foliage and crown and most common during cool and humid conditions. Brown ring patch has been reported occurring across a wide range of daytime temperatures around 16–35°C, with most infection and damage occurring between 18 and 29°C coinciding with high humidity and extended periods of leaf wetness (Wong and Kaminski, 2007).

Basal leaf blight caused by *W. prodiga* is a foliar disease of the lower canopy during warm and humid conditions. Knowledge of the ecology and epidemiology of basal leaf blight continues to evolve (Dant *et al.*, 2021). The high infection temperature of basal leaf blight overlaps with that with other diseases such as brown patch and leaf and sheath blight; and identification of the pathogen for proper management of the disease may be required.

Leaf and sheath spot (*W. zeae* and *W. oryzae*) is a disease of the foliage and crown during warmer seasons. Much of the ecology and epidemiology of leaf and sheath spot remains under investigation. Leaf and sheath rot, as opposed to brown patch (*R. solani*), is a hot-temperature disease because the pathogens cause disease at temperatures around 27°C (Kammerer and Harmon, 2008). Increased adoption of dwarf bermudagrass cultivars and the associated shallow-rooting characteristics of these grasses, exceptionally low mowing height, relatively low levels of sand-based rootzone fertility, and frequent surface-renovation (e.g. verticutting or scarification) and other practices have been proposed as possible explanations for emergence and continual occurrence of leaf and sheath rot on putting greens in the United States (Inguagiato and Martin, 2015).

General signs and symptoms

The presence of oval or roughly round (0.25–1.5 mm diameter), coloured (black, brown, and orange to red) survival structures is characteristic of *Rhizoctonia* and *Waitea* species and strains. Late-stage symptoms caused by *Rhizoctonia* and *Waitea* species may vary from entire blighted patches to coloured rings with unaffected centres (Table 7.15) depending on fungal species and environmental conditions.

Generalized integrated management of Rhizoctonia/Waitea diseases

Effective management of certain *Rhizoctonia*- and *Waitea*-incited diseases, once established, can be challenging. Response to mowing, nutrition and irrigation regimes and sensitivity to certain fungicides among *Rhizoctonia* species have been shown experimentally to be variable (Kerns and Tredway, 2013).

An integrated approach utilizing disease-resistant varieties, frequent monitoring of weather conditions, appropriate nitrogen fertilization,

Table 7.15. Key features of symptoms of *Rhizoctonia*- and *Waitea*-incited diseases.

Disease name	Late-stage symptom features
Brown patch	Roughly circular, straw- to brown-coloured patches 1–8 m in diameter, depending on turfgrass species and mowing height. Black-coloured smoke ring on cool-season grasses and orange-coloured smoke ring on warm-season grasses
Large patch	Spectacular, large-sized, roughly circular, straw- to brown-coloured patches up to 8 m in diameter, depending on turfgrass species and mowing height. Orange-coloured smoke ring
Yellow patch	Relatively circular-shaped rings, between 10 and 100 cm overall diameter, in clusters with some often coalescing forming larger areas. Foliage colour tends to fade during the day
Leaf and sheath spot	Roughly circular, straw- to brown-coloured patches 5–10 cm in diameter, depending on turfgrass species and mowing height. Some patches may coalesce forming larger doughnut-shaped patches up to 30 cm in diameter displaying bronze- to tan-coloured foliage
Brown ring patch	Clustered, roughly circular to irregularly shaped rings up to 30 cm in diameter, some often coalescing forming larger, irregular-shaped areas up to 1 m. Yellow-coloured foliage which may fade to a brownish colour during the day

---

Box 7.16. Generalized management factors of *Rhizoctonia*- and *Waitea*-incited diseases.

- Disease-resistant varieties
- Monitoring of weather conditions and disease symptoms
- Remove dew early morning where possible
- Avoid over-applications of nitrogen fertilizer
- Avoid mowing during wet surface conditions and collect leaf clippings
- Apply balanced levels of phosphorus and potassium
- Minimization of thatch-mat thickness
- Irrigate early morning and avoid irrigation during humid evenings and nights

---

scouting for symptoms in combination with proven cultural, biological and targeted application of registered fungicides and biofungicides is required (Box 7.16).

Southern Blight

The disease known as southern blight in the United States, or Rolf's disease in other countries where the disease occurs, can be a devastating patch disease once established on bentgrass golf greens during warmer conditions. The disease can be readily distinguished during conducive conditions by the spectacularly coloured, doughnut-shaped patches (Fig. 7.36). Southern blight occurs sporadically, and epidemics are particularly problematic to effectively manage.

Causal agent

Athelia rolfsii (formerly *Sclerotium rolfsii*) is now recognized the causal agent of southern blight (Stalpers *et al.*, 2021). The pathogen comprises different mycelium compatibility groups.

Geographic distribution

Southern blight has a wide distribution in warm temperate, subtropical and tropical regions (Table 7.16) and in legume crops from where the fungus may have originated (Lucas, 1976). The global distribution of southern blight on turfgrasses is unclear but probably under-reported in warmer regions.

Host range

Southern blight has an extensive host range of grasses and forbs. Host grasses are bluegrasses, bermudagrass, fescues, ryegrass and zoysia (Smith *et al.*, 1989; Couch, 1995).

Signs and symptoms

The presence of whitish-coloured mycelium and sclerotia during conducive conditions can be characteristic signs of southern blight. Massed, whitish-coloured mycelium during highly favourable

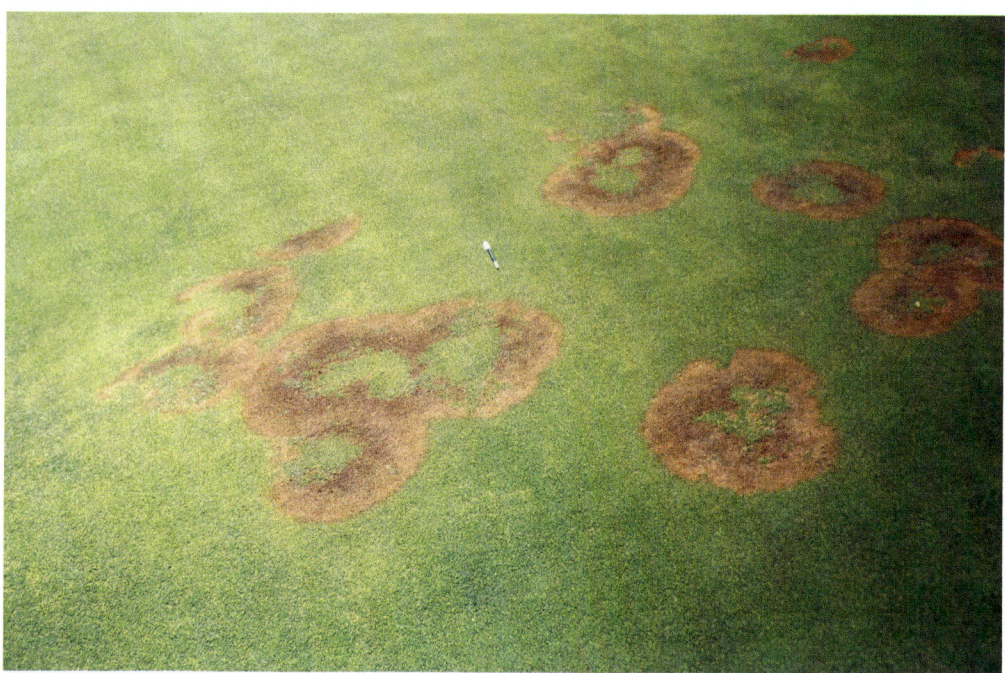

Fig. 7.36. Multiple southern blight symptoms on a creeping bentgrass golf green in southern California. (Courtesy of Gary W. Beehag.)

Table 7.16. Reported global distribution of southern blight on turfgrass.

Region/country	References
United States	Couch (1995); Tredway *et al.* (2023)
Southern Europe	Polizzi *et al.* (2006); Unal *et al.* (2019)
Australia	Woodcock (1983); Beehag (1986)
New Zealand	Cushnahan (2009)
East Asia (China)	Liu and Pu (2004)

temperature and humidity may be observed around the periphery of patches on golf greens (Fig. 7.37). Careful observation in the crown region of infected grass may reveal the presence of brown- to dark-coloured sclerotia. Mycelium and sclerotia are more readily observed with canopy moisture and high humidity during early mornings.

Symptoms of southern blight on bentgrass golf greens can vary depending on environmental conditions. Initial symptoms commence as yellow- to reddish-brown-coloured, almost indiscrete circles in isolation or randomly scattered across a green. In hotter conditions on greens with adequate moisture, symptoms become enlarged and more pronounced in size and colour. Individual rings and patches may expand as much as 10–20 cm and coalesce weekly during hot and humid conditions.

Late-stage symptoms may eventually enlarge up to 1.0 m or more in diameter displaying a bright orange to fiery red appearance (Fig. 7.38). Centres of enlarging rings and established circles remain unaffected. Infected turfgrass plants eventually become severely necrotic. On higher-mown aprons during early morning under high humidity, light white- to grey-coloured mycelium may be observed around the outer periphery of patches. Established dicotyledonous plants inside infected patches become themselves infected by the pathogen.

Disease profile

Southern blight is considered a monocyclic, soil-borne disease. Many key aspects of the biology, ecology and epidemiology of southern blight on turfgrass remain limited (Punja, 1985; Tredway *et al.*, 2023). A generalized disease profile only can be outlined given an absence of detailed

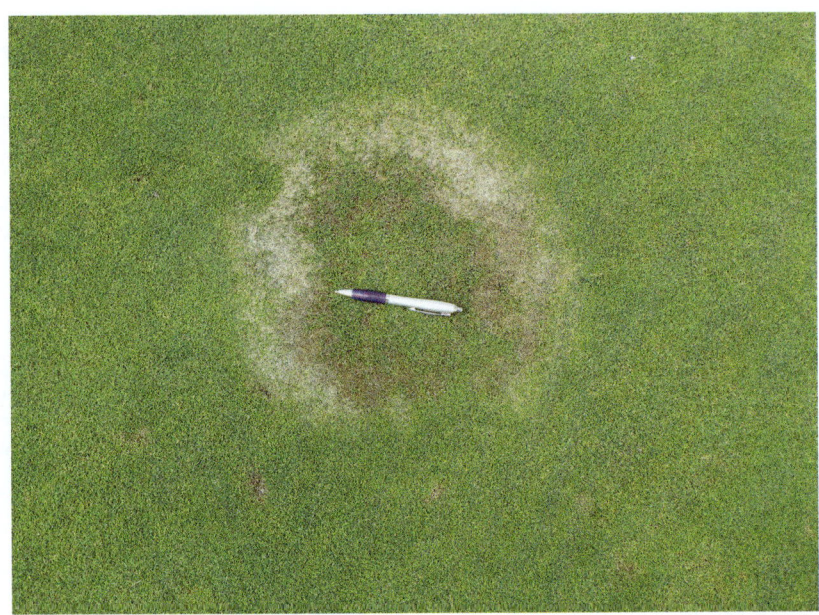

Fig. 7.37. White-coloured mycelium of *Athelia rolfsii* early morning on a creeping bentgrass golf green. (Courtesy of Gary W. Beehag.)

Fig. 7.38. Southern blight patches on a creeping bentgrass golf green in Australia. (Courtesy of Gary W. Beehag.)

understanding of the biology, ecology and epidemiology on turfgrass.

The southern blight fungus does not produce asexual spores and overwinters surviving as mycelium inside infected hosts or by producing sclerotia. Sclerotium is the primary source of inoculum; thus plays a key role in development and spread of the disease. Individual sclerotia are hardened, globular-shaped structures 0.3–3.0 mm in diameter, initially light-coloured and turning tan to dark brown at maturity, that can survive for extended periods. Sclerotia secrete a range of biochemical compounds aiding infection and pathogenesis.

Sclerotia are produced when atmospheric temperatures are above 24°C with adequate moisture. Under favourable warm and moist conditions in early summer, sclerotia produced the previous season undergo either 'hyphal germination' or 'eruptive germination' producing hyphae which grow radially to infect host plants. Surface-surviving sclerotia, as opposed to deeper ones, readily germinate suggestive of a high oxygen requirement.

Fungal penetration into host tissues is both intercellular and intracellular and pathogenicity is aided by production of biochemical compounds notably oxalic acid and cell-degrading enzymes. Hyphae penetrate intact crown and stem regions or through wounds. The fungus is capable of extensive saprophytic growth in the presence of decaying organic matter in the thatch-mat region. Physical dissemination of sclerotia aids in spreading the disease, causing secondary infections resulting in immature or satellite rings around late-stage symptoms.

Environmental factors

Several environmental and management-induced factors have been shown to be associated with southern blight (Box 7.17). Ambient temperature is a key factor of the disease.

TEMPERATURE AND MOISTURE Sclerotia formation and germination occur optimally at 27–35°C and host infection occurs at 27–30°C (Ayed *et al.*, 2018). Moisture levels of 50–75% in a sandy loam soil favour disease development. Non-limiting oxygen favours disease development as attested to greater germination of surface-resting sclerotia of the fungus.

HOST NUTRITION Acidic conditions favour mycelium growth and sclerotial germination. Soluble nitrogen compounds (e.g. calcium nitrate, ammonium nitrate and ammonium sulfate) showed no effect on sclerotial germination in fungal culture studies but did reduce incidence of southern blight on turfgrass (Punja *et al.*, 1982a,b; Unruh and Punja, 1982).

Biological management

Several different strains of antagonistic bacteria (*Bacillus*, *Paenibacillus*, *Pseudomonas* and *Stenotrophomonas*) were shown experimentally to suppress the disease to varying degrees under greenhouse-controlled conditions (Unal *et al.*, 2019). The authors are unaware of any commercially available, biological antagonist formulations for turfgrass application.

Integrated management

Presence and potential spread of sclerotia and limited registered fungicides make effective management of southern blight problematic, necessitating an integrated approach (Box 7.18). Monitoring for signs of fruiting bodies and cultural and fungicidal options are required for effective management of the disease. Aerification and thatch-management equipment may aid dispersal of fungal sclerotia.

Box 7.17. Key environmental factors of southern blight.

- Moderate to high temperature (30–35°C)
- High humidity (> 90%)
- Moderate moisture
- Low oxygen levels

Box 7.18. Integrated management factors of southern blight.

- Monitor weather conditions and turfgrass for signs
- Minimization of thatch-mat thickness
- Maintain growth using balanced NPK nutrition and check pH
- Judicious irrigation water applications
- Application of registered fungicides

References

Abernathy, S.D., White, R.H., Colbaugh, P.F., Engelke, M.C., Taylor, G.R. and Hale, T.C. (2001) Dollar spot resistance among blends of creeping bentgrass cultivars. *Crop Science* 41, 806–809.

Abler, S. and Jung, G. (2004) Leptosphaerulina leaf blight of turfgrasses ... lion or lamb? *The Grass Roots* (July/Aug) 15, 17.

Alizadeh, A., Javan-Nikkhah, M., Nazarian, R.N., Liu, F., Zare, R., et al. (2021) New species of *Colletotrichum* from wild Poaceae and Cyperaceae plants from Iran. *Mycologia* 114, 89–113. DOI: 10.1080/00275514.2021.2008765

Anon. (1959) *Diseases of Pasture Plants and Field Crops*. Annual Report. NSW Department of Agriculture, Sydney, Australia.

Ayanardi, B.A., Jimenez-Gasco, M.M. and Uddin, W. (2019) Effects of isolates of *Clarireedia jacksonii* and *Clarireedia monteithiana* on severity of dollar spot in turfgrasses by host type. *European Journal of Plant Pathology* 155, 817–829.

Ayed, F., Jabnoun-Khiareddine, H., Ben Abdullah, R.A. and Daami-Remadi, M. (2018) Effect of temperatures and culture media on *Sclerotium rolfsii* mycelial growth, sclerotial formation and germination. *Journal of Plant Pathology and Microbiology* 9, 8. DOI: 10.4172/2157-7471.1000446

Baxter, A.P., van der Westhuizen, G.C.A. and Elcker, A. (1983) Morphology and taxonomy of South African isolates of *Colletotrichum*. *South African Journal of Botany* 2, 259–289.

Beehag, G. (1986) Pest profile: Rolfs disease. *ATRI Turf Notes* 5, 8.

Beehag, G.W. (1999) Prescription surface development: lawn bowling greens, croquet lawns and tennis courts. In: Aldous, D.E. (ed.) *International Turf Management Handbook*. Inkata Press, Melbourne, Australia, pp. 265–280.

Belanger, F.C., Bonos, S. and Meyer, W.A. (2004) Dollar spot resistant hybrids between creeping bentgrass and colonial bentgrass. *Crop Science* 44, 581–586.

Bennett, F.T. (1937) Dollar spot disease of turf and its causal organism *Sclerotinia homoeocarpa*. *Annals of Applied Biology* 24, 236–257.

Boesewinkel, H.J. (1977) New plant disease records in New Zealand: records in the period 1969–76. *New Zealand Journal of Agricultural Research* 20, 583–589. DOI: 10.1080/00288233.1977.10427376

Bonos, S.A. and Huff, D.R. (2013) Cool-season grasses: biology and breeding. In: Stier, J.C., Horgan, B.P. and Bonos, S.A. (eds) *Turfgrass: Biology, Use, and Management*. Agronomy Monograph No. 56. American Society of Agronomy, Inc., Crop Science Society of America, Inc. and Soil Science Society of America, Inc., Madison, Wisconsin, pp. 591–660.

Bonos, S.A., Rush, D., Hignignt, K., Langlois, S. and Meyer, W.A. (2005) The effect of selection on gray leaf spot resistance in perennial ryegrass. *International Turfgrass Society Research Journal* 10, 501–507.

Bonos, S.A., Clarke, B.B. and Meyer, W.A. (2006) Breeding for disease resistance in the major cool-season turfgrasses. *Annual Review of Phytopathology* 44, 213–234. DOI: 10.1146/annual.phyto.44.070505.143338

Boulter, J.I., Boland, G.J. and Trevors, J.T. (2002) Evaluation of composts for suppression of dollar spot (*Sclerotinia homoeocarpa*) of turfgrass. *Plant Disease* 86, 405–410.

Bransgrove, K.L. (2005) Warm-season turfgrass diseases in Australia. *International Turfgrass Society Research Journal* 10, 152–155.

Braun, R.C., Patton, A.J., Watkins, E., Koch, P.L., Anderson, N.P., et al. (2020) Fine fescues: a review of the species, their management, production, establishment and management. *Crop Science* 60, 1142–1187. DOI: 10.10002/csc2.20122

Brecht, M.O., Stiles, C.M. and Datnoff, L.E. (2007) Evaluation of pathogenicity of *Bipolaris* and *Curvularia* spp. on dwarf and ultradwarf bermudagrasses in Florida. *Plant Protection Network*. University of Florida, Gainesville, Florida.

Brien, R.M. (1935) Three fungi causing 'brown-patch' of lawns in New Zealand. *New Zealand Journal of Agricultural Research* 51, 157–159.

Burpee, L.L. and Martin, B.S. (1996) Biology of turfgrass diseases incited by *Rhizoctonia* species. In: Sneh, B., Jabaji-Hare, S., Neate, S. and Dijst, G. (eds) *Rhizoctonia Species: Taxonomy, Molecular Biology, Ecology, Pathology and Disease Control*. Springer, Dordrecht, The Netherlands, pp. 359–368.

Charbonneau, P. and Hsiang, T. (eds) (2015) *Integrated Pest Management for Turf*. Publication 845. Ministry of Agriculture, Food and Rural Affairs, Toronto, Canada.

Choi, G.J., Kim, J.C., Shun, M.J., Kim, H.T. and Cho, K.Y. (2000) Phytotoxin production of *Nigrospora sphaerica* pathogenic on turfgrasses. *Plant Pathology Journal* 16, 137–141.

Christensen, M.J. (1992) *Rhizoctonia* fungi associated with turf diseases in New Zealand. *New Zealand Turf Management Journal* 6(February), 9–11.

Clarke, B.B. and Vaiciunas, S.S. (2001) Best management practices for the control of gray leaf spot. *Phytopathology* 91, S2.

Coelho, L., Borrero, C., Bueno-Pallero, F., Guerroro, C., Fonsceca, F., et al. (2009) First report of *Curvularia trifolli* causing Curvularia blight in *Agrostis stolonifera* in south of Portugal. *Plant Disease* 104, 292. DOI: 10.1094/PDIS-03-19-0517-PDN

Couch, H.B. (1995) *Diseases of Turfgrasses*, 3rd edn. Krieger Publishing Company, Malabar, Florida.

Crouch, J.A. (2014) *Colletotrichum caudatum* s.l. is a species complex. *IMA Fungus* 5, 17–30.

Crouch, J.A. and Beirn, L.A. (2009) Anthracnose of cereals and grasses. *Fungal Diversity* 39, 19–44.

Crouch, J.A. and Clarke, B.B. (2012) Biology and pathology of turfgrass anthracnose. *Golf Course Management* 80, 96–100.

Crouch, J.A. and Tomaso-Peterson, M. (2012) Anthracnose diseases of centipedegrass turf caused by *Colletotrichum eremochloae*, a new fungal species closely related to *Colletotrichum sublineola*. *Mycologia* 101, 1085–1096.

Crouch, J.A., Clarke, B.B., White, J.F. and Hillman, B.I. (2009) Systematic analysis of the falcate-spored graminicolous *Colletotrichum* and a description of six new species from warm-season grasses. *Mycologia* 101, 717–732.

Cushnahan, M. (2009) Rolf's disease (*Sclerotium rolfsii*) in golf greens. *New Zealand Turf Management Journal* 24, 13–14.

Dant, L.A., Martin, S.B., Kerns, J.P. and McCarty, L.B. (2021) Nitrogen source impacts on *Rhizoctonia* leaf and sheath spot severity in ultradwarf bermudagrass. *International Turfgrass Society Research Journal* 14, 940–950. DOI:10.1002/its2.15

Davis, D.B. and Dernoeden, P.H. (1991) Summer patch and Kentucky bluegrass quality as influenced by cultural practices. *Agronomy Journal* 83, 670–677. DOI: 10.2134/agronj1991.00021962008300040005x

Davis, J.G. and Dernoeden, P.H. (2002) Dollar spot severity, tissue nitrogen and soil microbial activity in bentgrass as influenced by nitrogen source. *Crop Science* 42, 480–488.

Datnoff, L.E., Brecht, M., Stiles, C. and Rutherford, B. (2005) The role of silicon in supressing foliar diseases in warm-season turf. *International Turfgrass Society Research Journal* 10, 175–179.

de la Cerda, K., Douhan, G.W. and Wong, F.P. (2007) Discovery and characterisation of *Waitea circinata* var. *circinata* affecting annual bluegrass from the Western United States. *Plant Disease* 91, 791–797.

Dernoeden, P.H. (2013) *Creeping Bentgrass Management*. CRC Press, Boca Raton, Florida.

Dick, M.W., Wong, P.T.W. and Clark, G. (1984) The identity of the oomycete causing 'kikuyu yellows', with reclassification of the downy mildews. *Botanical Journal of the Linnean Society* 89, 171–197.

Elliott, M.L. (1991) Determination of an etiological agent of bermudagrass decline. *Phytopathology* 81, 1380–1384. DOI: 10.1094/Phyto-81-1380.

Endo, R.M. (1963) Influence of temperature on rate of growth of five fungus pathogens of turfgrass and on rate of disease spread. *Phytopathology* 53, 857–861.

Endo, R.M. (1966) Control of dollar spot by nitrogen and its probable basis (abstract). *Phytopathology* 56, 877.

Espevig, T., Sundsdal, K., Aamlid, T.S., Crouch, J.A., Normann, K., et al. (2022) In vitro screening of turfgrass species and cultivars for resistance to dollar spot. *International Turfgrass Society Research Journal* 14, 873–882.

Falloon, R.E. (1975) *Curvularia trifolli* as a high-temperature turfgrass pathogen. *New Zealand Journal of Agricultural Research* 19, 243–248.

Fermanian, T.W., Shurtleff, M.C., Randell, R., Wilkinson, H.T. and Nixon, P.L. (2003) *Controlling Turfgrass Pests*, 3rd edn. Prentice-Hall, Upper Saddle River, New Jersey.

Fidanza, M.A. and Dernoeden, P.H. (1996) Brown patch severity in perennial ryegrass as influenced by irrigation, fungicide and fertiliser. *Crop Science* 36, 1620–1630.

Flor, N.C., Munoz, P., Harmon, P. and Kenworthy, K. (2013) Response of seashore paspalum genotypes to dollar spot disease. *International Turfgrass Society Research Journal* 12, 119–125.

Fuke, K., Naofumi, H., Enami, Y., Matsuura, K. and Tajimi, A. (2006) Anthcnose of centipede grass caused by *Colletotrichum caudatum*. *Journal of General Plant Pathology* 72, 74–75. DOI: 10.1007/s10327-005-0251-y

Fulkerson, W.J., Jennings, N.B., Callow, M., Harper, K.J., Wong, P.T.W. and Martin, P.M. (2021) Selection for resistance to fungal diseases and other desirable traits in kikuyu grass (*Cenchus clandestinus*). *Tropical Grasslands* 9, 60–69. DOI: 10.17138/tgft(9)60-69

Gilardi, G., Mocioni, M., Guillino, M.L. and Guarnaccia, V. (2022) *Curvularia americana* and *Curvularia tropicalis* cause leaf and crown necrosis on bermudagrass in Italy. *Phytopathologia Mediterranea* 61, 431–437. DOI: 10.36253/phyto-13825

Goh, T.K., Hyde, K.D. and Lee, D.K.L. (1998) Generic distinction in the *Helminthosporium*-complex based on restriction analysis of the ribosomal RNA gene. *Fungal Diversity* 1, 85–107.

Gomez de Barreda, D., de Luca, V., Ramon-Albalat, V., Leon, M. and Armengol, J. (2019) First report of dollar spot caused by *Clarireedia jacksonii* and brown ring patch caused by *Waitea circinata* var. *circinata* on *Agrostis stolonifera* in Spain. *Plant Disease* 103, 1771. DOI: 10/1094/PDIS-10-18-1816-PDN

Green, D.E., Fry, J.D., Pair, C. and Tisserat, N.A. (1993) Pathogenicity of *Rhizoctonia solani* AG-2-2 and *Ophiosphaella herpotricha* on zoysiagrass. *Plant Disease* 77, 1040–1044.

Grijalba, P.E., Palmucci, H.E. and Guillin, E. (2017) Identification and characterisation of *Pythium graminicola*, causal agent of kikuyu yellows in Argentina. *Tropical Plant Pathology* 42, 284–290. DOI: 10.1007/s40858-017-0149-1

Hall, R. (1984) Relationship between weather factors and dollar spot of creeping bentgrass. *Canadian Journal of Plant Science* 64, 167–174.

Han, J.H., Ahn, C.H., Lee, S.-Y., Back, C.-G., Kang, I.-K. and Jung, H.-Y. (2016) First report of summer patch caused by *Magnaporthiopsis poae* on cool season grass. *The Korean Journal of Mycology* 44, 196–200. DOI: 10.4489/KJM.2016.44.3.196

Hannan, B. and Cushnahan, M. (2010) Waitea patch in New Zealand. *New Zealand Turfgrass Management Journal* 25, 8–9.

Harmon, P.F. and Latin, R. (2005) Winter survival of the perennial ryegrass pathogen *Magnaporthe oryzae* in northern central Indiana. *Plant Disease* 89, 412–418. DOI: 10.1094/PD-89-0412

Hempfling, J.W., Schmidt, C.J., Wang, R., Clarke, B.B. and Murphy, J.A. (2017) Best management practice effects on anthracnose disease of annual bluegrass. *Crop Science* 57, 602–610. DOI: 10.2135/cropsci2016.06.0492

Hempfling, J.W., Murphy, J.A. and Clarke, B.B. (2020) Midseason cultivation effects on anthracnose of annual bluegrass turf. *Crop Science* 112, 3411–3417. DOI: 10.1002/agj2.20202

Hernández-Restrepo, M., Groenewald, J.Z. and Crous, P.W. (2016) Taxonomic and phylogenetic re-evaluation of *Microdochium*, *Monographella* and *Idriella*. *Persoonia* 36, 57–82.

Hodges, C.F. (1990) The microbiology of non-pathogens and minor root pathogens in high sand content greens. *Golf Course Management* 58, 60–75.

Hodges, C.F. and Campbell, D.A. (1995) Growth of *Agrostis palustris* in response to adventitious root infection by *Curvularia lunata*. *Journal of Phytopathology* 143, 639–642.

Hsiang, T., Wu, C., Yung, L. and Liu, L. (1995) Pythium root rot associated with cool-season dieback of turfgrass in Ontario and Quebec. *Canadian Plant Disease Survey* 72, 191–195.

Hu, J., Zhang, H., Dong, Y., Jiang, S., Lamour, K., et al. (2021) Global distributions of *Clarireedia* species and their *in vitro* sensitivity profiles to fungicides. *Agronomy Journal* 11, 2036. DOI: 10.3390/agronomy11102036

Huang, J., Zhang, L. and Hsiang, T. (2005) Turfgrass diseases of the central Yangtze River basin in China. *International Turfgrass Society Research Journal* 10, 202–205.

Hyakumachi, M. and Hayakawa, T. (2008) New isolates of *Rhizoctonia* diseases of turfgrasses. *Floriculture and Ornamental Biotechnology* 2, 14–24.

Inguagiato, J.C. and Martin, S.B. (2015) Diseases of cool- and warm-season putting greens. *USGA Green Section Record* 53, 1–19.

Inguagiato, J.C., Murphy, J.A. and Clarke, B.B. (2013) Topdressing sand particle shape and incorporation effects on anthracnose severity of an annual bluegrass putting green. *International Turfgrass Society Research Journal* 12, 127–133.

Jackson, N. (1993) Geographic distribution, host range, and symptomology of patch diseases caused by soilborne ectotrophic fungi. In: Clarke, B.B. and Gould, A.B. (eds) *Turfgrass Patch Diseases Caused by Ectotrophic Root-Infecting Fungi*. The American Phytopathological Society, St. Paul, Minnesota, pp. 17–40.

Jayawardena, R.S., Bhunjun, C.S., Hyde, K.D., Gentekaki, E. and Itthayakorn, P. (2021) *Colletotrichum*: lifestyles, biology, morpho-species, species complexes and accepted species. *Mycosphere* 12, 519–669. DOI: 10.5943/mycosphere/12/1/7

Kammerer, S.J. and Harmon, P.F. (2008) The importance of early and accurate diagnosis of *Rhizoctonia* diseases. *Golf Course Management* 11, 92–98.

Kammerer, S.J., Burpee, L.L. and Harmon, P.F. (2011) Identification of a new *Waitea circinata* variety causing basal leaf blight of seashore paspalum. *Plant Disease* 95, 515–522.

Kerns, J.P. and Tredway, L.P. (2013) Advances in turfgrass pathology since 1990. In: Stier, J.C., Horgan, B.P. and Bonos, S.A. (eds) *Turfgrass: Biology, Use and Management*. Agronomy Monograph No. 56.

American Society of Agronomy, Inc., Crop Science Society of America, Inc. and Soil Science Society of America, Inc., Madison, Wisconsin, pp. 733–776.

Khan, A. and Hsiang, T. (2003) The infection process of *Colletotrichum graminicola* and relative aggressiveness of four turfgrass species. *Canadian Journal of Microbiology* 49, 433–422. DOI: 10.1139/w03-059

Kim, J.H., Shim, G.-Y. and Kim, Y.H. (2010) Occurrence of Leptosphaerulina leaf blight on Kentucky bluegrass caused by *Leptosphaerulina trifolii*. *Research in Plant Disease* 16, 94–96.

Kim, J.-W. and Park, E.W. (1999) Occurrence and pathogenicity of *Pythium* species isolated from leaf blight symptoms of turfgrasses of golf courses in Korea. *Plant Pathology Journal* 15, 112–118.

Landschoot, P.J. (1993) Taxonomy and biology of ectotrophic root-infecting fungi associated with patch diseases of turfgrasses. In: Clarke, B.B. and Gould, A.B. (eds) *Turfgrass Patch Diseases Caused by Ectotrophic Root-Infecting Fungi*. The American Phytopathological Society, St. Paul, Minnesota, pp. 41–72.

Landschoot, P. and Hoyland, B. (1995) Shedding some light on anthracnose basal rot. *Golf Course Management* 63, 52–55.

Latin, R. (2021) *A Practical Guide to Turfgrass Fungicides*, 2nd edn. The American Phytopathological Society, St. Paul, Minnesota.

Lee, J.H., Shim, G.Y., Kim, J.H., Jeon, C.W. and Kwak, Y.S. (2017) Investigation of fungicides inhibitory effect of on summer patch disease, caused by *Magnaporthiopsis poae*, in Kentucky bluegrass. *Weed & Turfgrass Science* 6, 151–156. DOI: 10.5660/WTS.2017.6.2.151

Liu, F., May, Z.Y., Hou, L.W., Diao, Y.Z., Wu, W.P., et al. (2022) Updating species diversity of *Colletotrichum* with a phylogenetic overview. *Studies in Mycology* 101, 1–56. DOI: 10.3114/sim.2022.101.01

Liu, X.M. and and Jinji, Pu, J.J. (2004) Preliminary report on diseases in turf grasses. *Pratacultural Science* 21, 73–74.

Loch, D.S., McMaugh, P. and Scattini, W. (2013) A review of *Digitaria didactyla* Willd., a low-input warm-season turfgrass in Australia; biology, adaptation and management. *International Turfgrass Society Research Journal* 12, 1–14.

Lucas, L.T. (1976) *Sclerotium rolfsii* on bentgrass greens in North Carolina. *Plant Disease Reporter* 60, 820–822.

Luo, J. and Zhang, N. (2013) *Magnaporthiopsis*, a new genus in the Magnaporthaceae (Ascomycota). *Mycologia* 105, 1010–1029. DOI: 10.3852/12-359

Manamgoda, D.S., Cai, L., McKenzie, E.H.C., Crous, P.W., Madrid, H., et al. (2012) A phylogenetic and taxonomic re-evaluation of the *Bipolaris–Cochliobolus–Curvularia* complex. *Fungal Diversity* 56, 131–144.

Manamgoda, D.S., Rossman, A.Y., Castlebury, L.A., Crous, P.W., Madrid, H., et al. (2014) The genus *Bipolaris*. *Studies in Mycology* 79, 221–288.

Mann, R.L. (2004) A review of the main turfgrass diseases in Europe and their best management practices at present. *Journal of Turfgrass and Sports Surface Science* 80, 19–31.

Mann, R.L. and Newell, A.J. (2005) A survey to determine the incidence and severity of pests and diseases on golf course putting greens in England, Ireland, Scotland and Wales. *International Turfgrass Society Research Journal* 10, 224–229.

Mann, R. and Perris, J. (2005) Disease alert – dollar spot. *International Turfgrass Bulletin* 230, 21–22.

Massimo, M., Patrizia, T., Michele, C., Garibolodi, A. and Lodovica, G.M. (2005) Cool season grasses sensitivity to major grass diseases in Italy. Presented at the *10th International Turfgrass Research Society Conference, Llandudno, UK, 10–15 July 2005*.

Marvasti, F.B. and Banihashemi, Z. (1990) Identification and pathogenicity of turfgrass-infecting fungi in Shiraz landscape. *Iranian Journal of Plant Pathology* 47, 127–129.

Mei, S.S., Wang, Z.Y., Zhang, J. and Rong, W. (2019) First report of leaf blight on *Stenotaphrum secundatum* caused by *Nigrospora osmanthi* in China. *Plant Disease* 103, 1783. DOI: 10.1094/PDIS-02-19-0270-PDN

Milazzo, J., Pordel, A., Raven, S. and Didier, T. (2018) First scientific report of *Pyricularia oryzae* causing gray leaf spot disease *on* perennial ryegrass (*Lolium perenne*) in France. *Plant Disease* 103, 1024.

Mitkowski, N.A. and Browning, M. (2004) *Leptosphaerulina australis* associated with intensively managed stands of *Poa annua* and *Agrostis palustris*. *Canadian Journal of Plant Pathology* 26, 193–198.

Monteith, J. and Dahl, A.S. (1932) Turf diseases and their control. *Bulletin of the United States Golf Association* 12, 87–187.

Murphy, J.A., Clarke, B.B. and Inguagiato, J.A. (2018) Update on BMPs for anthracnose on annual bluegrass turf. *Golf Course Management* 86, 76–85.

Ni, X.X., Li, B.T. and Liu, X.L. (2012) First report of brown ring patch caused by *Waitea circinata* var. *circinata* on *Agrostis stolonifera* and *Poa pratensis* in China. *Plant Disease* 96, 1821.

Obasa, K., Fry, J., Bremer, D. and Kennelly, M. (2017) Evaluation of spring and fall fungicide applications for large patch management in zoysiagrass. *International Turfgrass Society Research Journal* 13, 191–197.

Ou, S.H. (1985) *Rice Diseases*, 2nd edn. Commonwealth Mycological Institute, Kew, UK.

Parmeter, J.R. (ed.) (1970) *Rhizoctonia solani, Biology and Pathology*. University of California Press, Los Angeles, California.

Polizzi, G., Vitale, A. and Castello, I. (2006) Southern blight of tall fescue and bluegrass caused by *Sclerotium rolfsii* in Italy. *Plant Disease* 90, 246.

Prusky, D., Freeman, S., Dickman, M.B. and Dickman, M. (eds) (2000) *Colletotrichum: Host Specificity, Pathology, and Host–Pathogen Interaction*. The American Phytopathological Society, St. Paul, Minnesota.

Punja, Z.K. (1985) The biology, ecology and control of *Sclerotium rolfsii*. *Annual Review of Phytopathology* 23, 97–127.

Punja, Z.K., Grogan, R.G. and Unruh, T. (1982a) Chemical control of *Sclerotium rolfsii* on golf greens in northern California. *Plant Disease* 66, 108–111.

Punja, Z.K., Grogan, R.G. and Unruh. T. (1982b) Comparative control of *Sclerotium rolfsii* on golf greens in northern California with fungicides, inorganic salts, and *Trichoderma* spp. *Plant Disease* 66, 1125-1128.

Rahman, A., Wallis, C.M. and Uddin, W. (2015) Silicon-induced systemic defence responses in perennial ryegrass against infection by *Magnaporthe oryzae*. *Phytopathology* 105, 748–757.

Rioux, R.A., Schultz, J., Garcia, M., Willis, D.K., Casler, M., et al. (2014) *Sclerotinia homoeocarpa* overwinters in turfgrass and is present in commercial seed. *PLoS ONE* 9, e110897. DOI: 10.1371/journal.pone.0110897

Roberts, J.A. and Tredway, L.P. (2008) First report of Curvularia blight of zoysiagrass caused by *Curvularia lunata* in the United States. *Plant Disease* 92, 173.

Roberts, J.A., Murphy, J.A. and Clarke, B.B. (2012) Lightweight rolling effects on anthracnose of annual bluegrass putting greens. *Agronomy Journal* 104, 1176–1181

Ryan, C.P., Dernoeden, P.H. and Grybauskas, A.P. (2012) Seasonal developments of dollar spot epidemics in six creeping bentgrass cultivars in Maryland. *HortScience* 47, 422–426.

Salgado-Salazar, C., Beirn, L.A., Ismaiel, A., Boehm, M.J., Carbone, I., et al. (2018) *Clarireedia*: a new fungal genus comprising four pathogenic species responsible for dollar spot disease of turfgrass. *Fungal Biology* 122, 761–773.

Schoeman, A.S., Brandenburg, R.L. and Martin, B. (2001) The effect of creeping bentgrass cultivars on the influence of dollar spot (*Sclerotinia homoeocarpa*) in South Africa. *International Turfgrass Society Research Journal* 9, 22.

Scott, J.B. (1937) Observations on turf pests and diseases. *The Australian Greenkeeper* 2, 3–34.

Shakesby, A. (1989) Anthracnose in wintergrass – epidemiology and control. In: Martin, P.M. (ed.) *Proceedings of the 1st Australasian Turf Researchers Seminar, Bermagui, NSW, Australia, 4–6 April 1989*. Australian Turfgrass Research Institute, Sydney. Australia, pp. 33–34.

Shelton, C.D., Askew, S.D., Ervin, E.H. and McCall, D.S. (2021) Impact of ferrous sulphate concentration on *Clarireedia* isolate growth and dollar spot development. *Crop Science* 61, 2148–2155.

Shim, G.-Y. and Lee, H. (2018) Identification and pathogenicity of *Rhizoctonia* spp. isolated from turfgrasses in golf courses in Korea (abstract in English). *Korean Journal of Turfgrass Science* 9, 235–252.

Shivas, R.G., Tan, Y.P., Edwards, J., Dinh, Q., Maxwell, A., et al. (2016) *Colletotrichum* species in Australia. *Australasian Plant Pathology* 45, 447–464. DOI: 10.1007/s13313-016-0443-2

Simmonds, J.H. (1966) *Host Index of Plant Diseases in Queensland*. Queensland Department of Primary Industries, Brisbane, Australia.

Sivanesan, A. (1987) *Graminicolous Species of 'Bipolaris', 'Curvularia', 'Drechslera', 'Exsorohilum' and Their Teleomorphs*. Mycological Paper 158. CAB International, Wallingford, UK.

Skorulski, J. (2014) Getting the upper hand on dollar-spot disease. *USGA Green Section Record* 52, 1–5.

Smiley, R.W., Fowler, M.C. and O'Knefski, R.C. (1985) Arsenate herbicide stress and incidence of summer patch on Kentucky bluegrass. *Plant Disease* 69, 44–48.

Smith, J.D., Jackson, N. and Woolhouse, A.R. (1989) *Fungal Diseases of Amenity Turf Grasses*, 3rd edn. E. & F.N. Spon, London.

Sneh, B., Jabaji-Hare, S., Neale, S. and Dijst, G. (eds) (1996) *Rhizoctonia Species: Taxonomy, Molecular Biology, Ecology, Pathology and Disease Control*. Springer, Dordrecht, The Netherlands.

Stalpers, J.A., Redhead, S.A., May, T.W., Rossman, A.Y., Crouch, J.A., et al. (2021) Competing sexual–asexual generic names in *Agaricomycotina* (*Basidiomycota*) with recommendations for use. *IMA Fungus* 12, 22. DOI: 10.1186/s43008-021-00061-3

Steketee, C.J., Martinez-Espinoza, A.D., Harris-Shultz, K.R., Henry. G.M. and Raymer, P.L. (2017) Evaluation of seashore paspalum germplasm for resistance to dollar spot. *International Turfgrass Society Research Journal* 13, 175–184.

Tan, Y.P., Crous, P.W. and Shivas, R.G. (2016) Eight novel *Bipolaris* species identified from John L. Alcorn's collections at the Queensland Plant Pathology Herbarium (BRIP). *Mycological Progress* 15, 1203–1214. DOI: 10.1007/s11557-016-1240-6

Tanaka, A., Sato, T. and Tani, T. (1999) Anthracnose of *Zoysia matrella* caused by *Colletotrichum caudatum*. *Journal of Japanese Society of Turfgrass Science* 28(Suppl. 1), 70–71. DOI: 10.11275/turfgrass1972.28.supplement1_70

Tani, T. and Beard, J.B. (1997) *Color Atlas of Turfgrass Diseases*. Wiley, Hoboken, New Jersey.

Teuton, T.C., Sorochan, J.C., Main, C.L., Samples, J.C., Parham, J.M. and Mueller, T.C. (2007) Hybrid bluegrass, Kentucky bluegrass and tall fescue response to nitrogen fertilisation in the transition zone. *HortScience* 42, 369–372.

Thompson, C., Zhang, Q., Kennelly, M., Stier, J., Blume, C., et al. (2019) The dollar spot susceptibility of 25 bentgrasses is consistent across five states in the central USA. *Crop, Forage & Turfgrass Management* 5, 1–4. DOI: 10.2134/cftm2018.09.0075

Thompson, D.C. and Clarke, B.B. (1992) Evalution of bacteria for biological control of summer patch of Kentucky bluegrass caused by *Magnaporthe poae* (abstract). *Phytopathology* 82, 1123.

Thompson, D.C., Fowler, M.C. and Smiley, R.W. (1982) Recognising Nigrospora blight. *Plant Disease* 66, 265.

Toda, T., Mushika, T., Hayakawa, T., Tanaka, A., Tani, T. and Hyakumachi, M. (2005) Brown ring patch: a new disease on bentgrass caused by *Waitea circinata* var. *circinata*. *Plant Disease* 89, 536–542.

Tomaso-Peterson, M. (2018) Ink spot of warm-season golf course turf. *Crop, Forage & Turfgrass Management* 4, 1–5. DOI: 10.2134/cftm2018.06.0044

Tomaso-Peterson, M. and Young, J. (2010) Cultivar response of seeded bermudagrass to leaf spot and the influence of nitrogen on disease severity. *Applied Turfgrass Science* 7, 1–8. DOI: 10.1094/ATS-2010-0326-01-RS

Tomaso-Peterson, M., Jo, Y.K., Vines, P.L. and Hoffman, F.G. (2016) *Curvularia malina* sp. nov. incites a new disease of warm-season turfgrasses in the southeastern United States. *Mycologia* 108, 915–924.

Tran, N.T., Geering, A.D.W., Tan, Y.P. and Shivas, R.G. (2021) *Curvularia stenotaphri*. *Fungal Planet* 1346 (24 December 2021).

Tredway, L.P., Tomaso-Peterson, M., Kerns, J.P. and Clarke, B.B. (2023) *Compendium of Turfgrass Diseases*, 4th edn. The American Phytopathological Society. St. Paul, Minnesota.

Turgeon, A.J. and Kaminski, J.E. (2019) *Turfgrass Management, Edition 1.0*. TurfPath LLC, State College, Pennsylvania.

Uddin, W. and Soika, M.D. (2000) Effects of plant growth regulators, herbicides, and fungicides on development of blast disease (gray leaf spot) on perennial ryegrass turf. *Phytopathology* 92, S82.

Uddin, W., Viji, G. and Vincelli, P. (2003) Gray leaf spot (blast) of perennial ryegrass turf: an emerging problem for the turfgrass industry. *Plant Disease* 87, 880–889.

Unal, F., Askin, A., Koca, E., Yildirir, M. and Bingol, M.U. (2019) Mycelial compatibility groups, pathogenic diversity and biological control of *Sclerotium rolfsii* on turfgrass. *Egyptian Journal of Biological Pest Control* 29, 44–51. DOI: 10.1186/s41938-019-0144-6

Unruh, T. and Punja, Z. (1982) I've never seen that before! *Sclerotium rolfsii* blight on golf greens. *USGA Green Section Record* 20, 8–10.

Vargas, J.M. (2005) *Management of Turfgrass Diseases*, 3rd edn. Wiley, Hoboken, New Jersey.

Vines, P.L., Hoffman, F.G., Meyer, F., Allen, T.W., Luo, J., et al. (2020) *Magnaporthiopsis cynodontis*, a novel turfgrass pathogen with widespread distribution in the United States. *Mycologia* 112, 52–63.

Wang, Z.Y., Xie, S.N., Wang, Y., Wu, H.Y. and Zhang, M. (2016) First report of *Bipolaris peregianensis* causing leaf spot of *Cynodon dactylon* in China. *Plant Disease* 96, 917. DOI: 10.1094/PDIS-12-11-1066-PDN

Watkins, J.E., Gaussoin, R.E., Frank, K.W. and Wit, L.A. (2001) Brown patch severity and perennial ryegrass quality as influenced by nitrogen rate and source and cultivar. *International Turfgrass Society Research Journal* 9, 723–728.

Watschke, T.L., Dernoeden, P.H. and Shetlar, D.J. (2013) *Managing Turfgrass Pests*, 2nd edn. CRC Press. Boca Raton, Florida.

Williams, D.W., Burrus, P.B. and Vincelli, P. (2001) Severity of gray leaf spot in perennial ryegrass as influenced by mowing height and nitrogen level. *Crop Science* 41, 1207–1211.

Wong, F. (2006) Gray leaf spot in the West. *Golf Course Management* 74, 97–101.

Wong, F.P. and Kaminski, J.E. (2007) A new *Rhizoctonia* disease of bluegrass putting greens. *Golf Course Management* 75, 98–103.

Wong, P.T.W. (1975) Kikuyu yellows – a disease caused by an undescribed phycomycete. *Plant Disease Reporter* 59, 800–801.

Wong, P.T.W. (1982) Kikuyu yellows – a potential hazard to kikuyu production in NSW. *Plant Disease Survey*. NSW Agriculture and Fisheries, Sydney, Australia, pp. 21–22.

Wong, P. (2011) *Screening of Kikuyu Selections and Hybrids for Resistance to Kikuyu Yellows*. Project No. TU09006. Horticulture Australia Ltd, Sydney, Australia.

Wong, P.T.W. and Tesoriero, L.A. (1990) Evaluation of fungicides for the control of kikuyu yellows (*Verrucalvus flavofaciens*). *Plant Protection Quarterly* 5, 76–77.

Wong, P.T.W. and Wilson, G.P.M. (1983) Reaction of kikuyu cultivars and breeding lines to kikuyu yellows disease. *Australasian Plant Pathology* 12, 47–48.

Wong, P.T.W., Khemmuk, W., Geering, A.D.W. and Shivas, R.G. (2015) *Magnaporthiopsis agrostidis*. *Persoonia* 35, 322–323.

Wong, P.T.W., Tan, Y.P., Weese, T.L. and Shivas, R.G. (2022) *Magnaporthiopsis* species associated with patch diseases in turfgrasses in Australia. *Mycosphere* 13, 602–611. DOI 10.5943/mycosphere/13/1/5

Woodcock, T. (1983) Patch diseases. In: *Proceedings of Turfgrass Disease Seminar, 12–13 July 1983*. Victorian College of Agriculture and Horticulture (Burnley) and Turf Research and Advisory Institute (Frankston), Victoria, Australia, pp. 13–29.

Yamanaka, S. (1982) A consideration of classification of *Pyricularia* spp. isolated from various gramineous plants in Japan (in Japanese). *Annals of the Phytopathological Society of Japan* 48, 245–248.

Yildiz, F., Yildiz, M. and Delen, N. (1990) The preliminary studies on the turfgrass diseases in Turkey. *Journal of Turkish Phytopathology* 19, 21–29.

Zhang, H., Dong, Y., Jin, P., Hu, J., Lamour, K. and Yang, Z. (2022) Genome resources for four *Clarireedia* species causing dollar spot on diverse turfgrass. *Plant Disease* 107, 929–934. DOI: 10.1094/PDIS-08-22-1921-A

Zhang, W., Nan, Z.B., Liu, G.D., Gao, Z.Y. and Li, M. (2014) First report of leaf and sheath spot caused by *Waitea circinata* var. *zeae* on *Paspalum vaginatum* and *Zoysia tenuifolia* in China. *Plant Disease* 98, 1436. DOI: 10.1094/PDIS-04-14-0404-PDN

Zhang, W., Liu, J., Pinghui, H. and Zhenchi, H. (2018) *Curvularia malina* causes a foliar disease on hybrid bermudagrass. *European Journal of Plant Pathology* 2, 557–562.

Zhang, W., Damm, U., Crous, P.W., Groenewald, J.Z., Niu, X., et al. (2020) Anthcnose disease of carpetgrass (*Axonopus compressus*) caused by *Colletotrichum hainanense* sp. nov. *Plant Disease* 104, 1744–1750.

Zheng, L., Shi, F., Kelly, D. and Hsiang, T. (2012) First report of leaf spot on Kentucky bluegrass (*Poa pratensis*) by *Nigrospora oryzae* in Ontario. *Plant Disease* 96, 909.

8

Fungal Diseases of Mown Forbs

Abstract

This chapter focuses on mown forbs and their fungal diseases. Forbs are broad-leafed, decumbent plants and several small-leafed species capable of tolerating low mowing heights are selectively managed for amenity purposes in many countries. Forbs, wherever managed, are hosts of numerous fungal diseases some of which also occur on cool-season and warm-season grasses.

Forbs are members of one of several dicotyledonous plant families and possess a fibrous root system and a variety of leaf shapes, depending on the plant species. A select number of perennial low-growing forbs, capable of tolerating and recovering from low mowing to greater or lesser degrees, are managed as grass alternatives in many countries, as outlined in Chapter 1 (this volume). Certain forb species (e.g. chamomile, clover and others) have a long historical use in lawn landscapes on large country estates of England and Western Europe (Gilbert, 1991).

Forbs and grasses are hosts to certain fungal diseases (e.g. brown patch and Rolf's disease) while other diseases (e.g. Phytophthora root rot) are restricted to forbs. Fungal diseases of forbs have not generally attracted the attention of plant pathologists. Consequently, the identity and biology of many fungal diseases of mown forbs remain unclear or unknown. The implementation of specific integrated management programmes of many fungal diseases of mown forbs is problematic given the lack of understanding of much of their ecology.

Mown Forbs

A select number of forb plants capable of tolerating low mowing are managed, to varying cultural intensities, as either domestic or estate lawns (e.g. chamomile and dichondra) while very few are managed as playing surfaces (e.g. cotula bowling greens in New Zealand).

Chamomile

English or Roman chamomile, *Chamaemelum nobile* (formerly *Anthemis nobilis*), is well known throughout Western Europe (Aldous, 1991). Chamomile still forms part of a lawn area in Buckingham Palace in London, England (Gilbert,

1991). *C. nobile* 'Treneague' is a non-flowering selection more capable of tolerating low mowing. The susceptibility of chamomile to fungal diseases is unclear.

Clover

Clovers (*Trifolium* spp.) are widespread legumes and several species such as strawberry clover (*Trifolium fragiferum*) and microclover (*Trifolium repens* var. Pirouette) are managed to a limited extent in certain countries (Baltensperger and Gaussoin, 1985; Turner and Caroll, 2015). As a group, clovers are susceptible to a range of foliar and root diseases (Johnstone and Barbetti, 1987) depending on the host–fungi combination.

Colobanthus

Colobanthus (*Colobanthus apetalus* and *Colobanthus muelleri*) is considered a colonizing plant common on mixed-cotula bowling greens in New Zealand and susceptible to golden bracelet and Sclerotinia patch diseases (Ormsby, 2012).

Cotula

Cotula (*Leptinella dioica* and *Leptinella maniototo*) is exclusively managed as a near-monoculture on bowling greens throughout New Zealand (Evans, 1984; Way, 1991). Cotula is widely managed on bowling greens and capable of surviving very low mowing (Fig. 8.1). 'Grasslands Pahia' and 'Pahia' are improved selections of cotula (Hickey *et al.*, 1993). Cotula is susceptible to brown patch, Phytophora root rot, Pythium blight, Sclerotium patch, southern blight and gold bracelet disease (Christensen, 1989; NZSTI, 2008).

Dichondra

Dichondra (*Dichondra repens* and *Dichondra micrantha*) has kidney-shaped leaves and is managed in subtropical regions in certain countries (Fig. 8.2). Dichondra is susceptible to Sclerotium blight,

Fig. 8.1. A cotula bowling green surface in New Zealand. (Courtesy of Gary W. Beehag.)

rust and leaf spot diseases (Cardin *et al.*, 2005; Garibaldi *et al.*, 2005).

Pinto peanut

Pinto peanut (*Arachis pintoi*) is a low-growing, tropical legume managed as low-intensity turfgrass in the warmer regions of the Americas. Pinto peanut is susceptible to several leaf spot and rust diseases (Kerridge and Hardy, 1994; Christians, 2004).

Starweed

Starweed (*Plantago triandra*) occurs in mixed-cotula bowling greens in New Zealand and is susceptible to Phytophthora root rot, Sclerotium patch, and southern blight or Rolf's disease (NZSTI, 2008).

Fungal Diseases of Mown Forbs

Fungal diseases of forbs may occur throughout the year during either cooler or warmer conditions, depending on the host–fungi combination and environmental conditions. Certain fungal pathogens may cause disease during cooler or warmer seasons under environmental conditions favouring fungal growth and infection at the expense of their host.

The most well-known fungal diseases of mown forbs occur on cotula and dichondra (Table 8.1). The causal agents of several fungal

Fig. 8.2. A dichondra domestic lawn in Australia. (Courtesy of Gary W. Beehag.)

Table 8.1. Fungal diseases of cotula and dichondra.

Forb	Disease name	Fungal pathogen	References
Cotula	Alternaria leaf spot	*Alternaria* sp.	Christensen (1989); NZSTI
	Brown patch	*Rhizoctonia solani*	(2008); Howard (2012);
	Fairy ring	Basidiomycetes	Ormsby and Howard (2021)
	Gold bracelet	*Rhizoctonia* sp.	
	Phytophthora root rot	*Phytophthora cryptogea*	
	Rolf's disease	*Athelia rolfsii*	
	Sclerotinia patch	*Sclerotinia minor*	
	Winter Pythium patch	*Pythium* sp.	
	White patch	Unknown	
Dichondra	Alternaria leaf spot	*Alternaria dichondrae*	Cardin *et al.* (2005); NZSTI
	Cercospora leaf spot	*Cercospora dichondrae*	(2008); Sarrocco *et al.* (2010)
	Rolf's disease	*Athelia rolfsii*	
	Rust	*Puccinia dichondrae*	

diseases of mown forbs are well established in those countries or regions where specific forb species have long been managed as domestic lawns (e.g. United States, South Africa, the United Kingdom and Western Europe) and on bowling greens (i.e. New Zealand). Under highly favourable conditions for infection and pathogenicity, certain fungal diseases of forbs cause unacceptable damage requiring remedial action.

The most common and potentially damaging fungal diseases on mixed cotula–starweed bowling greens in the North Island of New Zealand are brown patch and Rolf's disease (called southern blight in the United States), Sclerotinia patch, gold bracelet disease and white patch. Common fungal diseases on mixed-cotula bowling greens in the South Island of New Zealand are Sclerotinia patch, brown patch, gold bracelet disease, fairy rings and Rolf's disease (southern blight). Phytophthora root rot has become a lesser reported disease throughout New Zealand.

Symptomatology of certain fungal diseases on mown forbs and grasses may differ even though the same fungal pathogen may be involved. Structural, physiological and biochemical differences between grasses and forbs and their respective defence mechanisms account partly for these differences. Fungicide phytomobility in foliage and roots, thus their efficacy, may differ because of anatomical and biochemical differences between grasses and forbs.

Alternaria leaf spot

Alternaria leaf spot on managed dichondra has been reported in Western Europe, Australia, East Asia and New Zealand. In Western Europe, Australia and New Zealand, the disease is caused by *Alternaria dichondrae* and *Alternaria solani* (Endo, 1965; Laundon, 1979; Sivapalan and Pascoe, 1994; Cardin *et al.*, 2005). In New Zealand on cotula bowling greens the *Alternaria* species is unclear (Christensen, 1989) and in China the causal agent is *Pyrenophora dichondrae* (Huang *et al.*, 2008).

Alternaria pathogens grow over a wide temperature range (18–25°C) and the disease is generally favoured during cool conditions. *Alternaria* species survive unfavourable conditions as mycelium or spores on decomposing plant tissues or as latent infection in host seeds. Fungal spores are spread by wind to infect plants. *Alternaria* species generally primarily infect hosts already subjected to some form of environmental stress, senescence or wounding.

Symptoms appear initially as small-sized, irregular to roughly round spots or lesions 1.0–10.0 mm in diameter and light brown in colour. Lesions may extend downwards to the leaf petioles and coalesce as infection progresses, showing dark brown-coloured edges. Under a severely diseased condition, the sward displays a wilted and thinned appearance depending on conditions. In mixed-cotula greens the irregular symptoms are similar in pattern but the damaged cotula leaves tend to take on a yellow appearance.

Cercospora leaf spot

Cercospora leaf spot caused by fungi in the genus *Cercospora* is widespread worldwide. An unnamed *Cercospora* sp. has been isolated on an infected dichondra lawn in Italy (Cardin *et al.*, 2005).

Fairy ring

Fairy ring is a soil-borne formation incited by numerous species of basidiomycete fungi. Fairy ring formation may occur on numerous species of mown forbs. The biology, ecology and epidemiology and cultural management of fairy ring disease have been extensively documented and are discussed separately in Chapter 6 (this volume).

Gold bracelet disease

Gold bracelet describes unusual and variable symptoms associated with mixed-cotula bowling greens in New Zealand. Unknown disease symptoms were initially reported in the mid-1980s on infected cotula bowling greens on lower regions of the North Island. A slow-growing *Rhizoctonia*-like fungus is considered the causal agent, but the true identity and virulence of fungal isolates remain untested.

Geographic distribution, host range and susceptibility

Gold bracelet disease occurs on bowling greens throughout New Zealand. The host range is cotula (*Leptinella* spp.), colobanthus (*Colobanthus* spp.) and starweed (*Plantago* spp.). Cotula (*L. dioica* and *L. maniototo*) is more susceptible to this disease (Hannan and Christensen, 1999; Howard, 2012).

Signs and symptoms

The presence of small-diameter, dark-coloured hyphae and blackened sclerotia or fruiting bodes approximately 1.0 mm in diameter along infected stems and roots are signs of gold bracelet disease. Symptom development associated with gold bracelet disease varies widely depending on season of the year and other factors not fully understood. Initial symptoms appear as small patches 5–10 mm in diameter displaying yellow- to golden-coloured leaves. Patches enlarge and coalesce forming irregular shaped rings 300 cm to 1 m in diameter with the distinctive coloured foliage (Fig. 8.3). Infected perimeter leaves eventually turn brown and die.

Fig. 8.3. Early-stage symptoms of gold bracelet disease on a cotula bowling green in New Zealand. (Courtesy of the NZSTI.)

Late-stage symptoms of gold bracelet disease are characterized by roughly circular-shaped patches with multicoloured, mosaic centres displaying chlorosis and necrosis (Fig. 8.4). Infected patches may display an orange- to gold-coloured perimeter, depending on environmental conditions, hence the common name for this disease. Infected plants inside the ring generally show poor growth resulting in an irregular and bumpy surface. Additional symptoms observed in the South Island of New Zealand considered to be caused by the same fungus as gold bracelet are given the names 'summer spotting' and 'spring small patch'. Summer spotting symptoms occur from late spring (November) to summer (February).

Disease scars may persist throughout summer. Patches may reoccur in the same location, increasing in size each year. Differences in symptomatology between locations and seasons raise the possibility of a disease complex of more than one fungal pathogen.

Disease profile

The true identity, biology and epidemiology of the causal agent remain unknown. Fungal sclerotia may overwinter in the thatch-mat region. Gold bracelet disease can occur year-round (NZSTI, 2008) depending on regional weather conditions but primarily causes damage any time from late spring (November) through to autumn (June) contingent on the incidence of frost and rainfall (Hannan and Christensen, 1999; Howard, 2012).

Gold bracelet disease is favoured during periods of reduced plant growth, relatively dry soil conditions followed by light rainfall and high humidity. The disease may appear earlier coinciding with dry conditions in spring in certain regions. The disease is generally more severe when moisture level in the upper 25–30 mm soil profile is adequate but it is relatively dry below this layer. Gold bracelet disease can also be found on newly established cotula bowling greens using both bulbils and stolons, raising the potential of disease transmission by infected material.

Integrated management

Predictive management of gold bracelet disease remains problematic (Box 8.1) in regions where the incidence, severity and duration of the disease warrant treatment. There are no proven

Fig. 8.4. Late-stage symptoms of golden bracelet disease on a cotula bowling green in New Zealand. (Courtesy of the NZSTI.)

Box 8.1. Integrated management of gold bracelet disease.

- Monitor plants for signs of pathogen structures
- Maintain some plant growth, preventing plant dormancy with judicious use of light rates of N or NP nutrition particularly on newer greens
- Adequate irrigation water to maintain even moisture in the rootzone
- Use of registered fungicides

science-based cultural or biological measures providing effective management against the disease. However, maintaining even moisture levels in the profile over the summer period does help to reduce disease susceptibility and severity.

Leaf rust

The leaf rust pathogen *Puccinia dichondrae* is widespread on dichondra and occurs worldwide (Hickman and Holcomb, 1979; Cardin *et al.*, 2005; Sarrocco *et al.*, 2010). The biology, ecology and epidemiology and cultural management of rust diseases have been documented, as discussed separately in Chapter 6 (this volume).

Phytophthora root rot

Phytophthora is a large, cosmopolitan genus of soil-borne, fungal-like microbes or oomycetes with an extensive host range. Phytophthora root rot is caused by *Phytophthora cryptogea* which has an optimum temperature for growth of 22–25°C but is capable of growth at temperatures of less than 1°C to 31–33°C (Erwin and Ribeiro, 1996).

In New Zealand, Phytophthora root rot became problematic during the 1990s in the upper North Island (Anon., 2020) but is now considered generally uncommon. Phytophthora root rot can occur from August to October, depending on environmental conditions. The disease is favoured under moist soil and cool conditions. When present, overall symptoms are irregularly shaped, large-sized areas of unthrifty swards. Infected leaves range from a brownish-yellow to brown-black depending on stage of infection and plant species. Late-stage infected leaves are easily removed from the crown region and roots are brown and rotted (Christensen, 1989; NZSTI, 2008).

P. cryptogea is one of the water moulds related to *Pythium*, thus management of the oxygen–moisture balance is a key factor in its management. There are no specific, science-based cultural management practices proven against Phytophthora root rot on cotula bowling greens.

Rhizoctonia brown patch

Rhizoctonia brown patch caused by *Rhizoctonia solani* is a widespread soil-borne, fungal disease worldwide. Brown patch was isolated from a pearlwort (*Sagina procumbens*) lawn in South Australia (Kerr, 1956). In New Zealand, brown patch on bowling greens (Fig. 8.5) came to prominence during the 1970s coinciding with adoption of cotula *L. maniototo* in the North Island. Rhizoctonia brown patch is widely distributed on cotula and starweed bowling greens in New Zealand (Ormsby, 1990; NZSTI, 2008). *L. maniototo* is more susceptible than *L. dioica* (Christensen, 1989).

The biology, ecology and epidemiology and cultural management of brown patch have been extensively documented, as discussed in Chapter 7 (this volume). Several fungicides are available in New Zealand for application against brown patch, as discussed separately in Chapter 11 (this volume).

Rolf's disease (southern blight)

Rolf's disease, called southern blight on grasses in the United States, is caused by the soil-borne fungal pathogen *Athelia rolfsii* (formerly *Sclerotium rolfsii*) which has an extensive host range of grasses and forbs. The disease came to prominence on mixed-cotula bowling greens in New Zealand during the early 1970s. Rolf's disease remains problematic, capable of causing significant loss of surface to cotula and starweed bowling greens.

Rolf's disease also occurs on dichondra lawns in southern Europe (Garibaldi *et al.*, 2005). Rolf's disease occurs during the warmer seasons and is characterized by rapidly developing and reddish-coloured, roughly circular patches (Fig. 8.6). Patches vary widely in distribution, size and shape depending on the host plant and climatic conditions.

Fruiting bodies or sclerotia 1–3 mm in diameter, initially white (Fig. 8.7) and becoming dark brown when mature, may be observed adjacent to infected stems.

In New Zealand, Rolf's disease on mixed-cotula bowling greens is found throughout the North Island and in the upper South Island as

Fig. 8.5. Brown patch on a cotula bowling green in New Zealand. (Courtesy of the NZSTI.)

Fig. 8.6. Late-stage symptoms of Rolf's disease on a starweed bowling green in New Zealand. (Courtesy of the NZSTI.)

Fig. 8.7. White-coloured sclerotia of *Athelia rolfsii* on a starweed bowling green in New Zealand. (Courtesy of the NZSTI.)

far south as Nelson (Christensen, 1989). Rolf's disease occurs during December to March when daily temperatures exceed 25°C (Ormsby, 1990; Howard, 2012). Rolf's disease has been identified on dichondra lawns in New Zealand (Munro, 1997; Ormsby, 2005) and in Italy (Garibaldi et al., 2005).

The biology, ecology and cultural management of Rolf's disease have been extensively documented on true grasses, as covered in Chapter 7 (this volume). Several fungicides are available in New Zealand for application against Rolf's disease on bowling greens, as discussed in Chapter 11 (this volume).

Sclerotinia patch

Sclerotinia patch represents the most widespread and common fungal disease on mixed-cotula bowling greens in New Zealand. *Sclerotinia* patch is caused by *Sclerotinia minor*. *Sclerotinia* is a large, cosmopolitan fungal genus with an extensive plant host range. Sclerotinia patch can be effectively managed on cotula (*L. dioica*) bowling greens by adopting cultural practices and chemical products in an integrated approach.

Geographic distribution, host range and susceptibility

Sclerotinia patch occurs on bowling greens throughout New Zealand. Host forbs are cotula (*Leptinella* spp.), pennywort (*Hydrocotyle* spp.) and starweed (*Plantago* spp.) (Harrington, 1990). Starweed, pennywort and cotula *L. maniototo* are highly susceptible, while cotula *L. dioica* is the most resistant to Sclerotinia patch. Observations indicate risk of Sclerotinia patch is reduced in cotula *L. maniototo* plants older than 2 years (Ormsby, 2012; Ormsby and Howard, 2021). The pathogen *S. minor* was isolated from pennywort

(*Hydrocotyle* spp.) on a bowling green in 1940 in Canberra (Australia) (Ludbrook, 1940).

Signs, symptoms and occurrence

Early-morning presence of fungal mycelium, initially white fading to light grey, enveloping the foliage can be a characteristic sign of Sclerotinia patch. Late-stage symptoms and susceptibility of forbs to Sclerotinia patch vary depending on the growth stage of the host forb (Ormsby, 2012; Ormsby and Howard, 2021). Infected leaves on cotula (*L. maniototo*) go mushy and collapse resulting in the patch becoming greasy.

Sclerotinia patch symptoms vary depending on depending on host species and environmental conditions. Symptoms can be variable depending on weather conditions. The dark patches occur in wet conditions due to algal slime forming on the dying tissue. Under dry conditions the algal slime does not form, so the patches tend to be light brown in colour.

Symptoms on newly infected plants develop rapidly, appearing as roughly circular, larger-sized patches (Fig. 8.8). Infected starweed leaves turn yellow to yellow pink during dry surface conditions with generally no mycelium seen. However, mycelium is prominent during wetter conditions with the leaves turning yellow. Sclerotinia patch can result in death of individual starweed plants, normally the older plants. Superficial damage can also occur with mycelium present and death of outer leaves, but the crown remains intact.

Sclerotinia patch on starweed plants is more common from summer (December) to autumn (March). Sclerotinia patch occurs throughout the year in the upper North Island and during autumn and spring in cooler locations, depending on regional conditions.

Late-stage symptoms on pennywort (*Hydrocotyle* spp.) are roughly circular, larger-sized patches up to 200 mm in diameter. Mycelium is present and the infected leaves turn a mushy brown and die. Sclerotinia patch on pennywort plants is most common during autumn but may occur from late spring to early summer, depending on regional conditions. Symptoms on infected cotula (*L. maniototo*) plants appear as roughly circular, small-sized patches up to 50–75 mm in diameter. Patches may enlarge to coalesce forming larger-sized patches 0.5–1.0 m in diameter with infected leaves turning mushy brown and dying (Fig. 8.9).

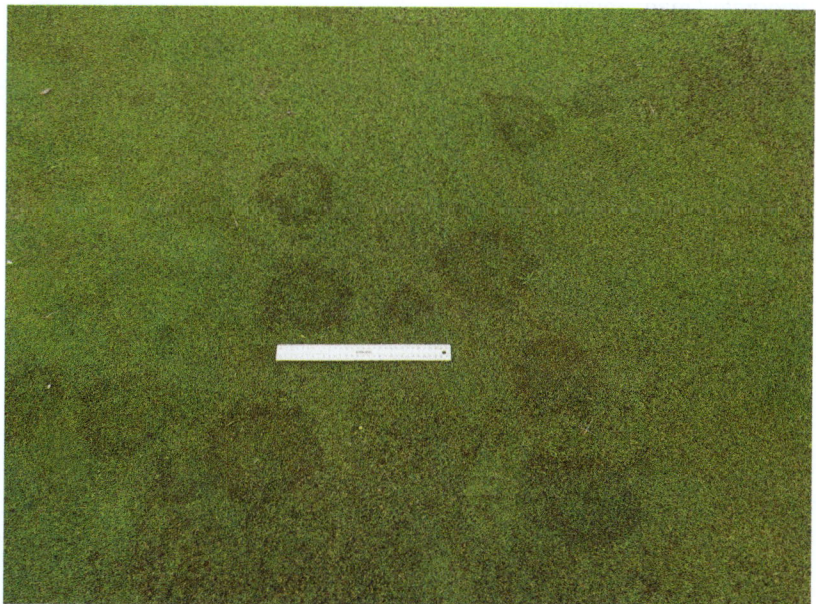

Fig. 8.8. Late-stage symptoms of Sclerotinia patch on a bowling green in New Zealand. (Courtesy of the NZSTI.)

Sclerotinia patch on infected, immature cotula is most common during spring but can occur in autumn, depending on conditions. Sclerotinia patch symptoms on *Leptinella dispersa* cotula plants is uncommon. Diseased patches are largely cosmetic and are mid-sized (50–75 mm diameter) with brownish-coloured leaves. Disease symptoms on *L. dioica* are most uncommon but may form mid-sized patches less than 100 mm in diameter during autumn (Ormsby, 2012; Ormsby and Howard, 2021) if the plants are lush or infrequently mown.

Recovery of diseased cotula plants is longer when infected during the cooler months, as opposed to damage during warmer periods.

Disease profile

Detailed understanding of Sclerotinia patch and the biology and epidemiology of the causal agent remains unclear. However, Sclerotinia patch is favoured by lush plant growth brought about by soil topdressing and nitrogen fertilizer applications especially in the spring. Locations of greens having extended periods of leaf wetness, in excess of 12 h, particularly in shade-affected places, also will encourage this disease.

Integrated management

Effective management of Sclerotinia patch disease can be problematic (Box 8.2) particularly on greens comprising either starweed or cotula (*L. maniototo*). Several cultural practices have proved effective.

Winter Pythium patch

Pythium is a cosmopolitan genus of fungal-like microbes collectively called oomycetes which have an extensive host range. In New Zealand, winter Pythium patch occurs on cotula bowling greens during July and August and ascribed the descriptive name of elephant's foot by some (Fig. 8.10). Damage observed on some bowling greens has been attributed to *Pythium* sp. and *Phytophthora* sp. (Christensen, 1989).

The biology, ecology and epidemiology and cultural management of *Pythium*-incited diseases

---

Box 8.2. Integrated management of Sclerotinia patch.

- Avoid spring nitrogen applications
- Promote regular mowing to avoid lush and long leaves
- Avoid topdressing in the spring and applying excess amounts at renovation time
- Remove dew from shaded areas of green
- Groom starweed greens to improve drying between the overlapping leaves of individual starweed plants
- Use of registered fungicides

Fig. 8.9. Scarred and dead plants caused by sclerotinia patch on a mixed-cotula bowling green in New Zealand. (Courtesy of the NZSTI.)

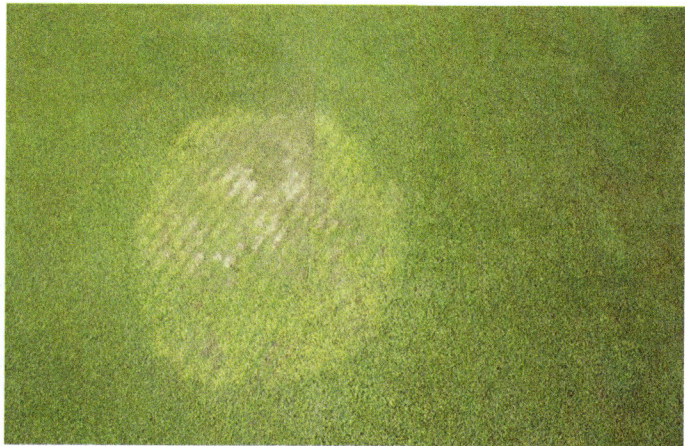

Fig. 8.10. Late-stage symptoms of winter Pythium patch on a cotula bowling green in New Zealand. (Courtesy of the NZSTI.)

on grasses have been extensively documented, as discussed in Chapter 7 (this volume). Several fungicides are available in New Zealand for application against winter Pythium patch, as discussed in Chapter 11 (this volume).

White patch disease

The malady known as white patch occurs on cotula bowling greens in New Zealand. The causal agent remains unknown. Symptoms appear as roughly circular, bleached patches primarily during autumn (March to June). Individual infected leaves display whitish-coloured tips with pinkish-coloured margins. Infected plants generally recover under favourable conditions for growth (Christensen, 1989; NZSTI, 2008). There are no specific, science-based cultural management practices proven against white patch disease on bowling greens. However the condition has responded to applications of magnesium sulfate, indicating that it could be related to magnesium deficiency in the plant.

References

Aldous, D.E. (1991) *Lawn Care and Lawn Alternatives*. Lothian Publishing Company Pty Ltd, Melbourne, Australia.
Anon. (2020) Timeline of turf diseases, plants and species. *New Zealand Turf Management Journal* 37, 2021.
Baltensperger, A.A. and Gaussoin, R.E. (1985) 'Fresa' strawberry clover, *Trifolium fragiferum* L. in reduced maintenance polystands and as monostand ground cover. In: Lemaire, F. (ed.) *Proceedings of the Fifth International Turfgrass Research Conference, Avignon, France, 1–5 July 1985*. INRA, Paris, pp. 311–316.
Cardin, L., Delecolle, B. and Moury, B. (2005) Occurrence of *Alternaria dichondrae*, *Cercospora* sp., and *Puccinia* sp. on *Dichondrae repens* in France and Italy. *Plant Disease* 89, 1012. DOI: 10.1094/PD-89-1012A
Christensen, M.J. (1989) Fungal diseases of cotula bowling greens. *New Zealand Turf Management Journal* 3, 9–11.
Christians, N. (2004) *Fundamentals of Turfgrass Management*, 2nd edn. Wiley, Hoboken, New Jersey.
Endo, R.M. (1965) Alternaria leaf spot and petiole blight of dichondra. *California Turfgrass Culture* 18, 32.
Erwin, D.C. and Ribeiro, O.K. (1996) *Phytophthora Diseases Worldwide*. The American Phytopathological Society, St. Paul, Minnesota.
Evans, P.S. (1984) The use of cotula for bowling greens in New Zealand. *Journal of the Sports Turf Research Institute* 60, 37–44.
Garibaldi, A., Minuto, A. and Gullino, M.L. (2005) First report of southern blight incited by *Sclerotium rolfsii* on *Dichondra repens* in Italy. *Plant Disease* 89, 203. DOI: 10.1094/PD-89-0203B
Gilbert, O.L. (1991) Gardens. *The Ecology of Urban Habitats*. Chapman and Hall, London, pp. 239–263.

Hannan, B. and Christensen, M. (1999) Golden bracelet: a disease with many masks. *New Zealand Turf Management Journal* 13, 17–20.

Harrington, K. (1990) Understanding hydrocotyle. In: *Proceedings of the Fourth New Zealand Sports Turf Convention, 13–16 August 1990*. Massey University, Palmerston North, New Zealand, pp. 62–63.

Hickey, M.J., Rumball, W., Murphy, J.W. and Field, T.R.O. (1993) The selection of 'Grasslands Pahia' (*Leptinella dioica* Hook. f.) for bowling greens. *New Zealand Journal of Agricultural Research* 40, 379–381.

Hickman, G.W. and Holcomb, J. (1979) Dichondra rust control: research progress report. *California Turfgrass Culture* 29, 18.

Howard, D. (2012) Common summer diseases on bowling greens. *New Zealand Turf Management Journal* 29, 40–42.

Huang, Y., Fang, F., Gong, G.-S. and Yun, Q. (2008) Symptoms of dichondra brown leaf spot and a new species of *Pyrenophora* (abstract in English). *Acta Phytopathologica Sinica* 38, 13–16.

Johnstone, G.R. and Barbetti, M.J. (1987) Impact of fungal and virus diseases on pastures. In: Wheeler, J.L., Pearson, C.J. and Robards, G.E. (eds) *Temperate Pastures: Their Production, Use and Management*. Australian Wool Corporation Technical Publication. CSIRO Publishing, Melbourne, Australia, pp. 235–248.

Kerr, A. (1956) Factors influencing the development of brown patch in lawns of *Sagina procumbens* L. *Australian Journal of Biological Science* 9, 323–338.

Kerridge, P.C. and Hardy, B. (eds) (1994) *Biology and Agronomy of Forage Arachis*. CIAF Publication No. 240. Centro International de Agricultura Tropical (CIAT), Cali, Columbia.

Laundon, G.F. (1979) New plant disease record in New Zealand: *Alternaria dichondrae* on *Dichondra repens*. *New Zealand Journal of Agricultural Research* 22, 647–648.

Ludbrook, W.V. (1940) Occurrence of *Sclerotinia minor* on a bowling green. *Journal of the Australian Institute of Agricultural Science* 6, 51–53.

Munro, P. (1997) The dichondra lawn (Mercury Bay weed): low maintenance or not? *New Zealand Turf Management Journal* 11, 5–7.

NZSTI (2008) *Establishment and Management of Natural Bowling Greens in New Zealand*. New Zealand Sports Turf Research Institute, Palmerston North, New Zealand.

Ormsby, D. (1990) Diseases of cotula in the North Island. In: *Proceedings of the Fourth New Zealand Sports Turf Convention, 13–16 August 1990*. Massey University, Palmerston North, New Zealand, pp. 55–57.

Ormsby, D. (2005) Starweed. *New Zealand Turf Management Journal* 20, 29–31.

Ormsby, D. (2012) *Sclerotinia minor* is a major. *New Zealand Turf Management Journal* 29, 20–23.

Ormsby, D. and Howard, D. (2021) *Sclerotinia minor*: the off-season scourge on bowling greens. *New Zealand Turf Management Journal* 40, 22–24.

Sarrocco, S., Vergara, M. and Vannacci, G. (2010) First conformed report of the leaf rust *Puccinia dichondrae* on *Dichondra repens* in Italy. *Plant Pathology* 59, 802. doi: 10.1111/j.1365-3059.2009.02236.x

Sivapalan, A. and Pascoe, I.G. (1994) Leaf blight of *Dichondra repens* in Australia caused by *Alternaria dichondrae*. *Australasian Plant Pathology* 23, 8–10.

Turner, T. and Carroll, M. (2015) *Microclover – Tall Fescue Lawns in the Mid-Atlantic Region*. Turfgrass Technical Update TT121–July 2015. University of Maryland, College Park, Maryland.

Way, B. (1991) New Zealand's unique bowling turf. *TurfCraft Australia* 24, 13–15.

9

Plant-Pathogenic Bacteria, Primitive Microbes and Viruses

Abstract

A multitude of lesser-known, microscopic and submicroscopic organisms are associated with different turfgrass diseases worldwide. Few of these organisms are recognized as turfgrass pathogens and for the most part the biology and the precise epidemiological factors influencing disease remain unclear. Once established, management options are limited largely to cultural practices.

An extensive range of microorganisms classified as either plant-pathogenic bacteria, phytoplasmas, primitive fungus-like microbes or viruses have long been known to be associated with grasses worldwide. Classification of many prokaryotes (e.g. bacteria) has been subjected to taxonomic revisions based on modern molecular DNA investigations and several groups of these organisms, some being true pathogens, have been associated with symptomatic turfgrasses. The global distribution of plant-pathogenic bacteria, phytoplasmas, primitive fungus-like microbes and viruses on turfgrass is undoubtedly under-reported.

The existing dogma is that the number of bacterial, phytoplasma, primitive microbe and viral pathogens and their associated diseases, as opposed to fungal diseases, currently acknowledged on turfgrass is relatively small. Knowledge of the biology and epidemiology of most plant-pathogenic bacteria, phytoplasmas, primitive fungus-like microbes and viruses is far from understood. The identity of plant-pathogenic bacterial, phytoplasma, primitive fungus-like microbe and viral diseases may often be misdiagnosed and attributed to other biotic causes or abiotic maladies.

The very small size or unculturable nature of some of these agents (i.e. phytoplasmas and viruses) can make detection difficult and often requires serological (antibody detection kits) or genetic-based technologies (e.g. DNA sequencing) for identification and highly specialized skills on the part of the diagnostician. Plant-pathogenic bacteria, phytoplasmas and viruses are microbes of potential quarantine and biosecurity risk given the difficulty to detect their presence and widespread transport of turfgrass. Many phytoplasmas and viruses of grasses commonly have insect vectors that feed on turfgrasses.

Most of these pathogenic organisms are obligate pathogens of the foliage or roots, resulting in chlorosis, but are capable of causing discernible symptoms and loss of host vigour during conducive environmental conditions. All species and cultivars of cool-season and warm-season grasses are potential hosts, whether managed in tropical, subtropical or temperate regions of the world.

Circumstantial evidence strongly suggests multiple species and subspecies of plant-pathogenic bacteria, phytoplasmas and viruses may occur simultaneously, raising the possibility of disease complexes with other turfgrass pathogens. Once established, effective management of plant-pathogenic bacterial, phytoplasma and viral diseases on turfgrass becomes highly problematic due to the lack of available and predictable management options.

Bacterial and Phytoplasma Diseases

Bacteria and phytoplasmas are prokaryotes and numerous plant-pathogenic species have long been important in cropping systems (Rosete and Jones, 2009; Borkar and Yumlembam, 2017). Phytoplasmas, some of which are plant-pathogenic, are similar to mycoplasma organisms and now are generally designated as *Candidatus* Phytoplasma (Bertaccini and Duduck, 2009).

Several plant-pathogenic bacteria and phytoplasmas cause named diseases on turfgrass (Table 9.1). Several species have been the subject of taxonomic reclassification. Bacterial infection of cool-season and warm-season turfgrasses can express various symptoms, depending on the host–pathogen interaction, which are far from completely understood. Plant-pathogenic bacteria gain entry into host leaves via natural openings such as stomates and hydathodes or through physical wounds.

Elongation and etiolation are common foliar symptoms associated with certain bacterial diseases (Fig. 9.1). Bacteria are implicated in etiolated tiller syndrome in the United States (Fidanza *et al.*, 2008; Roberts and Kerns, 2018), mad tiller disease in New Zealand (Stewart, 2002) and ghost grass disorder in the United Kingdom (Knowles, 2011). Plant-pathogenic bacteria can colonize the xylem or water-conducting tissues and infected leaves when carefully cut may display a mass streaming of bacterial cells when placed in water on a microscope slide. Mass streaming is not diagnostic for a specific bacterial disease.

Etiolated foliage may be a management-induced disease, as attested by being less problematic under lower cultural-management intensities. Limited investigations have indicated bacterial wilt and etiolated symptoms are impacted by temperature and hormones (Liu *et al.*, 2019) and are generally more pronounced with continued applications of certain commercial biostimulants and plant-growth regulators (Roberts *et al.*, 2015, 2016). The application of soluble nitrogen such as ammonium sulfate or applications of chlorothalonil plus acibenzolar-S-methyl may suppress disease symptoms (Vargas *et al.*, 2014).

Bacterial wilt

Bacterial wilt is the first bacterial disease verified on highly managed turfgrass that became prominent in the 1980s on golf greens in the United States (Roberts *et al.*, 1982). The disease was initially ascribed to infected 'Toronto' creeping bentgrass following isolation of an unknown bacterium and was given the name *Xanthomonas campestris*. Bacterial wilt is more commonly associated with certain highly intensively managed creeping bentgrass cultivars used on putting greens.

Xanthomonas translucens is the causal agent of bacterial wilt of annual bluegrass. Additional *X. translucens* pathovars occur on other cool-season grasses in Europe and New Zealand (Fermanian *et al.*, 2003). *X. translucens* has been more recently isolated in conjunction with another bacterium, *Pantoea ananatis*, from creeping bentgrass (Roberts *et al.*, 2018).

Distribution and host range

Bacterial wilt is widely distributed throughout the north-eastern United States and in Canada. Annual bluegrass is the principal host of the disease. Other turfgrass hosts associated with

Table 9.1. Bacterial diseases and their causal agents on turfgrass.

Bacterial disease	Causal agent	References
Bacterial decline	*Acidovorax avenae*	Dernoeden (2013); Mitrovic *et al.* (2015); Turgeon and Kaminski (2019)
Bacterial wilt	*Xanthomonas translucens*	
Bermudagrass white leaf	*Candidatus* Phytoplasma cynodontis	

Fig. 9.1. Bacterial etiolation on a creeping bentgrass putting green in the United States. (Courtesy of Paul Giordano.)

specific pathovars of *X. translucens* are creeping bentgrass cultivars, Kentucky bluegrass, fescues, perennial ryegrass and bermudagrass (Tredway *et al.*, 2023).

Signs and symptoms

On leaves, symptoms are typically a water soaked-like appearance. Foliar symptoms are highly variable depending on age of the host species; hence diagnosis based solely on field symptoms is unreliable. Initial symptoms appear as downward-progressing wilt from the leaf tip and water-soaked lesions. Indirect experimental evidence for the presence of bacterial pathogens can be obtained by treating symptomatic grass with antibiotics (e.g. oxytetracycline) but these chemicals are only registered for use for limited agricultural purposes in a few countries.

Epidemics are favoured by prolonged periods of rainfall followed by cloudless days and warm temperatures coinciding with spring and autumn. Bacterial wilt is more prevalent in turfgrass subject to shade and excess moisture. Infected annual bluegrass leaves may initially appear as faint light green, yellow or blue-green in colour (Fig. 9.2). Disease progression results in small-sized, faint reddish-brown- to copper-coloured spots (up to 2 cm in diameter) randomly scattered across a sward of grass. Symptoms rapidly advance within 24–48 h under highly favourable conditions. Immature leaves may become chlorotic and elongated (i.e. etiolated) growing slightly higher, curved, and above unaffected turfgrasses.

Infected creeping bentgrass leaves appear faint reddish-brown and may exhibit copper-coloured spots within 24–48 h under highly favourable conditions. Affected areas may coalesce forming larger symptoms of speckled or mottled diseased plants resulting in death within a few days. Plant death is caused by severe wilting due to colonization by the bacterium throughout the xylem tissues of infected plants which interrupts the internal movement of water in the plant.

Fig. 9.2. Bacterial wilt (*Xanthomonas* sp.) on an annual bluegrass green in the United States. (Courtesy of Paul Giordano.)

Bacterial decline

Bacterial decline is a relatively recent disease on creeping bentgrass in the United States (Giordano et al., 2010). Investigations of infected creeping bentgrass collected throughout numerous states resulted in the isolation of a bacterium under the former name *Acidovorax avenae* subsp. *avenae*. The disease name bacterial etiolation or decline was used to distinguish it from different symptoms caused by the same pathogen in Japan. *A. avenae* is a common seed-borne pathogen. Much of the biology and epidemiology of bacterial decline remains unknown.

Signs and symptoms

Observed symptoms of bacterial decline on creeping bentgrasses range from random chlorotic, etiolated growth and twisting of individual leaves, advancing to randomly scattered severe chlorosis through to necrosis and gradual declining patches of various sizes and shapes. The presence of mass streaming of bacterial cells may be observed microscopically (Fig. 9.3). Less obvious symptoms may be expressed on creeping bentgrass swards maintained at higher height/less frequently mowed.

Bacterial decline may begin in late spring and is enhanced when daytime temperatures are above 31°C and night-time temperatures above 26.7°C (Dernoeden, 2013). *A. avenae* in culture has been shown to produce phytohormones with increasing temperatures of 35–40°C (Liu et al., 2019) and can be associated with foliage etiolation on creeping bentgrass.

Bermudagrass white leaf

Bermudagrass white leaf is the result of a foliar infection primarily occurring on non-cultivated, common bermudagrasses and is associated with the phytoplasma *Candidatus cynodontis*. The degree of infection and symptoms vary widely depending on circumstances not fully understood but are usually mild and patchy in occurrence. Phytoplasma-like organisms have been reported on bermudagrass lawns in southern Europe (e.g. Mitrovic et al., 2015) and in South-East Asia (Stem, 1992; Koh et al., 2008).

Phytoplasmas are transmitted by stylet-feeding insect vectors (e.g. planthoppers and leafhoppers) and once inside move internally within the plant phloem tissues. Symptoms range from slight to severe yellowing to a bleached white appearance, stunted leaves, excessive tillering resulting in witches' brooming, and a general decline. There are no specific or predictable management strategies against phytoplasma-like diseases. Care should be taken when making a

Fig. 9.3. Mass streaming of *Acidovorax* sp. bacterial cells from a severed bentgrass leaf. (Courtesy of Paul Giordano.)

diagnosis as eriophyid mites (e.g. *Aceria* spp.) are a common cause of witches' broom symptoms in bermudagrass in many locations worldwide.

Cyanobacteria and yellow spot disease

Cyanobacteria are a diverse and ubiquitous group of microbes inhabiting both terrestrial and water environments worldwide (Rushford *et al.*, 2022). Classified as prokaryotes, these microbes conduct a plant-like photosynthesis. Cyanobacteria in the genera *Anabaena*, *Anacystis*, *Lyngbya*, *Oscillatoria*, *Nostoc* and *Phormidium* have been identified as occurring in turfgrass systems (Maddox *et al.*, 1997; Tredway *et al.*, 2023). Cyanobacteria microbes are often but incorrectly called 'blue-green' algae.

Cyanobacteria and algae are often co-inhabitants in turfgrass systems subject to anaerobic conditions that are unable to maintain an acceptable oxygen–moisture balance. However, the impacts of the presence and effects between cyanobacteria and algae are quite different. Management problems long associated with algae are the formation of surface mucilaginous slime during wet periods and surface crusts during dry conditions, a reduction in water infiltration and the formation of black-layering from the interaction with nitrogen-fixing bacteria (Baldwin and Whitton, 1992; Hodges, 1993). All of these problems contribute to low-quality playing surfaces.

The name yellow spot describes the chlorotic appearance of infected foliage and the size and shape of symptoms (Tredway *et al.*, 2023). Small-sized, roughly round symptoms were reported on golf greens during the late 1990s in the United States. Members of *Oscillatoria* and *Phormidium* are associated with yellow spot in the United States (Tredway *et al.*, 2006).

Cyanobacteria are generally not considered to directly parasitize turfgrass and therefore are not considered to cause infectious disease. There does not appear to be documented proof of the completion of Koch's postulates, demonstrating true diseases. The predisposing factors resulting in the emergence and possible occurrence of these microbes in turfgrass systems remain debated.

Geographic distribution and host range

Yellow spot was reported in California and the south-eastern region of the United States during the late 1990s. The global distribution and host range of yellow spot outside the United States remain unknown. Cyanobacteria causing yellow spot symptoms have been confirmed in Australia (New South Wales) on a mixed bentgrass–annual bluegrass golf green (A. Daly, New South Wales, 2023, personal communication; P.T. Wong, New South Wales, 2023, personal communication).

Symptoms mimicking those of yellow spot have been reported in Japan (Tani and Beard, 1997). Annual bluegrass, creeping bentgrass and bermudagrass are hosts of the disease (Gelernter and Stowell, 2000; Tredway et al., 2006).

Signs and symptoms

Cyanobacteria possess individual, hair-like filaments and require sample preparation and microscopy for identification. Evidence of the filamentous growth on infected foliage is not readily visible during daylight hours. Submitted samples are best placed in plastic bags overnight before examination. Suspect turfgrass samples require dissection into small-sized sections and placement into Petri dishes partially filled with water.

Under high magnification (400×), individual hair-like filaments of cyanobacteria are light green in colour (Fig. 9.4). When in mass, the presence of dark-coloured, mesh-like networks of joined filaments observed in the crown and sheath regions of chlorotic plants can be indicative of the presence of various cyanobacteria species. Initial symptoms of yellow spots often rapidly appearing overnight during summer include chlorotic or yellow-coloured, roughly circular-shaped areas 2.0–3.0 cm in diameter. Symptoms are more readily observed during early morning.

During severe outbreaks, the spots may display a fluorescent yellow–green-coloured foliage, appearing rapidly and randomly throughout a location. Certain spots may enlarge slightly up to 4.0 cm, and some may coalesce forming unevenly sized, chlorotic spots or small-sized patches (Fig. 9.5). The morphology of infected plants within the spot appears relatively normal aside from the chlorotic leaves.

Close observation shows infected plants remain chlorotic with some thinning, but this does not result in plant death (Fig. 9.6). The chlorotic spots may gradually disappear coinciding with reduced night temperatures around 18°C (Dernoeden, 2013) depending on environmental conditions. The chlorotic leaves may appear similar to those of yellow tuft (downy mildew) and can be easily misdiagnosed.

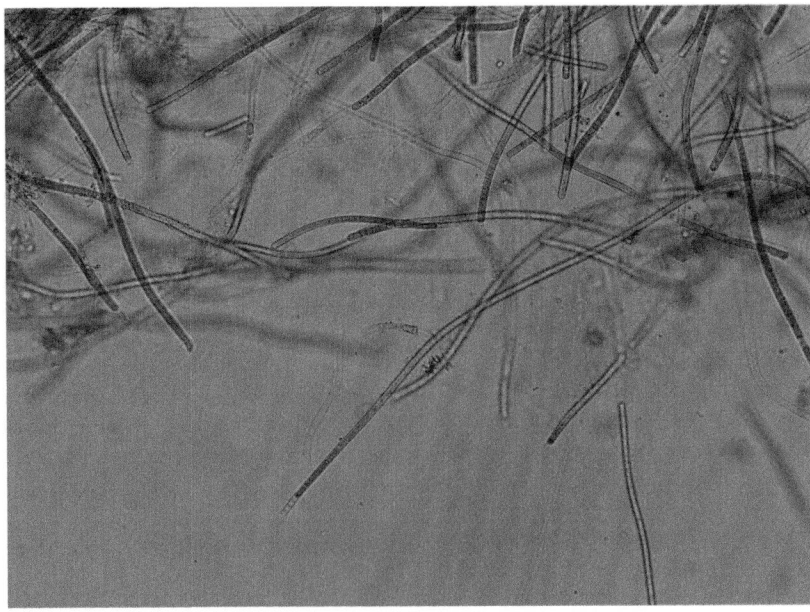

Fig. 9.4. Microscopic image of cyanobacteria filaments. (Courtesy of Andrew Daly.)

Fig. 9.5. Advanced symptoms of yellow spot on a mixed annual bluegrass–creeping bentgrass golf green in Australia. (Courtesy of Gary W. Beehag.)

Fig. 9.6. Close-up of chlorotic leaves caused by yellow spot. (Courtesy of the NZSTI.)

Disease profile

The biological nature of cyanobacteria–turfgrass interactions is imprecise. Yellow spot is best profiled as a cosmetic, foliar condition on established golf greens observed primarily during periods of prolonged moisture and high temperatures.

The chlorotic response of the symptomatic host has been generally attributed to the secretion of secondary metabolites called cyanotoxins (Chauhan *et al.*, 1992) or iron-chelating compounds (Brown and Trick, 1992). In other words, the symptomatic host suffers from reduced photosynthesis. Secretion of the toxins occurs from their hair-like filaments as the bacteria move throughout the plant's vascular system presumably by hydathodes or cut leaf ends. The production of cyanotoxins is influenced by numerous environmental interactions (e.g. temperature, light and nutrients).

Environmental influences

Several environmental and management-induced factors have been attributed to the occurrence of yellow spot on golf greens (Box 9.1). Excessive moisture and high temperatures are considered the key contributing factors of the condition.

Prolonged rain events or frequent irrigation are conducive for cyanobacteria activity because of the resultant anaerobic condition in the upper soil profile. Canopy temperatures exceeding 16°C

> **Box 9.1.** Key environmental factors of yellow spot.
>
> - Prolonged periods of excessive moisture and high humidity
> - Ambient temperatures around 20°C
> - Light intensity and duration
> - Exceptionally low and frequently mown turfgrass
> - Application of lime products and alkaline pH
> - Excessive application of plant-growth regulators
> - Application of QoI-based fungicides

> **Box 9.2.** Integrated management of yellow spot.
>
> - Monitor turfgrass for signs and early symptoms
> - Maintain an appropriate mowing regime of height and frequency
> - Implementation of narrow-diameter hollow or solid tines
> - Consider light application of topdressing sand or 'dusting'
> - Careful use of a balanced NPK nutrition
> - Minimization of thatch-mat accumulation

favour the development of cyanobacteria and their associated toxins. That low light intensity and duration affect yellow spot is because cyanobacteria have a low light-intensity requirement for photosynthesis and 'move' downwards during early morning; hence why they are not easily observed during daylight hours.

Cyanobacteria must be inside the plant and the significance of mowing height lies in the fact that cyanobacteria have a 'gliding mobility', thus are able to more easily gain entry into cut leaf ends closer to the crown. Greater time while at the end of leaves potentially results in greater damage. The newer bentgrass and bermudagrass cultivars may pose greater risk of infection because of their tolerance to extreme low mowing heights.

Alkaline soil pH possibly from using calcareous sands, excessive application of liming products or treated irrigation water generally favours cyanobacteria activity. Circumstantial evidence suggests excessive use of quinone outside inhibiting (QoI) (i.e. strobilurin) fungicides may intensify yellow spot symptoms (Tredway *et al.*, 2006).

Integrated management

Yellow spot occurrence is difficult to predict and problematic once established on golf greens having inadequate drainage. Proven management strategies for yellow spot on greens remain imprecise (Box 9.2). Minimizing soil moisture by careful irrigation programming in combination with aerification practices are key remedial strategies. Low populations of cyanobacteria will undoubtedly persist on golf greens subsequent to the disappearance of symptoms.

Monitoring for early symptoms needs to commence on greens having inadequate drainage at the commencement of prolonged rainfall events coinciding with increased humidity and temperatures around 15°C. Short-term implementation of narrow-diameter hollow or solid tines in combination with light sand topdressing aids to reduce the profile moisture content. Longer-term adoption of more vigorous thatch-mat minimization strategies should aid in reducing the damage caused by yellow spot.

Light applications of ammonium sulfate (Dernoeden, 2013) and certain phosphate, copper or zinc fungicides (Inguagiato *et al.*, 2018; Tredway *et al.*, 2023) may suppress yellow spot symptoms. Application of QoI fungicides should be limited.

Primitive Foliar- and Root-Infecting Microbes

Primitive foliar- and root-infecting, fungus-like organisms colonize host roots often without disfiguring symptoms. Morphologically simple or primitive root-infecting, fungus-like microbes are associated with grassland systems worldwide. *Cladochytrium*, *Olpidium*, *Physoderma* and *Polymyxa* are among the principal genera (Smith *et al.*, 1989). The taxonomy of many genera and species is subject to reclassification. Most species are obligate parasites of roots generally causing minimal damage or discernible symptoms and many are generally categorized as minor diseases (Smith *et al.*, 1989; Tredway *et al.*, 2023). Collectively, these primitive root-infecting microbes have a wide host range in grassland systems.

Many species infect seedlings or are root pathogens surviving as spores during unfavourable environmental conditions. Certain species are known to be vectors of viruses while others form disease complexes with bacteria and plant-parasitic nematodes. Much of their ecology and significance on turfgrass systems remains unknown.

Polymyxa root rot

Polymyxa root rot is caused by *Polymyxa graminis*, a root-infecting pathogen on cool-season and warm-season grasses (Couch, 1995). Polymyxa root rot has been documented on turfgrass in the United States (Tredway *et al.*, 2023), Europe (Smith *et al.*, 1989) and Australia (Forrest, 2014). The pathogen has temperate and tropical origins and can be a vector of several viruses.

Infected creeping bentgrass and hybrid bermudagrass on bowling and golf greens may display an overall loss of normal foliage colour, sward density and plant vigour. Under conducive environmental conditions infected plants may display a mixture of normal, chlorotic and necrotic plants (Fig. 9.7). Infected root hairs are typically brown in colour. A combination of increased soil moisture and ambient temperatures of 15–18°C is optimum for root infection and colonization (Couch, 1995) depending on the fungal isolate.

There are no specific remedial actions for Polymyxa root rot other than reducing environmental stress factors by moderating cultural practices (e.g. mowing height, adjusting irrigation and appropriate fertilization regimes).

Rapid blight

Rapid blight, caused by *Labyrinthula terrestris*, is a turfgrass disease that can affect bluegrasses (*Poa annua*, *Poa trivialis*) and perennial ryegrass (*Lolium perenne*) and is a relatively recent and unique disease. The pathogen is a single-celled organism collectively called marine net slime moulds that are generally saprophytic and associated with seagrasses and algae and not related to true slime moulds. Identification of the microbe requires a highly skilled diagnostician. The causal agent can be a pathogen of a marine grasses and is not a fungus or bacterium and is associated with highly saline environments.

The emergence of rapid blight on mown grass is ecologically significant because the pathogen was previously only known to cause diseases of marine grasses; however, it is now associated with turfgrass managed in the presence of high salinity. Equally intriguing is the disease's apparent decline on turfgrass in many regions of the United States and its emergence in the United Kingdom in recent years.

The name rapid blight was ascribed because of the spread of the disease throughout the mainland United States (Martin *et al.*, 2002). The pathogen represents the first species of the

Fig. 9.7. A bermudagrass bowling green in Australia infected by Polymyxa root rot. (Courtesy of John Forrest.)

genus known to inhabit terrestrial ecosystems. In view of increased utilization of non-potable water sources with elevated salinity, the potential exists for the disease to occur in other countries especially where rainfall levels are low.

Geographic distribution

The documented global distribution of rapid blight on turfgrass remains limited and unclear. Rapid blight was first reported in the United States initially in California in the mid-1990s then was found in the south and eastern coastal regions (Kerns and Tredway, 2013). Rapid blight was initially diagnosed in the United Kingdom in 1995 specifically in Scotland (Entwistle, 2005) and has now been confirmed in Ireland, Portugal and Spain (Entwistle, 2012).

Host range and susceptibility

Rapid blight has been observed on annual bluegrass, rough bluegrass, colonial, creeping and velvet bentgrasses, and chewing's fescue (Kerrigan et al., 2012). Perennial ryegrass cultivars appear to be equally highly susceptible to rapid blight (Petelewicz et al., 2019). Suffice to say, a range of environmental and management-induced factors over the long term will further impact the resistance of grass species and cultivars to rapid blight.

Signs and symptoms

Accurate diagnosis of field symptoms of rapid blight is problematic. Indicative signs are absent and late-stage symptoms can resemble other turfgrass diseases and abnormalities. Microscopic examination may reveal fusiform-shaped cells inside infected leaf and root tissues. Symptoms on closely mown grass appear as random, highly irregularly shaped, blighted patches ranging in overall size from 2.0 to 30.0 cm. The colour of infected areas varies from yellow-brown to bronze depending on age of turfgrass species. Certain patches may have dark-coloured outer edges. Larger-sized patches may coalesce as the disease progresses.

Late-stage symptoms may have a collapsed centre region. Close examination using a handheld lens may reveal mottled and water-soaked leaves. Symptoms may be differentially observed on mixed annual bluegrass–bentgrass greens.

Disease profile

The biology of the rapid blight pathogen and the ecology, emerging significance and possible management of the disease have been extensively documented in the United States (e.g. Kerrigan et al., 2012; Kerns and Tredway, 2013).

Rapid blight can be a highly devastating disease under favourable conditions primarily during warm and dry conditions coinciding with autumn and spring. The streaming of vegetative cells is believed to be the mechanism by which the pathogen moves and makes contact with adjacent plants. Dissemination may also occur via machinery and foot traffic. The pathogen can infect roots and leaves through stomates or severed leaf ends. Once inside, the pathogen multiplies within the host. Extensive damage can occur within a short period after infection under favourable conditions.

Reproductive structures of *L. terrestris* have not been found on turfgrass. Experimentally, the microbe has the ability to survive inside dry-stored leaf tissue and as a saprophyte during unfavourable conditions. In addition, observations indicate it may survive in symptomless warm-season grasses.

Environmental influences

Several key environmental and management-induced factors have been shown to specifically influence development of rapid blight disease (Box 9.3).

TEMPERATURE AND SALINITY Rapid blight disease can occur over a relatively wide temperature range from below 4°C to above 40°C (Olsen et al., 2004). The pathogen survives over a wide range of water salinities. Results have generally shown experimentally that cool-season grass species having low salinity tolerance are more susceptible to rapid blight, and severity of the disease increases with soil and water salinity. As a group, fine fescue, creeping bentgrass and alkali grasses were the most tolerant to high salinity and thus rapid blight disease (Peterson et al., 2005).

Box 9.3. Key environmental factors of rapid blight.

- Moderate temperatures
- High salinity levels (2.0–10.0 dS/m)
- Extended periods of leaf wetness (36–40 h)
- Warm and dry conditions

Integrated management

Predictable management against rapid blight on golf greens is limited but necessitates an integrated approach (Box 9.4). Generally, reduction in soil salinity levels reduces the severity of rapid blight.

Disease-resistant varieties

Degrees of genetic resistance have been shown experimentally among cultivars of creeping bentgrass, fescue and alkali grass (Peterson et al., 2005). Bermudagrass may be a non-symptomatic host acting as a reservoir of the pathogen for cool-season grasses.

Slime moulds

Slime moulds do not cause infectious disease but when present give an unsightly appearance on otherwise high-quality turfgrass. Slime moulds are primitive microbes lacking cell walls and classified as obligate saprophytes gaining nutrition from bacteria, fungi and other microbes. The presence of slime moulds does not indicate a disease but rather a cosmetic concern. Slime moulds can be readily managed by physical removal or mowing.

Causal agents

Didymium, Fuligo, Mucilago, Physarium and *Stemonitis* are the most common species of slime moulds occurring on turfgrass (Fermanian et al., 2003; Tredway et al., 2023).

Box 9.4. Generalized management of rapid blight.

- Monitoring of rainfall and temperature conditions
- Maintain consistent growth with use of a balanced NPK nutrition
- Frequent monitoring of recycled or treated irrigation water salinity
- Judicious irrigation water application and consider other sources
- Consider disease-resistant varieties where available

Geographic distribution and host range

Slime moulds may potentially appear on grasses worldwide.

Signs and symptoms

The presence of coloured fruiting bodies (sporangia) in masses along covered foliage is a characteristic sign of slime mould (Fig. 9.8). Coloured sporangia appear randomly scattered in irregularly shaped regions within a stand of turfgrass. Slime moulds may appear regularly in the same regions each year.

Disease profile

Slime moulds are not pathogenic but occur on turfgrass during warm, overcast and humid conditions. The biology and ecology of slime moulds on turfgrass have not been extensively documented (Smith et al., 1989; Fermanian et al., 2003).

Viruses and Viral Diseases

A multitude of plant viruses and viroids have long been recognized as economically important agents of transmissible diseases in agricultural and horticultural crop systems worldwide (Harris, 2000; Lapierre and Signoret, 2004; Geering and Randles, 2012; Fermin et al., 2015). Viral diseases are common in tropical and subtropical regions that are conducive for the multiplication of viruses and their respective vectors. Viruses are associated with wild grasses and a select few have been isolated and identified on mown turfgrasses (Vargas, 2005; Tredway et al., 2023).

Viruses are submicroscopic entities grouped according to their size, shape and other factors (e.g. RNA ribonucleic acid or DNA deoxyribonucleic acid). Viruses are obligate parasites of a wide range of plant species and require a living host in which to multiply. All plant viruses are pathogenic, and many are transmitted by sap-sucking insects, phytophagous mites, plant-parasitic nematodes or fungal vectors. Viroids comprise single-stranded RNA.

Virus nomenclature is complex, using a *Latinized* system analogous to genera and families ending in the suffix *–virus* for genera (e.g. *Sobemovirus*)

Fig. 9.8. Slime mould sporangia on a grass leaf. (Courtesy of Ben Evans.)

and –*viridae* for families (e.g. *Potyviridae*) as recognized by the International Committee on Taxonomy of Viruses (ICTV). However, it is commonplace in the turfgrass literature for a virus to be assigned an English name (e.g. panicum mosaic virus and sugarcane mosaic virus) acknowledging the name of the wild or cultivated host grass from which each was initially isolated and identified. This simple system is generally not recognized by the ICTV.

Viruses may be seed-borne or enter their hosts by physical wounds (natural or anthropogenic) or by vectors (e.g. insects). Once inside the plant, viruses cannot be managed by conventional pesticides and host plants remain permanently infected, displaying varying degrees of leaf mottling or streaking (Fig. 9.9). Application of insecticides does not provide predictable protection against insect viral vectors.

Viruses have generally been considered not commonly associated with infectious diseases on highly managed turfgrass. Several virus diseases occur on managed turfgrasses most notably on warm-season species in North America, Europe, East Asia and Australasia (Table 9.2). Viruses of the family *Potyviridae* are among the most common on grasses and are transmitted by aphids but their importance as a vector on turfgrass systems is unknown. New viruses and virus-like organisms continue to be isolated and identified on uncultivated grasses and unthrifty turfgrasses in certain countries (Tran *et al.*, 2022).

Fig. 9.9. Leaf streaking caused by a viral infection. (Courtesy of Ben Evans.)

The true identity, biology and global distribution of many viruses and viroids with turfgrass systems have not been intensively documented and many key aspects remain unclear or unknown. All turfgrass species and cultivars are potential virus hosts. Several viruses have been isolated from symptomatic and non-symptomatic turfgrasses and viral infections are not always clearly obvious on foliage. Viral-induced disease symptoms are generally distinct from those caused by other plant pathogens (e.g. fungi and nematodes) and often go unnoticed or are misdiagnosed (e.g. nutrient deficiency) until the host becomes severely chlorotic. Careful examination for signs of mottling or streaks (Fig. 9.10) can be characteristic of a viral disease.

Table 9.2. Virus diseases and their causal agents on turfgrass.

Turfgrass viral disease	Virus name	References
St. Augustine decline	Panicum mosaic virus	Geering (2020)
Mosaic disease of St. Augustinegrass	Sugarcane mosaic virus	Harmon (2019)
Centipede grass mosaic disease	Panicum mosaic virus	Holcomb et al. (1989)
Mosaic disease of Queensland blue couch	Digitaria didactyla striate mosaic virus	Briddon et al. (2010)
Zoysia virus	Wheat mottle dwarf virus	Tani and Beard (1997)
	Zoysia mosaic virus	

Fig. 9.10. Mottled St. Augustinegrass (buffalo grass) leaves infected with Panicum mosaic virus in Australia. (Courtesy of Andrew Geering.)

Certain plant viruses may act alone causing cosmetic or relatively mild symptoms. In other cases, chlorosis may be severe or infection of highly susceptible turfgrass species (e.g. St Augustine decline of cv. Floratam St. Augustinegrass) may result in plant death under highly favourable conditions. Co-infection by more than one virus species or strain may act in combination causing symptoms ranging from slightly chlorotic to necrosis on many common turfgrasses (Fig. 9.11). Importantly, a plant cannot be cured of a viral infection and the only options are to prevent infection in the first place or destroy the plant to prevent further spread. Plant viruses are also readily spread by movement of infected turfgrass.

Several virus diseases are important on St. Augustinegrass in the United States (Tredway et al., 2023). In Australia, several viruses have been isolated from Durban grass, Queensland blue couch, buffalo grass (St. Augustinegrass) and weeping grass (Greber, 1989; Pares and Gillings, 1990; Tran et al., 2020, 2022). The identification of Spartina mottle virus on hybrid bermudagrass imported into the country highlights the potential significance of plant-virus transmission (Thomas et al., 2021).

St. Augustine decline (SAD)

St. Augustine decline (SAD) is the name ascribed to a well-known viral disease on St. Augustinegrass in the southern United States. St. Augustine decline represents the first turfgrass viral disease recognized on managed turfgrass and was first described during the late 1960s in Texas (United States) but more recently has been found in many Gulf States.

Distribution, host range and susceptibility

St. Augustine decline is widely distributed throughout the southern United States and Mexico. Panicum mosaic virus (PMV) is the causal agent. In the United States, St. Augustinegrass cultivars Bitterblue, Floralawn, Floratam, Raleigh and

Fig. 9.11. Severe chlorosis on St Augustinegrass (buffalo grass) caused by Sugarcane mosaic virus in Australia. (Courtesy of Andrew Geering.)

Seville possess increased tolerance to St. Augustine decline (Fermanian et al., 2003; McCarty, 2010).

Signs and symptoms

St. Augustine decline symptoms commence with slight chlorosis, progressing to more distinct mottling and streaking of individual infected leaves. Late-stage symptoms are severe chlorosis, stunting and necrosis eventually leading to death of entire sections of turfgrass over several years. The symptoms tend to be exacerbated by shade.

Mosaic disease of St. Augustinegrass

Mosaic disease of St. Augustinegrass is a more recent virus-like disease having similar symptoms to St. Augustine decline. Mosaic disease of St. Augustinegrass is caused by sugarcane mosaic virus (SCMV). In Florida, the strain of sugarcane mosaic virus affecting St. Augustinegrass is related to that occurring in sugarcane. In Australia the strain is the same as that infecting Queensland blue couch (*Digitaria didactyla*) (A.D.W. Geering, Queensland, 2023, personal communication). Sugarcane mosaic virus causes lethal necrosis of cv. Floratam in central and southern Florida in the United States, and very severe symptoms appear to be induced by cool evening temperatures that occur in autumn (Harmon et al., 2015).

Outside the Americas, several viruses causing mosaic symptoms have been reported on St. Augustinegrass. In eastern Australia, Johnson grass mosaic virus (JGMV), panicum mosaic virus from infected 'Palmetto' and sugarcane mosaic virus from 'Sir Walter' buffalo grasses have been isolated and identified from samples collected from commercial sod farms (Pares and Gillings, 1990; Thomas and Steele, 2011; Tran et al., 2020).

Centipede grass mosaic disease (CGMD)

Centipede grass mosaic disease (CGMD) was initially reported on a chlorotic, centipede grass lawn in mid-1980 in the southern United States. Centipede grass mosaic disease is caused by a strain of panicum mosaic virus (Holcomb, 1985) and distributed throughout the southern United States where a strain of the virus also occurs on St. Augustinegrass (Haygood and Barnett, 1992).

Mosaic disease of Queensland blue couch

A strain of sugarcane mosaic virus has been known to occur naturally on Queensland blue couch in Australia (Teakle and Grylls, 1973). A virus species initially isolated from chlorotic Queensland blue couch collected in Brisbane (Australia) in the late 1980s (Greber, 1989) was ascribed the name Digitaria didactyla striate mosaic virus (Briddon et al., 2010).

Zoysia dwarf virus/mosaic virus

Zoysia dwarf virus (wheat mottle dwarf virus (WMDV)) and zoysia mosaic virus (ZMV) occur on Japanese zoysiagrass in Japan (Tani and Beard, 1997). Severe damage has not been reported and information on the disease remains unclear.

Management of Plant-Pathogenic Bacterial, Phytoplasma and Virus Diseases

Bacterial, phytoplasma and virus diseases have many common aspects making effective management problematic (Box 9.5). Initial infections by bacteria, phytoplasmas and viruses in host plants may go undetected for some time, being partly disguised or mistaken for other biological causes (e.g. fungal disease) or management-induced (e.g. imbalanced nutrition) inputs. Severe infection resulting in turfgrass loss may warrant partial or total replacement of the sward by reseeding or vegetative-propagation practices, depending on circumstances.

Plant-associated bacteria are abundant on decomposing organic matter in the thatch-mat layer. Bacteria occupy the water-conducting (xylem) tissues of the hosts; thus presenting significant physiological challenges in the potential management of the disease. Phytoplasmas and viruses are intracellular organisms infecting hosts systemically. Bacteria, phytoplasmas and viruses are potentially distributed within and between sites physically by anything that can move infected plant tissues, i.e. equipment, plant sap or plant material.

Many viruses associated with uncultivated grasses are transmitted by specific species of sap-sucking insect vectors where they occur naturally (or otherwise) in certain regions or countries.

Box 9.5. Management limitations of plant-parasitic microscopic organisms.

- Subcellular, microscopic size
- Cryptic location inside host plant
- Difficulty in their isolation and detection
- Non-symptomatic hosts
- Mechanical and/or vector transmission
- Limited options of disease-resistant cultivars
- Lack of proven antibiotics for turfgrass application

Thus, the potential of bacteria, phytoplasma and virus transmission and infection on managed turfgrass remains ever present. Management of unwanted plants decreases the potential of a virus host reservoir.

The use of disease-resistant turfgrasses is a significant limiting factor given a lack of cultivars with documented resistance. Seeded, creeping bentgrass cultivars (e.g. Penncross and Penneagle) have been shown experimentally to possess a degree of resistance against bacterial wilt and cv. Floratam St. Augustinegrass has resistance against the viral disease St. Augustine decline (Vargas, 2005). 'Tifdwarf' bermudagrass was purported to be less susceptible as opposed to 'Tifton 419' hybrid bermudagrass against bermudagrass white leaf disease (Stem, 1992).

Stated cultural practices (e.g. increased mowing height, reduced mowing frequency, mowing only during dry conditions and avoidance of sand topdressing) may prove impractical to effectively implement on high-quality bowling and golf greens. Bacteria, phytoplasmas and viruses are not managed by conventional pesticides or biopesticides. Symptom suppression of bacterial wilt has been demonstrated experimentally using the antibiotic products oxytetracycline, streptomycin and copper hydroxide (Roberts et al., 1982).

Disinfestation of mowing equipment using water-based solutions and avoiding products containing gibberellic acid are advised when managing bacterial-incited diseases. Management of most virus-like diseases by attempting to reduce the populations of possible insect vectors has been proven experimentally to be ineffective because of the specific timing required to kill the insects prior to transmission of the pathogens (Geering and Randles, 2012).

References

Baldwin, N.A. and Whitton, B.A. (1992) Cyanobacteria and eukaryotic algae in sports turf and amenity grasslands: a review. *Journal of Applied Phycology* 4, 39–47.

Bertaccini, A. and Duduk, B. (2009) Phytoplasma and phytoplasma diseases: a review of recent research. *Phytopathology Mediterranea* 48, 355–378.

Borkar, S.G. and Yumlembam, R.A. (2017) *Bacterial Diseases of Crop Plants*. CRC Press, Boca Raton, Florida.

Briddon, R.W., Martin, D.P., Owor, B.E., Donaldson, L., Markham, P.G., et al. (2010) A novel species of mastrevirus (family *Geminiviridae*) isolated from *Digitaria didactyla* grass from Australia. *Archives of Virology* 155, 1529–1534. DOI: 10.1007/s00705-010-0759-0

Brown, C.M. and Trick, C.G. (1992) Response of cyanobacterium, *Oscillatoria tenuis*, to low iron environments: the effect on growth rate and evidence for siderophore production. *Archives of Microbiology* 157, 349–354.

Chauhan, V.S., Marwah, J.B. and Bagchi, S.N. (1992) Effect of an antibiotic from *Oscillatoria* sp., on phytoplankers, higher plants and mice. *New Phytologist* 120, 251–257.

Couch, H.B. (1995) *Diseases of Turfgrasses*, 3rd edn. Krieger Publishing Company, Malabar, Florida.

Dernoeden, P.H. (2013) *Creeping Bentgrass Management*. CRC Press, Boca Raton, Florida.

Entwistle, C.A. (2005) First report of *Labyrinthula* sp. causing rapid blight of *Agrostis capillaris* and *Poa annua* on amenity turfgrass in the UK. *Plant Pathology* 55, 306.

Entwistle, C.A. (2012) Course disease alert. *Greenkeeper International* (April), 30–33.

Fermanian, T.W., Shurtleff, M.C., Randell, R., Wilkinson, H.T. and Nixon, P.L. (2003) *Controlling Turfgrass Pests*, 3rd edn. Prentice-Hall, Upper Saddle River, New Jersey.

Fermin, G., Verchot, J., Azizi, A. and Tennant, P. (2015) Viruses affecting tropical and subtropical crops: biology, diversity and management. In: Tennant, P. and Fermin, G. (eds) *Viruses Affecting Tropical and Subtropical Crops*. CABI Protection Series. CAB International, Wallingford, UK, pp. 1–16.

Fidanza, M., Gregos, J. and Brickley, D. (2008) Are etiolated tillers a visual nuisance or something else? *Turfgrass Trends* (October), 60–64.

Forrest, J. (2014) New disease hits bowls club. *TurfCraft International* 154, 38–42.

Gelernter, W. and Stowell, L.J. (2000) Cyanobacteria (AKA blue-green algae): wanted for serious damage to turf. *PACE Insights* 6, 1–4.

Geering, A. (2020) *Identification and Management of Mosaic Viruses and Secondary Pathogens in Buffalo Turf*. Final Report No. TU9000. Hort Innovation, Sydney, Australia.

Geering, A.D.W. and Randles, J.W. (2012) *Virus Diseases of Tropical Crops*. Wiley, Chichester, UK. DOI: 10.1002/9780470015902.a0000767.pub2

Giordano, P.R., Vargas, J.M. and Detweller, A.R. (2010) First report of a bacterial disease on creeping bentgrass (*Agrostis stolonifera*) caused by *Acidovorax* spp. in the United States. *Plant Disease* 94, 922.

Greber, R.S. (1989) Biological characteristics of grass geminiviruses from eastern Australia. *Annals of Applied Biology* 114, 471–480.

Harmon, P. (2019) *Mosaic Disease of St. Augustinegrass Caused by Sugarcane Mosaic Virus*. Document No. PP313. University of Florida IFAS Extension, Gainesville, Florida.

Harmon, P.F., Alcala-Briseno, R.I. and Poiston, J.E. (2015) Severe symptoms of mosaic and necrosis in cv. Floratam St Augustinegrass associated with *sugarcane mosaic virus* in neighborhoods of St. Petersburg, Florida. *Plant Disease* 99, 557. DOI: 10.1094/PDIS-11-14-1140-PDN

Harris, A. (2000) *Viruses, Phytoplasmas and Spiroplasmas of Clonal Grasses and their Diagnosis*. Plant Biosecurity, Biosecurity Australia, Department of Agriculture, Fisheries and Forestry, Canberra.

Haygood, R.A. and Barnett, G.W. (1992) Widespread occurrence of centipede mosaic in South Carolina. *Plant Disease* 76, 46–49.

Hodges, C.F. (1993) The biology of algae in turf. *Golf Course Management* 61, 44–56.

Holcomb, G.E. (1985) A mosaic disease of centipedegrass and crowsfootgrass (abstract). *Phytopathology* 75, 500.

Holcomb, G.E., Liu, T.-Z.L. and Derrick, K.S. (1989) Comparison of isolates of panicum mosaic virus from St. Augustinegrass and centipedegrass. *Plant Disease* 73, 355–358.

Inguagiato, J.C., Kaminski, J.E. and Lulis, T.T. (2018) effect of phosphate rate and source on cyanobacteria colonisation on putting green turf. *Golf Course Management* 3, 66–71.

Kerns, J.P. and Tredway, L.P. (2013) Advances in turfgrass pathology since 1990. In: Stier, J.C., Horgan, B.P. and Bonos, S.A. (eds) *Turfgrasses: Biology, Use, and Management*. Agronomy Monograph No. 56.

American Society of Agronomy, Inc., Crop Science Society of America, Inc. and Soil Science Society of America, Inc., Madison, Wisconsin, pp. 733–776.

Kerrigan, J.L., Olsen, M.W. and Martin, S.B. (2012) Rapid blight of turfgrass. *The Plant Health Instructor.* DOI: 10.1094/PHI-I-2012-0621-01

Knowles, J. (2011) Ghost grass. *Greenkeeper International* (July), 24–25.

Koh, L.H., Yap, M.L., Yik, C.P., Niu, N. and Wong, S.M. (2008) First report of phytoplasma infection of grasses in Singapore. *Plant Disease* 92, 317. DOI: 10.1094/PDIS-92-2-0317C

Lapierre, H. and Signoret, P.-A. (eds) (2004) *Viruses and Virus Diseases of Poaceae (Gramineae).* INRA, Paris.

Liu, S., Vargas, J. and Merewitz, M. (2019) Temperature and hormones associated with bacterial etiolation symptoms of creeping bentgrass and annual bluegrass. *Journal of Plant Growth Regulation* 38, 249–261.

McCarty, L.B. (2010) *Best Golf Course Management Practices: Construction, Watering, Fertilizing, Cultural Practices and Pest Management Strategies to Maintain Golf Course Turf with Minimal Environmental Impact*, 3rd edn. Prentice-Hall, Upper Saddle River, New Jersey.

Maddox, V.L., Krans, J.V. and Sullivan, M.J. (1997) Survey of algal and cyanobacterial species on golf putting greens in Mississippi, USA. *International Turfgrass Society Research Journal* 8, 495–505.

Martin, B., Stowell, L. and Gelernter, W. (2002) Rough bluegrass, annual bluegrass and perennial ryegrass hit by new disease. *Golf Course Management* 70, 61–65.

Mitrovic, J., Smiljkovic, M., Seemuller, E., Reinhardt, R., Huttel, B., et al. (2015) Differentiation of 'Candidatus Phytoplasma cynodontis' based on 16S rRNA and *groEL* genes and identification of a new subgroup 16SrXIV-c. *Plant Disease* 99, 1578–1583.

Olsen, M.W., Bigelow, D.M., Kohout, M.J., Gilbert, J. and Kopec, D. (2004) Rapid blight: a new disease of cool-season turf. *Golf Course Management* 72, 87–91.

Pares, R.D. and Gillings, M.R. (1990) Two new records of diseases caused by potyviruses in Australia. *Australasian Plant Pathology* 19, 36–37.

Petelewicz, P., Glegola, S., Gomez, D. and Baird, J. (2019) *Management of Salinity and Rapid Blight Disease on Annual Bluegrass Putting Greens. 2019 Report.* University of California, Riverside, California.

Peterson, P.D., Martin, S.B. and Camberato, J.J. (2005) Tolerance of cool-season turfgrasses to rapid blight disease. *Applied Turfgrass Science* 2, 1–8. DOI: 10.1094/ATS-2005-0328-01-RS

Roberts, D.L., Vargas, J.M., Detweller, R. and Baker, K.K. (1982) Symptom expression with oxytetracycline of a Toronto creeping bentgrass disease presumed bacterial etiology. *Plant Disease* 66, 804–806

Roberts, J.A. and Kerns, J.P. (2018) Etiolation and decline: an enigmatic condition in creeping bentgrass putting greens. In: Burman, S. and Walcott, R.R. (eds) *Plant-Pathogenic Acidovorax Species.* The American Phytopathological Society, St. Paul, Minnesota, pp. 89–100.

Roberts, J.A., Kerns, J.P. and Ritchie, D.F. (2015) Bacterial etiolation of creeping bentgrass as influenced by biostimulants and trinexapac-ethyl. *Crop Protection* 72, 119–126. DOI: 10.1016/j.cropro.2015.03.009

Roberts, J.A., Ritchie, D.F. and Kerns, J.P. (2016) Plant growth regulator effects on bacterial etiolation of creeping bentgrass putting greens turf caused by *Acidovorax avenae*. *Plant Disease* 100, 577–582.

Roberts, J.A., Ma, B., Tredway, L.P., Ritchie, D.F. and Kerns, J.P. (2018) Identification and pathogenicity of bacteria associated with etiolation and decline of creeping bentgrass golf course putting greens. *Phytopathology* 108, 23–30. DOI: 10.1094/PHYTO-01-17-0015-R

Rosete, Y.A. and Jones, P. (2009) Phytoplasma diseases of the Gramineae. In: Weintraub, P.G. and Jones, P. (eds) *Phytoplasmas: Genomes, Plant Hosts and Vectors.* CAB International, Wallingford, UK, pp. 170–187.

Rushford, C.A., North, R.L. and Miller, G.L. (2022) Detection of cyanotoxins in irrigation water and potential impact on putting green health. *International Turfgrass Society Research Journal* 14, 994–996. DOI: 10.1002/its2.40

Smith, J.D., Jackson, N. and Woolhouse, A.R. (1989) *Fungal Diseases of Amenity Turfgrasses*, 3rd edn. E. & F.N. Spon, London.

Stem, M. (1992) White leaf. *Golf Course Management* 60, 86–89.

Stewart, A. (2002) Mad tiller disease: a common but only recently named condition in turf. *New Zealand Turf Management Journal* 17, 25–26.

Tani, T. and Beard, J.B. (1997) *Color Atlas of Turfgrass Diseases.* Wiley, Hoboken, New Jersey.

Teakle, D.S. and Grylls, N.E. (1973) Four strains of sugarcane mosaic virus infecting cereals and other grasses in Australia. *Australian Journal of Agricultural Research* 24, 465–477.

Thomas, J.E. and Steele, V. (2011) First report of panicum mosaic virus in buffalo grass (*Stenotaphrum secundatum*) from Australia. *Australasian Plant Disease Notes* 6, 16–17.

Thomas, J.E., Raymond, M., Tran, N.T., Crew, K.S., Teo, A.C. and Geering, A.D.W. (2021) Complete genome sequences and properties of Spartina mottle virus isolates from hybrid bermudagrass (*Cynodon dactylon* × *Cynodon tranvaalensis*). *Plant Pathology* 70, 1062–1071.

Tran, N.T., Teo, A.C., Thomas, J.E., Crew, K.S. and Geering, A.D.W. (2020) Sugarcane mosaic virus infects *Stenotaphrum secundatum* in Australia. *Australasian Plant Disease Notes* 15, 41. DOI: 10.1007/s13314-020-00410-y

Tran, N.T., Chin, A.C., Crew, K.S., Thomas, J.E., Campbell, P.R. and Geering, A.D.W. (2022) Bermuda grass latent virus in Australia: genome sequence, sequence variation, and new hosts. *Archives of Virology* 167, 1317–1323. DOI: 10.1007/s00705-022-05434-6

Tredway, L.P., Stowell, L.J. and Gelernter, W.D. (2006) Yellow spot and the potential role of cyanobacteria as turfgrass pathogens. *Golf Course Management* 74, 83–86.

Tredway, L.P., Tomaso-Peterson, M., Kerns, J.P. and Clarke, B.B. (2023) *Compendium of Turfgrass Diseases*, 4th edn. The American Phytopathological Society. St. Paul, Minnesota.

Turgeon, A.J. and Kaminski, J.E. (2019) *Turfgrass Management, Edition 1.0*. Turfpath LLC, State College, Pennsylvania.

Vargas, J.M. (2005) *Management of Turfgrass Diseases*, 3rd edn. Wiley, Hoboken, New Jersey.

Vargas, J., Giordano, P., Detweiler, R. and Dykema, N. (2014) Occurrence and identification of an emerging bacterial pathogen of creeping bentgrass. *USGA Turfgrass and Environmental Research Online* 13, 24–26.

10

Plant-Parasitic and Beneficial Nematodes

Abstract

Nematodes, free-living, plant-parasitic and predacious species, are natural inhabitants in ecosystems worldwide. Nematode populations are influenced by food availability, management-driven inputs and environmental conditions. Numerous genera of root-feeding nematodes are widely acknowledged as important turfgrass pests capable of causing significant host damage during highly favourable environmental conditions. Potential damaging effects of high populations of root-feeding nematodes can be mitigated only using an integrated systems approach.

Nematodes are ubiquitous and natural co-inhabitants of grasslands and turfgrass systems worldwide. Free-living nematodes survive on decomposing organic matter or the microbes degrading the organic matter or as predators feeding on other nematodes or invertebrates and are collectively the most abundant species in soil environments. Phytophagous or plant-feeding nematodes encompass both foliar-feeding and root-feeding species with the latter group being the more important plant pathogens. Most species of plant-parasitic nematodes are obligate pathogens that require living plants to feed on and have populations synchronized to and driven by the seasonal growth of host plants.

Root-feeding nematodes pose the greatest threat when populations exceed suggested threshold levels when feeding on turfgrasses having dysfunctional root systems under stress-related conditions. Turfgrass damage by high populations of one more species of root-feeding nematodes can result in a random mosaic pattern of chlorotic and necrotic plants (Fig. 10.1) which may be misdiagnosed and attributed to other causes (e.g. heat and drought stress).

The species distribution and populations of root-feeding nematodes vary seasonally and spatially, both horizontally and vertically, in the soil and are strongly influenced by host root health and underlying management-induced, environmental or mitigating biological factors. The subterranean lifestyle of root-feeding nematodes presents significant challenges for their diagnosis and effective management. Effective, long-term management of root-feeding nematodes can be gained only by an integrated systems approach that encourages and maintains a highly functional plant root system.

Significance of Turfgrass-Parasitic Nematodes

The significance of root-feeding nematodes as turfgrass pests relates to their feeding activity by

Fig. 10.1. Plant-parasitic nematode damage on bermudagrass (couchgrass) in the United States. (Courtesy of Nathan R. Walker.)

decreasing the capacity of the host to acquire soil water and nutrient resources because of a reduced functional root system.

The host's inability to acquire these resources becomes critical during periods coinciding with increasing temperatures and reduced moisture. Most species of root-feeding nematodes have a body length ranging 0.25–3.0 mm or more at maturity (Crow, 2008).

The association between root-feeding nematodes and unthrifty turfgrass was acknowledged from the late 1900s on golf greens in North America (Perry *et al.*, 1970; Fushtey and McElroy, 1977) and later on bowling and golf greens in Australia (Stynes, 1968; Siviour and McLeod, 1979) and in New Zealand (Yeates, 1980) primarily on mixed-cotula bowling greens.

Generally, populations of root-feeding nematodes in highly managed, monocultured turfgrass, as opposed to natural plant systems, may not be suppressed by high populations of antagonistic and parasitic microorganisms. Anecdotal evidence suggests populations of root-feeding nematodes may increase over time and persist on newly established turfgrass subsequent to reshaping and uncontrolled mixing of natural sands in specific coastal sites as a possible consequence of loss of naturally occurring microbial antagonists.

Turfgrasses possessing dysfunctional root systems, for whatever reason, inevitably deteriorate from nematode feeding damage from moderate to high populations when under management-induced (e.g. close mowing) or environmental stresses. Once established in turfgrass systems, root-feeding nematode populations can be reduced but never completely eliminated.

Nematode Feeding Mechanisms

Nematodes gain their nutrition based on their mouthpart structure (Fig. 10.2). Free-living nematodes can feed on bacteria, fungi or other nematodes and are part of nutrient recycling. Plant-parasitic nematodes are generally characterized by a highly specialized, hypodermic needle-like device or stylet in their head. Plant-parasitic nematodes feed on root hairs; however other nematodes such as fungal-feeding,

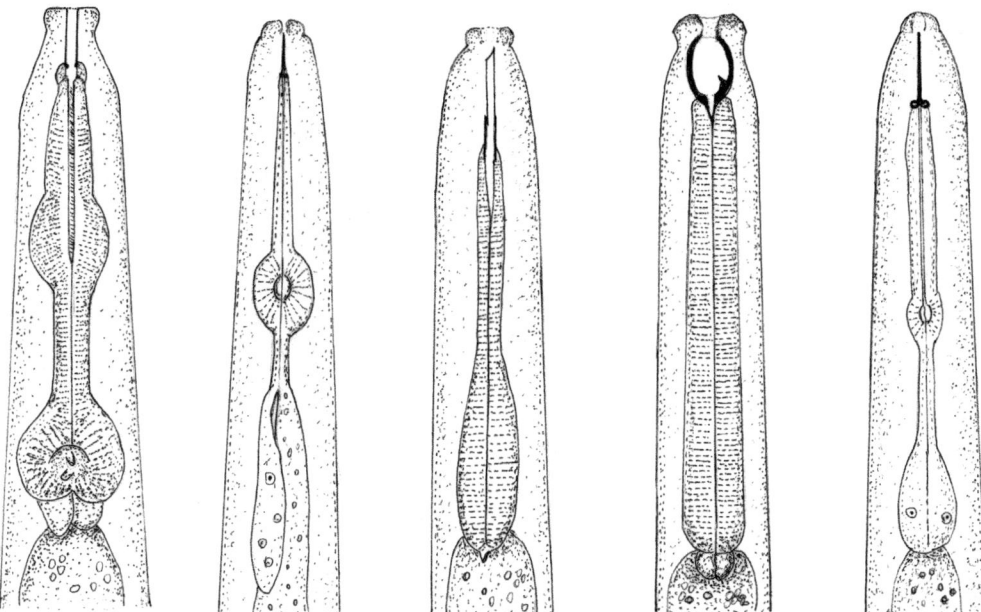

Fig. 10.2. Categorization of soil-borne nematodes based on mouthpart structure. Left to right: bacterial-feeding, fungal-feeding, omnivorous, predatory and plant-parasitic nematode types. (Courtesy of Marcelle Stirling.)

omnivorous and some predatory nematodes may possess a similar tooth or stylet. The presence of a stylet does not always mean the nematode is a plant parasite.

The stylet of plant-parasitic nematodes is inserted into plant cells and digestive enzymes may be excreted with the end result being the withdrawal of cell contents. During feeding, injected compounds (e.g. enzymes) and root-surface wounding may cause morphological changes to the host, potentially rending the plant more susceptible to both pathogenic and opportunistic soil microbes. Certain species of plant-parasitic nematodes (e.g. *Longidorus*, *Trichodorus* or *Xiphinema*) are virus vectors (Tredway *et al.*, 2023). Larger and longer stylet length partly explains the degree of physical damage to the host plant by certain species of plant-parasitic nematodes.

Genera of Turfgrass-Parasitic Nematodes

Numerous genera of plant-parasitic nematodes are common on turfgrass systems (Table 10.1).

The most common nematode genera are not necessarily the most damaging. The identity of the most common and highly damaging, root-feeding nematodes in turfgrass systems emanates from surveys conducted primarily throughout North America (e.g. Yu *et al.*, 1998; Jordan and Mitkowski, 2006; Simard *et al.*, 2008; Zeng *et al.*, 2012) mainly on golf courses.

Outside North America, limited surveys of the identity of root-feeding nematodes have been published in Australia (Neylan, 1996; Stirling, 2008) and in New Zealand (Knight *et al.*, 1997; Howard and Ormsby, 2017) focusing on cotula bowling greens. More recent information about nematodes infesting turfgrasses has been published in Western Europe and the United Kingdom (Fleming *et al.*, 2010; Vandenbossche *et al.*, 2011; Entwistle *et al.*, 2014), South Africa (Swart *et al.*, 2000) and East Asia (Onoda *et al.*, 1998; Mwamula and Dong, 2021; Dong *et al.*, 2022).

Parasitic Nematode Feeding Groups

Root-feeding nematodes are aquatic organisms requiring moisture for their development and

Table 10.1. Root-feeding nematodes by common name and genera.

Nematode	Genera	Nematode	Genera
Awl	*Dolichodorus*	Ring	*Criconemella, Mesocriconema*
Burrowing	*Radopholus*	Root-knot	*Meloidogyne*
Cyst	*Heterodera, Punctodera*	Root-lesion	*Pratylenchus*
Dagger	*Xiphinema*	Sheath	*Caloosia, Hemicycliophora*
False root-lesion	*Pratylenchoides*	Spiral	*Helicotylenchus, Rotylenchus*
Lance	*Hoplolaimus*	Stem	*Ditylenchus*
Lesion	*Pratylenchus*	Sting	*Belonolaimus, Ibipora*
Needle	*Longidorus, Paralongidorus*	Stubby-root	*Paratrichodorus, Trichodorus*
Pin	*Paratylenchus*	Stunt	*Tylenchorhynchus*
Reniform	*Rotylenchulus*		

survival. Soil-dwelling, root-feeding nematode species move in the microscopic films of moisture surrounding soil particles. Depending on the species, some live fully inside the host, partially inside or outside the host or mostly outside the host. Root-feeding nematodes are grouped according to their feeding lifestyle and most nematologists generally recognize four broad groups: migratory or sedentary endoparasites and migratory or sedentary ectoparasites; however there are always intermediatory categories (Table 10.2). An ectoparasitic feeding habit represents the most common form of nematode feeding on turfgrass.

Sedentary endoparasites completely enter their host where females become enlarged inducing physical change to their hosts (root-feeding types). Sedentary semi-endoparasites establish a permanent feeding site inside the host but with a significant portion of their body remaining outside the host (e.g. cyst nematodes). Migratory endoparasites remain worm-like (i.e. vermiform) and mobile throughout their life cycle, feeding on host foliage, stem and/or roots (Fig. 10.3), depending on nematode species.

Root-feeding, migratory endoparasites feed and move internally within the host and generally only move outside to the rootzone to infect new roots on which to feed. Certain migratory endoparasites (*Anguina* spp., *Ditylenchus* spp.) are seed-, leaf- or stem-feeders causing superficial to significant damage, depending on host–nematode combination. Ectoparasites feed externally at the end of the root (i.e. root cap) and their body remains mostly or entirely in the rhizosphere.

Economically Important Turfgrass-Parasitic Nematodes

Members of certain genera of root-feeding nematodes are generally recognized as more important on turfgrass systems (Table 10.3). Their importance varies widely between countries and among regions depending on nematode–host species combination and environmental conditions (Crow, 2008).

Ectoparasitic

Sting nematode (*Belonolaimus longicaudatus*)

The sting nematode is one of the most destructive and well-studied plant-parasitic nematodes in the Americas. The sting nematode is native to the southern United States and occurs along the regions of the Gulf of Mexico and Atlantic coasts from California to Virginia as far north as Ohio and in the Caribbean islands (Crow, 2008).

The sting nematode has an extensive host range of cool-season and warm-season grasses and has a relatively large body up to 3.0 mm in length and 0.35 mm wide at maturity with a stylet being approximately 100 µm long (Fig. 10.4). The relatively large size explains partly why sting nematodes require high-sand-content rootzones and can cause severe physical damage to turfgrass roots when feeding. Sting nematodes can complete their life cycle in 18–24 days under highly favourable conditions.

Table 10.2. Feeding groups of plant-parasitic nematodes.

Feeding group	Nematode genera	Reference
Sedentary endoparasites	*Heterodera, Meloidogyne, Punctodera*	Eisenback (2002)
Sedentary semi-endoparasites	*Rotylenchulus, Tylenchulus*	
Migratory endoparasites	*Hoplolaimus, Pratylenchus, Pratylenchoides, Radopholus*	
Ectoparasites	*Belonolaimus, Caloosia, Criconemella, Dolichodorus, Helicotylenchus, Hemicycliophora, Hoplolaimus, Ibipora, Longidorus, Mesocriconema, Paratrichodorus, Paratylenchus, Peltamigratus, Rotylenchus, Scutellonema, Trichodorus, Tylenchorhynchus, Xiphinema*	
Foliar parasites	*Anguina, Ditylenchus*	

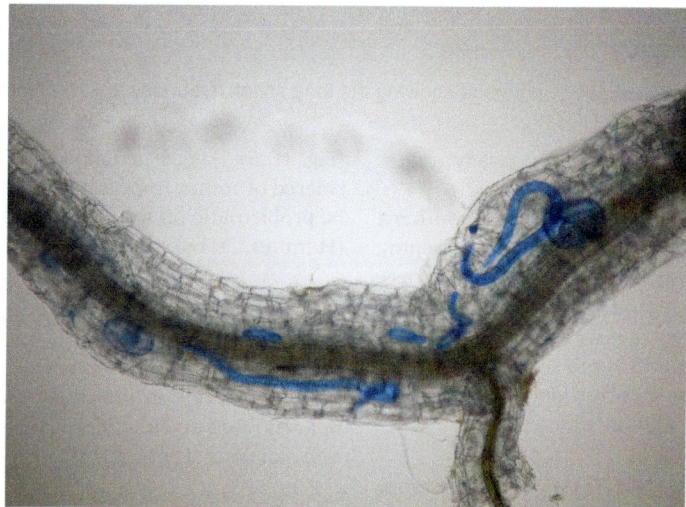

Fig. 10.3. Stained *Pratylenchoides* spp. adults and juveniles feeding inside a bentgrass root. (Courtesy of Colin Fleming.)

Table 10.3. Economically important root-feeding nematode names and genera (listed in order of importance).

Ectoparasites	Genera	Endoparasites	Genera
Awl	*Dolichodorus*	Root-knot	*Meloidogyne*
Sting	*Belonolaimus, Ibipora*	Lesion	*Pratylenchus*
Stubby-root	*Paratrichodorus, Trichodorus*	Cyst	*Heterodera*
Needle	*Longidorus, Paralongidorus*		
Dagger	*Xiphinema*		

Southern sting nematode (*Ibipora lolii*)

The southern sting nematode is the most investigated and damaging nematode of turfgrass in Australia. Southern sting nematode is distributed along coastal regions of New South Wales, Victoria and the southern coast of south-western Western Australia. The southern sting nematode is relatively thin but long, with males and females measuring 2.0 mm or more in length at maturity

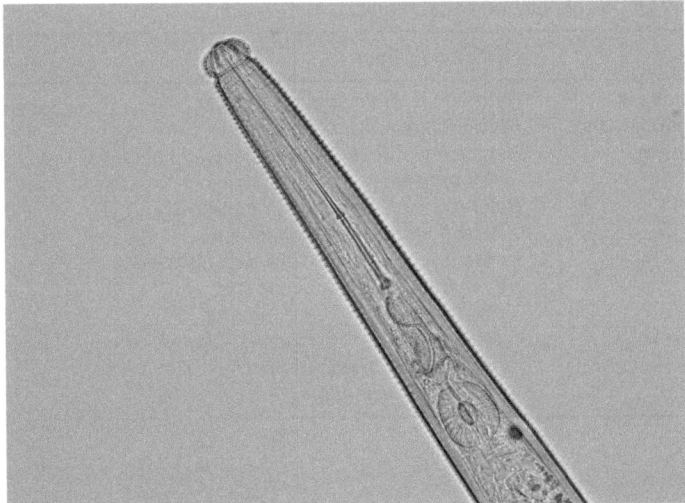

Fig. 10.4. Head section of sting nematode showing the long stylet. (Courtesy of Colin Fleming.)

and having a relatively long stylet typically greater than 100 μm (Ruscoe and Stirling, 2020).

Known turfgrass hosts of the southern sting nematode are bermudagrass, creeping bentgrass, kikuyugrass and perennial ryegrass. Southern sting nematode populations generally increase during cool and wet seasons occurring during late autumn through to mid-spring and decline during the hot and dry period of late spring through to mid-autumn.

Stubby-root nematode (*Paratrichodorus* spp. and *Trichodorus* spp.)

Stubby-root nematodes are 'cigar-shaped' and gain their common name by the root symptoms being 'stubby' in appearance.

Needle nematode (*Longidorus* spp. and *Paralongidorus* spp.)

Needle nematodes are among the largest plant-parasitic nematodes with some species having a body over 1.0 cm in length. They feed near the root tip and cause curved galls at the tips of roots which eventually get colonized by soil microbes and the galls become rotted.

Dagger nematode (*Xiphinema* spp.)

Dagger nematodes are relatively large nematodes but still require microscopic means to be visible (Fig. 10.5). Dagger nematodes are generally considered of minor importance on grasses but can be problematic on mixed-cotula bowling greens (Hannan, 2018).

Awl nematode (*Dolichodorus* spp.)

Awl nematodes gain their name from their 'awl-like' shaped tail on males. The nematode feeds near the root tip often reducing fine root growth and can cause root necrosis.

Endoparasitic

Root-knot nematode (*Meloidogyne* spp.)

Root-knot nematodes are among the most important plant-parasitic nematodes worldwide. The nematodes' collective name characterizes the galling or knot formed on host roots, but the galls are not always readily observed on infected turfgrass roots. Root-knot nematodes have a large host range and to feed, inject salivary secretions into their host resulting in altered physiological functioning and root swelling or galling. *Meloidogyne minor* has become problematic on putting greens in the United Kingdom (Fleming et al., 2008) and *Meloidogyne graminis* and *Meloidogyne marylandi* are a common problem on dwarf bermudagrass greens in the United States (Crow, 2021).

Lance nematode (*Hoplolaimus* spp.)

Lance nematodes are relatively large nematodes and can reach up to 1.0 mm at maturity (Fig. 10.6). Due to their size, lance nematodes (like sting nematodes) are restricted to sandy soils. Lance nematodes are among the most important nematodes parasitizing turfgrasses in the United States (Crow, 2008).

Lesion nematode (*Pratylenchus* spp.)

Lesion nematodes have a wide turfgrass host range and gain their common name by the formation of root lesions as a result of their feeding and physical damage to root tissues. Generally, lesion nematodes complete their life cycle in 6–9 weeks under optimal conditions.

Cyst nematode (*Heterodera* spp.)

Cyst nematodes are a relatively large species when females reach maturity. Cyst nematodes are characterized by the female cyst that is formed at death. The nematode cysts (0.5–1.0 mm in diameter) are lemon-shaped and the cyst contains eggs which can remain viable for years awaiting favourable conditions to hatch.

False root-lesion nematode (*Pratylenchoides* spp.)

The false root-lesion nematode (*Pratylenchoides crenicauda*) is an emerging pest species on golf courses in Great Britain (Fleming *et al.*, 2008).

Root-gall nematode (*Subanguina radicicola*)

The root-gall nematode occurs on annual bluegrass in Great Britain, Europe and North America (Anon., 2007; Mitkowski and Jackson, 2003).

Foliar nematode parasites

Stem-gall nematode (*Anguina* spp.)

The Pacific stem-gall nematode (*Anguina pacificae*) occurs on mixed annual bluegrass–bentgrass

Fig. 10.5. Dagger nematodes on cotula (*Leptinella* sp.) roots of a bowling green in New Zealand. (Courtesy of the NZSTI.)

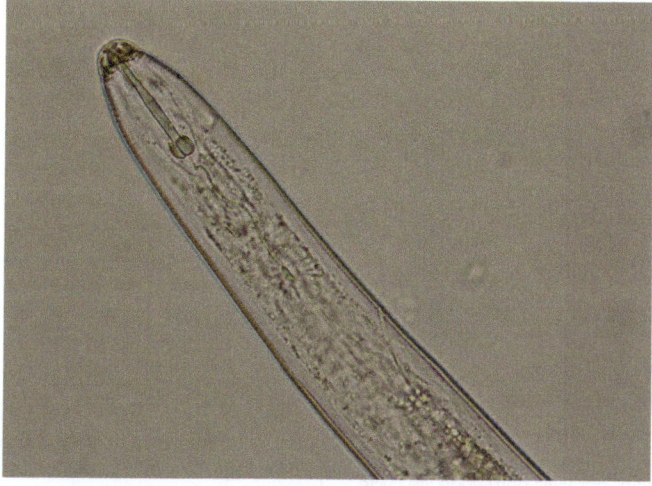

Fig. 10.6. Head section of a female lance nematode. (Courtesy of Colin Fleming.)

golf greens in California, USA (McClure et al., 2008) where extended periods of fog occur. *A. pacificae* causes stem galls on individual plants and yellow-coloured patches and has recently been identified on a golf green in Ireland (Entwistle et al., 2014).

Basic Nematode Biology

The life cycle of root-feeding nematodes is relatively simple (Fig. 10.7). Nematodes have an egg stage, four immature (juvenile) stages and a reproductive adult stage. Most nematodes develop inside the eggs as first-stage juveniles (J1) undergoing one moult then hatch as second-stage juveniles (J2). The hatched nematode undergoes three additional moults developing into third (J3) and fourth (J4) juvenile stages and finally into an adult stage.

Nematodes reproduce either sexually or asexually (i.e. parthenogenesis or reproduction by development of an unfertilized egg without males) depending on nematode species. Life-cycle length from egg to adult varies from weeks to years depending on nematode species and environmental conditions. The life cycle of root-feeding nematodes is generally shorter in tropical and subtropical regions and longer in cooler, temperate regions.

Nematode population dynamics

Root-feeding nematode populations fluctuate seasonally and spatially, both horizontally and vertically (Jordan and Mitkowski, 2006; Kaapro, 2020; Stirling et al., 2013). Nematode populations increase or decrease in response to seasonal changes in turfgrass root growth and soil temperatures. There may be several nematode generations occurring per season with population peaks that generally coincide during warmer soil conditions. Nematode populations generally tend to peak on cool-season grasses in spring and autumn and in summer on warm-season grasses, provided plant health and soil moisture are not limiting.

Population development of root-feeding nematodes and the capacity to cause physical damage to turfgrasses are governed by numerous factors (Box 10.1). Most root-feeding nematodes are found in the regions of most active root growth and the availability of functioning root systems is a primary factor.

Certain nematode species may undergo rapid changes in population increase, having relatively short life cycles, high colonization capacity and wide tolerance to environmental disturbances. On the other hand, stable populations are often characterized by low reproduction rate, long life cycles, low colonization capacity and sensitivity to disturbances. The former nematode types are typically numerically dominant in samples; thus pose a significant risk in highly stressed turfgrass systems.

Rootzone texture

High sand content of the rootzone generally favours root-feeding nematodes (Fig. 10.8). Sand particle size (0.12–0.37 mm diameter) appears to favour sting nematode (*B. longicaudatus*)

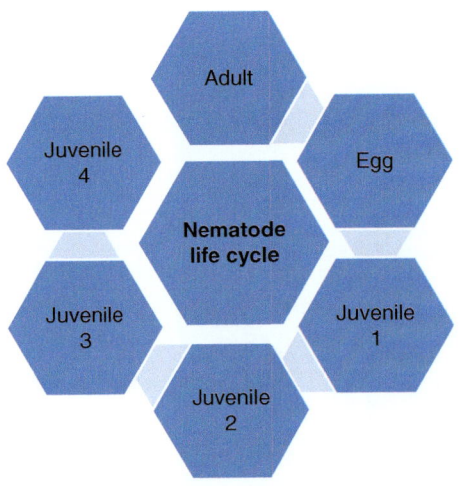

Fig. 10.7. Simplified plant-parasitic nematode life cycle.

Box 10.1. Nematode population factors.

- Rootzone texture
- Moderate temperature
- Moisture–oxygen balance
- Root growth and host food availability
- Rhizosphere chemical compounds
- Antagonists and parasites

populations (Robbins and Barker, 1974) while populations of ring nematode (*Criconemella* sp.) are favoured in fine sand (< 1.0 mm diameter) and low in coarse sands (1–2 mm diameter) (Walker and Martin, 2002).

Temperature and moisture

The optimum temperature for development and feeding of most root-feeding nematode species is around 20–30°C (Tredway *et al.*, 2023). As rootzone temperatures increase above the minimum up to the maximum for each species, nematode development, feeding and reproduction progress. Rootzone moisture content between field capacity and wilting point generally favours root-feeding nematodes. A rootzone moisture content averaging 15–25% (by volume) favours *Belonolaimus*, *Pratylenchus* and *Trichodorus* nematode populations (Brodie and Quattlebaum, 1970).

Food quantity and root growth

Turfgrass species, health and age all interact in determining the seasonal quantity and quality of food source for root-feeding nematodes. Developmental temperature of nematode populations generally coincides with temperatures favourable for turfgrass root growth. This point has practical implications in the timing of cultural practices designed to stimulate healthy root growth and application of nematicides and bionematicides, if and when required.

Lateral movement and feeding depth of ectoparasitic nematodes around turfgrass roots remain speculative. Sting nematode (*B. longicaudatus*) populations have been reported to a depth of 15–30 cm in California, USA (Bekal and Becker, 2000). Populations of the southern sting nematode (*I. lolii*) on a kikuyugrass sportsground in Perth (Western Australia) were reported to occur as deep as 70 cm (Ruscoe and Stirling, 2020).

Rhizosphere chemical compounds

Concentrations of rootzone chemical compounds (e.g. plant-root exudates or allelochemicals) may have a stimulatory or inhibitory effect on root-feeding nematodes, depending on the chemical and circumstances.

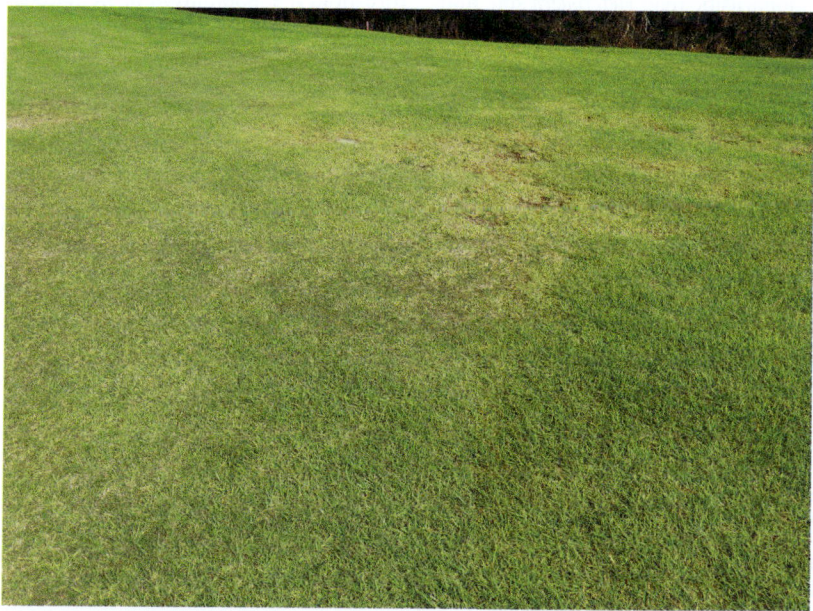

Fig. 10.8. Extensive chlorosis attributed primarily to damaging populations of southern sting nematode on a coastal golf fairway in Australia. (Courtesy of Jyri Kaapro.)

Antagonistic and parasitic organisms

A suite of antagonistic, suppressive or predatory microorganisms can in part regulate root-feeding nematode populations in natural plant systems (Table 10.4). Cultural manipulation of the rhizosphere in combination with the purposeful introduction of antagonistic or suppressive organisms using organic amendments, microbial inoculants or microbial metabolites having nematicidal properties offers the potential to suppress populations of root-feeding nematodes. Understanding of the exact nature of the biological interactions between nematode populations and microbial antagonists is far from complete. Use of entomopathogenic nematodes against root-feeding nematode populations in turfgrass systems has proven largely unsuccessful (Crow et al., 2006).

Parasitic nematode–soil-borne fungal interactions

Biological interactions between root-feeding nematodes and soil-borne pathogenic fungi resulting in disease complexes are known to occur in cropping systems (Back et al., 2002). Interactions are known to occur between certain ectoparasitic (e.g. *Longidorus* spp. and *Xiphinema* spp.) and endoparasitic (e.g. *Heterodera* spp. and *Meloidogyne* spp.) nematodes and fungal genera such as *Fusarium*, *Gaeumannomyces*, *Polymyxa*, *Pythium* and *Rhizoctonia*. The resultant nematode–fungal disease complex may manifest in the form of either additive or synergistic interactions (Back et al., 2002).

While a detailed understanding of the complex physical, biochemical or physiological mechanisms explaining parasitic nematode–pathogenic fungal interactions and outcomes in cropping systems is far from fully known, it is reasonable to presume similar root-feeding nematode and pathogenic soil-borne fungi complexes occur in turfgrass systems.

The reported mechanisms explaining these interactions are nematode-inflicted physical wounding to gall formation allowing for fungal and opportunistic microbe infection or chemical and physiological changes to plant roots. These changes may manifest as alterations in root exudates, overall reduction in host resistance to fungal pathogens or increased plant senescence (Back et al., 2002). Suffice to say, the activity of plant-parasitic nematodes and soil-borne fungal pathogens and the degree of synergism or antagonistic interactions are mediated by key environmental factors (e.g. temperature and moisture).

Nematode damage thresholds

The issue of damage thresholds raises the complex question of how many root-feeding nematodes turfgrasses can tolerate while remaining healthy. Determination of a direct cause-and-effect relationship between root-feeding nematodes and unhealthy turfgrass is problematic given all the interacting, biological and environmental factors (Box 10.2). Most biological and environmental variables cannot be accurately quantified and are subject to wide interpretation. A nematode damage threshold, as opposed to an economic threshold (i.e. population of nematodes present to determine the need for a management action), is simply the minimum population of root-feeding nematodes resulting in a response from or symptom to the infected plant.

The nematode damage threshold concept originated in cropping systems based on the premise that when nematode populations are below a certain number (i.e. threshold level), healthy non-stressed plants can tolerate a certain level of nematode predation. Exceedance of damage thresholds for a specific host–nematode combination(s) prompts management actions (Fig. 10.9). A simple concept but complex to determine precisely in turfgrass systems.

Table 10.4. Antagonistic microbes and parasitic organisms.

Microbial group	Organism example	References
Parasitic bacteria	*Pasteuria*	Viaene et al. (2013); Stirling (2014)
Plant-growth-promoting rhizobacteria	*Bacillus*, *Burkholderia*, *Pseudomonas*	
Arbuscular mycorrhizal fungi	*Funneliformis*, *Glomus*, *Rhizophagus*	
Endophytic fungi	*Acremonium*, *Fusarium*, *Neotyphodium*	
Nematophagous fungi	*Purpureocillium lilacinum*	

Nematode damage thresholds have been extensively published in the United States (e.g. Couch, 1995; Buckley et al., 2010; McCarty, 2010) and to a much lesser extent in Australia (Neylan, 1996; Stirling, 2008) and the United Kingdom (Fleming et al., 2008). Damage thresholds have only been established for single nematode genera for the most common grasses. Most published damage thresholds only apply to a specific region and do not account for multiple nematode species being present or turfgrass blends or mixtures.

The apparent limited development of stolons among the 'newer' creeping bentgrass cultivars (Beehag and McMaugh, 2018), propensity for shallower root growth among certain dwarf bermudagrass cultivars (Inguagiato and Martin, 2015) and observable differences in stolon and rhizome architecture raise speculation that their reduced tolerance may require longer time for recovery from nematode feeding.

The results of nematode assays conducted during a summer on numerous golf course greens in the north-eastern United States (Jordan and Mitkowski, 2006) reported nematode populations in excess of threshold levels without damage symptoms. The presence of multiple nematode species may have inhibitory effects to one species (Johnson, 1970; Crow et al., 2013) depending on host–nematode combination in ways far from understood.

Relative pathogenicity

Contradictory results for the degree of pathogenicity (i.e. capacity of a pathogen to cause disease) are common. Relative pathogenicity is an alternative way of assessing the likelihood of high populations of root-feeding nematodes causing unacceptable levels of plant damage (Table 10.5).

The foregoing discussion highlights the point that nematode damage thresholds or relative pathogenicity should not be interpreted as absolute for all nematode–turfgrass species combinations. The activity of a few root-feeding nematodes is most likely to be insignificant. Under highly favourable conditions of temperature, moisture and adequate food resources, root-feeding nematode reproduction, development and numbers can increase rapidly causing unacceptable damage, most notably on environmentally stressed turfgrass.

Box 10.2. Nematode damage threshold variables.

- Turfgrass species or cultivar
- Level of cultural management
- Stress and wear level
- Seasonal root growth and volume
- Rootzone biology and health
- Presence of chemical compounds
- Nematode species and population
- Co-infection by other root pathogens

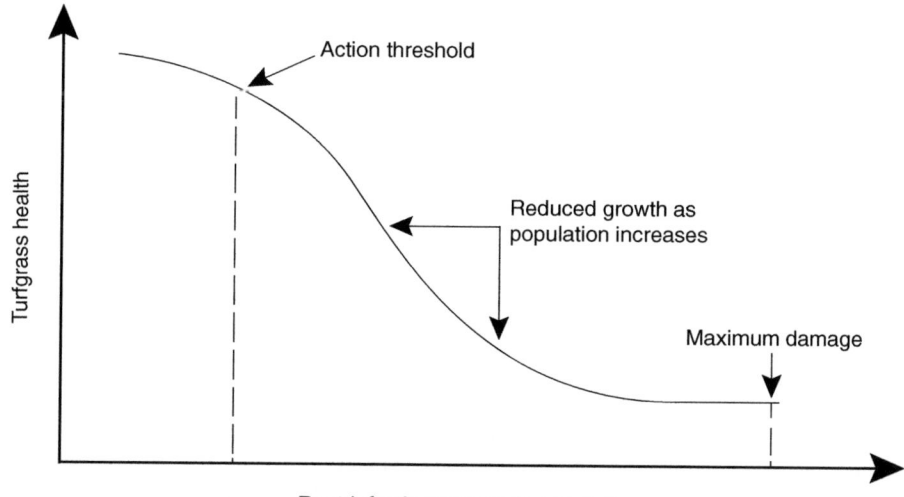

Fig. 10.9. Simplified hypothetical relationship between nematode populations and turfgrass damage.

Table 10.5. Relative pathogenicity of root-feeding turfgrass nematodes.

Relative pathogenicity	Common name and nematode genera	References
High	Sting (*Belonolaimus*)	Couch (1995); Fleming *et al.* (2008);
	Southern sting (*Ibipora*)	Stirling (2008); Buckley *et al.*
	Needle (*Paralongidorus*)	(2010); McCarty (2010)
	Dagger (*Xiphinema*)	
	Lance (*Hoplolaimus*)	
	Root-gall (*Subanguina*)	
Medium	Stubby-root (*Paratrichodorus*)	
	Cyst (*Heterodera*)	
	Sheath (*Hemicycliophora*)	
	Lesion (*Pratylenchus*)	
	Stunt (*Tylenchorhynchus*)	
	Root-knot (*Meloidogyne*)	
Low–minimal	Pin (*Paratylenchus*)	
	Ring (*Criconemella*, *Macroposthonia*)	
	Reniform (*Rotylenchulus*)	
	Spiral (*Helicotylenchus*, *Rotylenchus*)	

Root-Feeding Nematode Signs and Damage Symptoms

Root-feeding nematodes with few exceptions (e.g. dagger nematode) are not readily visible to the naked eye requiring microscopic techniques. Aboveground and belowground feeding signs and symptoms vary widely depending on host–nematode population combinations and environmental factors. The presence of other microbial pathogens (e.g. foliar and/or soil-borne fungi) may partly disguise obvious signs and symptoms characteristic of feeding nematodes, contributing to a possible misdiagnosis.

Aboveground symptoms

Aboveground plant symptoms due to the presence of relatively low to moderate numbers of root-feeding nematodes in the one turfgrass site may simply range from isolated areas of mildly chlorotic plants to a slow recovery from other injury (Fig. 10.10) to larger-sized, random areas of severely chlorotic plants (Fig. 10.11) depending on host–nematode population combinations and environmental factors. The playing surface may remain intact and nematode-affected plants may recover over time, depending on environmental conditions and cultural inputs. The degree of chlorosis and necrosis of some grass species in mixed swards may be due to differential plant nematode susceptibility between the turfgrass species and among cultivars.

Aboveground plant symptoms in the presence of moderate to relatively high numbers of root-feeding nematodes may appear as larger-sized, random areas of severely chlorotic and necrotic plants resembling turfgrass under summer stress (Fig. 10.12). The end effect is partial or total loss of the playing surface depending on host–nematode population combinations and environmental factors.

Belowground symptoms

Belowground symptoms of nematode-infected root systems may range from the presence of dark-coloured lesions, reduced root length, minimal branching and lack of root hairs to distorted root sections (e.g. stunting and galling) depending on nematode–host combination. Sunken red- to dark-coloured lesions observed along infected roots can be characteristic of the site of entry, feeding and reproduction of certain nematode species (e.g. *Pratylenchus* or lesion nematode) or root-infecting fungi (Davis *et al.*, 1994).

Signs of the sedentary root-knot nematode (*Meloidogyne* sp.) may be readily confirmed by the presence of nematode galls located along

Fig. 10.10. Slight chlorosis caused by root-feeding nematodes on a creeping bentgrass golf green. (Courtesy of Jyri Kaapro.)

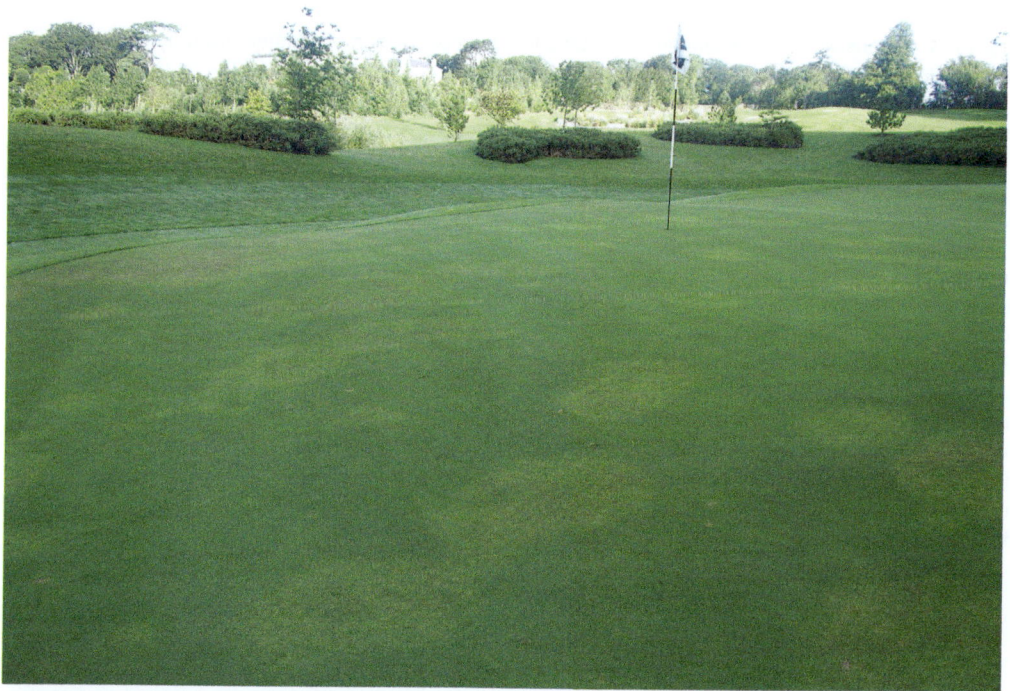

Fig. 10.11. Widespread and severely chlorotic and necrotic areas caused by the root-knot nematode (*Meloidogyne minor*) on a creeping bentgrass golf green. (Courtesy of Colin Fleming.)

Fig. 10.12. Extensive chlorosis primarily attributed to a high population of southern sting nematodes on a golf fairway in Australia. (Courtesy of Jyri Kaapro.)

infected roots (Fig. 10.13). Galls are swollen regions of abnormal plant cells acting as energy sinks in which the female nematode gains nutrition.

Similar root galling is characteristic of the root-gall nematode (*S. radicicola*) on annual bluegrass roots (Fig. 10.14). All root-feeding nematodes reduce the size and volume of roots and their normal functioning which can be expressed as plant wilting and chlorosis. Colonization of infected roots by relatively weak bacterial and fungal pathogens may occur because of physical damage to host roots, reduced turfgrass health and vigour; thus complicating symptom expression, diagnosis and management actions.

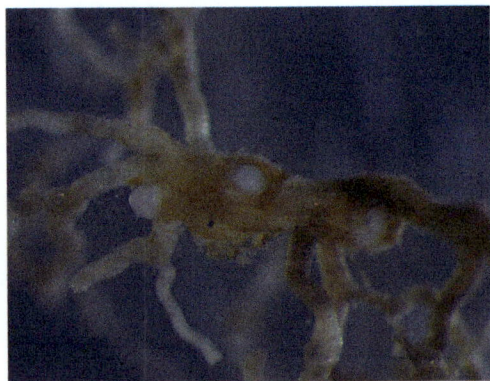

Fig. 10.13. Galling caused by the root-knot nematode (*Meloidogyne minor*) to a creeping bentgrass root. (Courtesy Colin Fleming.)

Confirmation of root-feeding nematodes

Confirmation of the presence of damaging levels of root-feeding nematodes requires a soil or plant nematode assay. A nematode assay is the process whereby plant or soil samples are prepared in the laboratory for nematode extraction, microscopic identification and then counting of numbers present.

Fig. 10.14. Damage caused by the root-gall nematode (*Subanguina radicicola*) to annual bluegrass roots. (Courtesy of Colin Fleming.)

Fig. 10.15. Example of a root-feeding nematode extraction system. (Courtesy of Colin Fleming.)

Diagnostic laboratories use one or more techniques (Fig. 10.15) when physically separating nematodes, soil or plant samples in a water solution. It is critical to understand all nematodes present are not extracted. Each extraction method (e.g. Baermann funnel, Whitehead tray or centrifugal sugar-flotation) has an associated extraction efficacy in recovering a percentage of the total number of plant-parasitic nematodes present. Another method called mist chamber extraction can be used to enumerate root-knot nematode populations more accurately (Crow, 2021).

A nematode assay normally reports multiple nematode genera and the numbers present based on a volume or weight of soil or roots. The validity of any nematode assay is predicated on the adopted extraction method and the accuracy in the identification of the nematode genera (species) and numbers of each present.

Most nematode assays conducted by commercial diagnostic laboratories for monitoring and problem-solving purposes currently assign nematodes only to a particular genus based on published morphological criteria (e.g. Eisenback, 2002). A high level of taxonomic skill, high-resolution microscopes and considerable time are required. Technological advances in the molecular taxonomy of nematodes using genetic-based methodologies (e.g. DNA sequencing and others) allows for greater accuracy to identify and quantify populations (e.g. Nisa *et al.*, 2022).

Differences in nematode population counts because of extraction efficiencies between the various nematode assay methods have practical significance when assessing nematode counts against published nematode damage thresholds if using different diagnostic laboratories. Population counts of free-living nematodes (i.e. bacterial and fungal feeders) are useful to assess those that may be antagonistic against root-feeding nematodes for the purpose of assessing the overall health of turfgrass rootzones. High populations of free-living nematodes can be indicative of a healthy rhizosphere.

Systems Approach to Parasitic Nematode Management

Key to effectively managing root-feeding nematodes is maintaining and encouraging a voluminous and functional root system (Fig. 10.16). Highly

Fig. 10.16. A highly functional root system on a mature creeping bentgrass golf green. (Courtesy of Gary W. Beehag.)

> **Box 10.3.** Integrated systems approach to nematode management.
>
> - Biosecurity and surveillance programme
> - Frequent monitoring and periodic sampling
> - Selection of resistant turfgrass varieties
> - Sanitation and biosecurity procedures
> - Cultural practices (mowing, nutrition, irrigation)
> - Proven organic-based products
> - Proven microbial antagonists
> - Chemical pesticides (nematicides, bionematicides)

functional, turfgrass root systems can tolerate relatively high populations of root-feeding nematodes and can be maintained in high-sand-content rootzones without significant injury or surface disfiguration.

Effective, long-term management of root-feeding nematode populations below damaging levels necessitates an integrated systems approach (Box 10.3) aimed at maintaining a functional root system.

Surveillance and biosecurity programmes

Inadvertent transport of infected turfgrass has raised the possibility of the introduction of plant-parasitic nematodes into several countries (Oka *et al.*, 2004; Fleming *et al.*, 2015; Stirling *et al.*, 2021). Implementation of a well-conceived biosecurity and surveillance programme against root-feeding nematodes and other plant pests would identify potential sources of introduction and limit their unintentional transfer to non-infested regions.

Consideration is required for thorough cleansing of aerification machinery having the ability to transfer turfgrass fragments, soil and root-feeding nematodes. Paying attention to the regions from where sod and construction and topdressing sand are purchased further forms part of biosecurity and surveillance effort.

Nematode population monitoring and sampling

Monitoring and sampling for the purpose of a nematode assay requires careful consideration given the seasonal variations and spatially clustered rather than random distribution of nematode populations. Routine sampling for the purpose of merely monitoring nematode numbers should commence in spring and continue into late summer, depending on nematode numbers and regional weather conditions.

Sampling in non-affected and in areas exhibiting decline should be conducted and placed in separate, labelled bags in order to compare nematode populations. Sampling in areas of dead plants will yield very low nematode populations. The procedure for sampling of root-feeding nematode populations involves extracting multiple cores (10–20) from the site to a uniform depth of the rootzone using an appropriate device (Fig. 10.17) having a core diameter of 1.0–2.0 cm. Samples from affected regions can be mixed and placed in a sealed, plastic bag. Sampling allows the opportunity to assess root depth and branching.

Wrapping the composite sample in moistened paper or adding a small volume of water in the bag minimizes sample desiccation. Collected samples must be kept in a cool environment and dispatched for next-day delivery to the diagnostic laboratory. Using the same laboratory and nematode assay method minimizes the risk of misinterpretation of nematode numbers.

Nematode-resistant varieties

Resistance is the ability of the plant host to reduce nematode reproduction and tolerance is ability of the host to withstand nematode feeding damage. On one hand, a turfgrass species (cultivar) is said to be a good host (tolerance) by supporting and withstanding high populations of feeding nematodes. On the other, a poor host (resistance) is a turfgrass species that only supports a relatively low nematode population.

Selection of improved turfgrasses to withstand high populations of root-feeding nematodes is problematic given multiple nematode species are normally present. Slight differences in susceptibility among cool-season and warm-season grass cultivars to specific root-feeding nematode species have been demonstrated experimentally under controlled conditions (Table 10.6). Evaluations have focused primarily on bermudagrass and St. Augustinegrass cultivars against lance, spiral and sting nematodes (Schwartz *et al.*, 2010; Pang *et al.*, 2011).

Possession of an extensive root system in terms of depth, diameter and degree of branching appear to be important morphological characteristics of turfgrass tolerance to root-feeding nematodes.

Cultural practices

Numerous cultural practices can be co-implemented to mitigate against high populations of root-feeding nematodes when establishing and maintaining turfgrass swards (Table 10.7) depending on the turfgrass site.

A healthy rhizosphere that encourages an adequate moisture–oxygen balance and an active, functional root system enables improved tolerance to the effects of root-feeding nematodes. The combination of increased mowing height and reduced mowing frequency has been widely stated as an important practice to prevent plant decline caused by root-feeding nematodes. Moderation of the mowing regime is practical only for relatively short periods on closely mown bowling and golf greens. More frequent rolling

Fig. 10.17. Examples of appropriate sampling devices, one having screw-in cores of multiple diameters (left) and the other having variable depth with cut-out side section (right). (Courtesy of Gary W. Beehag.)

Table 10.6. Reported nematode tolerance studies among turfgrass species.

Turfgrass	References
Bermudagrass Creeping bentgrass Seashore paspalum St. Augustinegrass Zoysiagrass	Johnson (1970); Henn and Dunn (1989); Busey *et al.* (1991); Giblin-Davis *et al.* (1992, 1995); Walker and Martin (2001); Hixson *et al.* (2004); Settle *et al.* (2007); Schwartz *et al.* (2010); Pang *et al.* (2011); Aryal *et al.* (2015); Choi *et al.* (2022)

in conjunction with mowing at a higher height is an option on bowling and golf greens.

Maintaining balanced NPK nutrition and adequate soil moisture are key to maximizing turfgrass root growth during conditions highly favourable for the specific cool-season or warm-season grass species. Minimizing or avoiding pre-emergent herbicides and plant-growth regulators in spring in areas known to have problematic nematode populations needs consideration to avoid phytotoxicity to turfgrass root growth.

Organic amendments and composts

The overall benefits of organic amendments through 'organic matter-mediated suppression' of plant-feeding nematodes has been extensively documented in cropping systems (e.g. McSorley, 2011; Stirling, 2014; Rosskopf *et al.*, 2020). Organic amendments and composts are widely used in turfgrass systems, and an ever-increasing number of formulations continue to be commercialized worldwide (Table 10.8). Applications of organic amendments must be targeted to the rootzone either being incorporated during pre-establishment or by thorough incorporation down open aerification holes post-establishment. The benefits of organic amendments for nematode suppression must be considered over the longer term.

Release of simple chemicals (e.g. ammonia-nitrogen, organic acids), pre-formed nematicidal compounds (e.g. alkaloids, phenolics) and stimulation of microbial antagonists (e.g. bacteria) as well as increased populations of free-living nematodes have been proposed explaining root-feeding nematode reductions following organic amendment application (e.g. Jones *et al.*, 2020). The biological and biochemical quality, application frequency and rate, effectiveness of incorporation into the rootzone and unknown impacts from turfgrass practices (e.g. chemical and pesticide regime) all potentially come into play leading to inconsistent and unpredictable outcomes.

Botanical and arthropod extracts

Secondary metabolites (allelochemicals) biosynthesized in certain plant families and having suppressive microbial activity continue to attract interest as potential compounds against plant-parasitic nematodes (Chitwood, 2002). Several purified, single-plant-based extracts have been demonstrated experimentally to possess degrees of antagonism or suppression towards plant-feeding nematodes (Table 10.9). Maintaining effective concentrations of these compounds following rootzone application remains challenging.

Many botanical-based products comprising two or more compounds are marketed as plant-growth promotants or 'plant biostimulants' implying suppression of plant-parasitic nematodes (Crow, 2008). Sequential applications of seaweed

Table 10.7. Pre- and post-establishment practices in managing root-feeding nematodes.

Phase	Cultural practices
Pre-establishment	Selection of rootzone media
	Sanitation procedures
	Organic amendments
	Plant extracts
	Microbial inoculants
Post-establishment	Mowing and rolling frequency
	Nutrition programme
	Irrigation and wetting agents
	Aerification, topdressing and dusting
	Organic amendments
	Plant extracts
	Microbial inoculants

Table 10.8. Common organic amendments.

Origin	Product examples	References
Farmyard manures	Cattle, horse, pigs and poultry	Stirling (2014); Rosskopf *et al.* (2020)
Plant residues	Peat moss, sawdust, sugarcane, woodchip, kelp	
Composts	Municipal green waste	
Industrial wastes	Biosolids, paper, abattoir by-products	
Seafood wastes	Molluscs and shellfish	

extracts have been demonstrated experimentally to provide suppression of root-feeding nematodes through increased creeping bentgrass root growth (Sun et al., 1997). The precise biological mode of action and nematicidal activity of most allelochemicals remain unclear.

Microbial inoculants and parasitic organisms

A suite of non-pathogenic microbes (e.g. plant-growth-promoting rhizobacteria, endophytic and mycorrhizal fungi) and associated nematicidal metabolites continue to be evaluated against root-feeding nematodes in cropping systems (Table 10.10). Limited species and strains of non-pathogenic microbes have been evaluated under controlled conditions against certain species of root-feeding nematodes in turfgrass systems (Crow, 2008; Groover et al., 2020).

Formulations may comprise single or multiple products based on bacterial and fungal species or their synthesized derivatives. Many products are combinations of simple chemicals and complex compounds. Utilization of endospore-forming bacteria (e.g. *Pasteuria* spp.) may offer the most promise to suppress certain species of root-feeding nematode populations (Rungrassamee et al., 2003).

Factors limiting the effectiveness of microbial inoculants against root-feeding nematode populations are the nematode species and population present, nematode feeding habit (e.g. endoparasitic lifestyle), life stages (e.g. cysts) and high reproduction rates. Efficacy of many proprietary products against root-feeding nematodes has been shown to be inconsistent. The challenge has been to encourage and maintain high populations of antagonists and parasites at sufficient populations in the long term to suppress root-feeding nematodes.

Nematicides and bionematicides

Chemical nematicides and an increasing range of bionematicides remain important components in managing high populations of root-feeding nematodes. Observable responses of turfgrass foliage and root growth following a timely nematicide treatment to small areas often provide the only clear indication of the presence of damaging nematode populations (Fig. 10.18). Measurable reductions in nematode populations do not always occur, especially for nematode species with relatively high reproductive rates and short life cycles.

Each nematicide and bionematicide product has strengths and limitations and no one product has absolute efficacy against all species of root-feeding nematodes. Certain formulations are applied preventively (i.e. pre-planting) and others curatively (i.e. post-planting).

How nematicides and bionematicides work

The exact biochemical mode of action of many nematicides and most bionematicides is

Table 10.9. Examples of botanical and animal extracts against root-feeding nematodes.

Biological origin	Plant/animal species	References
Indian neem	*Azadirachta indica*	Ntalli and Carboni (2012)
Lantana	*Lantana camara*	Cox et al. (2007)
Marine algae	*Ascophyllum nodosum*	Sun et al. (1997)
Wild mustard	*Brassica juncea*	Handiseni et al. (2017)
Shellfish	Crab, lobster	Stirling (2014)

Table 10.10. Examples of antagonistic and predatory microbes.

Microbial agent	Genera	References
Bacteria	*Bacillus, Pasteuria*	Luc et al. (2011); Crow (2014)
Fungi	*Myrothecium, Paecilomyces, Trichoderma*	Stirling (2014); Wilson and Jackson (2019); Poveda et al. (2020)

Fig. 10.18. Response to a nematicide application (left) to none (right) on a golf green in the United States. (Courtesy of Nathan R. Walker.)

imprecisely known. Exceptions are synthetic nematicides having a dual registration as fungicide, insecticide or miticide. Nematicide is a general term encompassing both chemical compounds that will directly kill nematodes (i.e. truly nematicidal) and other compounds that only disrupt nematode behaviour for a variable time period (i.e. nemastatic). True nematicides are quick-acting biocides and many are formulated for turfgrass application (Table 10.11) in certain countries and states.

Fumigants are broad-spectrum biocides formulated as liquefied gas, volatile liquids or granular solids. Liquefied gases vaporize on release, moving passively as a gas through the soil pore spaces. Volatile liquids (e.g. 1,3-dichloropropene) readily convert to a gaseous state once in the rootzone. Solid particles (e.g. dazomet) are microgranules which on application release methyl isothiocyanate under moist soil conditions. Fumigant nematicides penetrate the nematode body wall and once inside affect various physiological functions of the nematode. Nematode death generally occurs relatively quickly.

Non-fumigant nematicides are either granules or liquids. Non-fumigant nematicides also penetrate the body wall of nematodes and are assumed to affect their primitive nervous system. Ultimately, nematodes are affected by contact action.

Bionematicides are derived from plant extracts, bacteria or fungi or their metabolites (Table 10.12). These microorganisms or their nematicidal metabolites induce resistance through various mechanisms (e.g. parasitism or the production of enzymes) against root-feeding nematodes (Poveda *et al.*, 2020). Availability and registration of turfgrass nematicides and bionematicides continually change. Turfgrass managers are strongly advised to consult product distributors and product labels.

Factors affecting nematicide and bionematicide efficacy

Cessation of feeding, life-stage development and reproduction rate of root-feeding nematodes

Table 10.11. Categorization of nematicides.

Origin	Method of application	Chemical name	References
Synthetic	Fumigant	1,3-Dichloropropene Metam-sodium Dazomet Ethanedinitrile	Park *et al.* (2014); Crow *et al.* (2017)
	Non-fumigant	Abamectin Fluensulfone Fluopyram Fluazaindolizine Oxamyl	

Table 10.12. Categorization of bionematicides.

Origin	Example	References
Botanical	*Allium sativum* (garlic) *Brassica juncea* (mustard) *Eugenia caryophillata* (clove)	Migunova and Sasanelli (2021); Sasanelli *et al.* (2021)
Bacterial	*Azotobacter* spp. *Bacillus* spp. *Burkholderia* spp. *Pasteuria* spp. *Pseudomonas* spp. *Rhizobacteria* spp. *Serratia* spp.	
Fungal	*Myrothecium verrucaria* *Pochonia chlamydospora* *Purpureocillium lilacinum*	

effected by a nematicide or bionematicide treatment is governed by specific circumstances (Box 10.4). The lethal effect of nematicides is primarily determined by two components, concentration and exposure time.

The residual concentration and exposure time of nematodes to non-fumigant nematicides are critical for product efficacy. Under conditions of low rootzone concentrations and short exposure time, effects of non-fumigant nematicides may be short-lived, allowing the affected nematode to recover partly or fully over time, depending on circumstances.

Concentrations of nematicides and bionematicides decrease from application over time, eventually falling below the lethal or effective levels. Not all nematode species are equally susceptible to a given nematicide, nor are all life stages of a given nematode equally sensitive given the same exposure concentration and/or time (Noling, 2019). Bionematicides are slow-acting products best applied as a preventive application.

Box 10.4. Chemical nematicide efficacy factors.

- Nematicide selectivity and water solubility
- Exposure period
- Nematicide concentration
- Nematode species and life stage

Maximizing nematicide and bionematicide efficacy

Fumigants used as pre-plant nematicides are applied with the objective of reducing nematode populations prior to turfgrass establishment. Fumigants are prone to atmospheric loss (i.e. volatilization) and the treated surface must be covered following application. Most fumigants have maximum efficiency in the temperature range of 7–27°C (Haydock *et al.*, 2006). Vertical penetration of fumigants is maximized in sand-based rootzones with minimal moisture.

Non-fumigant or post-planting nematicides are applied to existing turfgrass and are subject to

Table 10.13. Factors, potential issues and management practices influencing nematicide outcomes.

Factor	Potential issue	Management practice
Nematicide formulation	Fumigant – volatilization	Post-application coverage
	Non-fumigant – photodecomposition	Post-application irrigation
Nematicide incorporation	Non-uniform distribution	Even and correct application
Thatch-mat accumulation	Adsorption of non-fumigants	Organic matter minimization
Sand hydrophobicity	Water-repellent rootzone	Proven soil-wetting agents
Rootzone porosity	Fumigant concentration/exposure	Aerification technologies
Irrigation water volume	Nematicide vertical movement	Correct water volume
Developing resistance	Reduced efficacy	Rotation of products
Enhanced biodegradation	Reduced efficacy	Correct application

Table 10.14. Solubility of turfgrass nematicides.

Name of nematicide	Water solubility (mg/l) at 20°C	Reference
1,3-Dichloropropene	2,450	PPDB (2021)
Dazomet	3,500	
Metam-sodium	578,290	
Abamectin	10	
Fluensulfone	545	
Fluopyram	16	
Furfural	78,1000	
Oxamyl	14,100	

numerous variables (Table 10.13). Minimization of thatch-mat density and thickness and adequate volume of post-application water are critical to ensure successful vertical movement of a nematicide or bionematicide treatment to the rootzone.

Effective rootzone concentration of applied nematicides and bionematicides will be governed partly by post-application water and the solubility of the product applied (Table 10.14). Post-application water volumes and need for additional wetting agents are often stated on nematicide and bionematicide labels. Avoid application of highly soluble formulations if rainfall events are expected. Water volume has been shown experimentally to impact the vertical movement of applied *Pasteuria* sp. endospores (Luc *et al.*, 2011).

Nematicide resistance and biodegradation

Development of nematicide resistance against certain root-feeding nematode species remains real under conditions of excessive application regimes and using products having a dual registration (i.e. nematicide and miticide or fungicide). Development of enhanced biodegradation (i.e. pesticides or their metabolites being degraded by soil-borne microbes at an accelerated rate resulting in loss of pesticidal activity) has been documented in turfgrass systems for the deregistered nematicide fenamiphos (Beehag, 1995). Minimization or avoidance of potential development of nematicide resistance and enhanced biodegradation can be achieved through appropriate management strategies, based on nematicide label statements.

References

Anon. (2007) *The Root Gall Nematode (Subanguina radiciola)*. ASBINI Fact Sheet No. 1, Version March 2007. Agri-Food and Bioscience Institute, Belfast, UK.

Aryal, S.K., Crow, W.T., McSorley, R., Giblin-Davis, R.M., Rowland. D.L., *et al.* (2015) Effects of infection by *Belonolaimus longicaudatus* on rooting dynamics among St. Augustinegrass and bermudagrass genotypes. *Journal of Nematology* 47, 322–331.

Back, M.A., Haydock, P.P.J. and Jenkinson, P. (2002) Disease complexes involving plant parasitic nematodes and soilborne pathogens. *Plant Pathology* 51, 683–697.

Bekal, S. and Becker, J.O. (2000) Population dynamics of the sting nematode in California turfgrass. *Plant Disease* 84, 1081–1084.

Beehag, G.W. (1995) A review of enhanced biodegradation of fenamiphos in turf in Australia. In: *Proceedings of the 2nd Turf Research Conference, Sydney, Australia*. Australian Turfgrass Research Institute, Sydney, Australia, pp. 61–65.

Beehag, G.W. and McMaugh, P. (2018) An investigation of thatch-mat and root architecture beneath new bentgrass (*Agrostis* spp.) putting greens. In: *Proceedings of the Australasian Turfgrass Conference & Trade Exhibition 2018, Wellington, New Zealand, 25–28 June 2018* [USB stick]. Australian Golf Course Superintendents Association, Melbourne, Australia and New Zealand Golf Course Superintendents Association, Auckland, New Zealand.

Brodie, B.B. and Quattlebaum, B.H. (1970) Vertical distribution and population fluctuations on three nematode species as correlated with soil temperature, moisture and texture (abstract). *Phytopathology* 60, 1286.

Buckley, R.J., Koppenhofer, A.M. and Tirpak, S. (2010) *An Integrated Approach to Pest Management in Turfgrass: Nematodes*. Fact Sheet No. FS1014. Rutgers University Cooperative Extension, New Brunswick, New Jersey.

Busey, P., Giblin-Davis, R.M., Riger, C.W. and Zaenker, E.J. (1991) Susceptibility of diploid St Augustinegrasses to *Belonolaimus longicaudatus*. *Journal of Nematology* 23(4S), 604–610.

Chitwood, D.J. (2002) Phytochemical based strategies for nematode control. *Annual Review of Phytopathology* 40, 221–249.

Choi, C.J., Valiente, J., Schiavon, M., Dhilion, B. and Crow, W.T. (2022) Bermudagrass cultivars with different tolerance to nematode damage are characterised by distinct fungal but similar bacterial and archaeal microbiomes. *Microorganisms* 10, 457. DOI: 10.3390/microorganisms10020457

Couch, H.B. (1995) *Diseases of Turfgrasses*, 3rd edn. Krieger Publishing Company, Malabar, Florida.

Cox, C.J., McCarty, B. and Martin, S.B. (2007) Suppressing sting nematodes using botanical extracts. *Golf Course Management* 79, 94–97.

Crow, W.T. (2008) Understanding and managing plant-parasitic nematodes on turfgrass. In: Pessarakli, M. (ed.) *Handbook of Turfgrass Management and Physiology*. CRC Press, Boca Raton, Florida, pp. 351–388.

Crow, W.T. (2014) Turfgrass nematicide and bionematicide research in Florida. *Outlooks on Pest Management* 25, 222–225. doi: 10.1564/v25_jun_08

Crow, W.T. (2018) *Sting Nematode Belonolaimus longicaudatus Rau (Nematoda: Secernentea: Tylenchida: Tylenchina: Belonolaimidae: Belonolaiminae)*. Leaflet No. EENY618. University of Florida IFAS Extension, Gainesville, Florida.

Crow, W.T. (2021) Root-knot nematode on warm-season turf: chasing the ghost. *Golf Course Management* 89, 88–93.

Crow, W.T., Porazinska, D.L., Giblin-Davis, R.M. and Grewal, P.S. (2006) Entomopathogenic nematodes are not an alternative to fenamiphos for management of plant-parasitic nematodes on golf courses in Florida. *Journal of Nematology* 38, 52–58.

Crow, W.T., Luc, J.E., Sekora, N.S. and Pang, W. (2013) Interaction between *Belonolaimus longicaudatus* and *Helicotylenchus pseudorobustus* on bermudagrass and seashore paspalum hosts. *Journal of Nematology* 45, 17–20.

Crow, W.T., Becker, J.O. and Baird, J.H. (2017) New golf course nematicides. *Golf Course Management* 7, 66–71.

Davis, R.F., Wilkinson, H.T. and Noel, G.R. (1994) Root growth of bentgrass and annual bluegrass as influenced by coinfection of *Tylenchorhynchus nudus* and *Magnaporthe poae*. *Journal of Nematology* 26, 86–99.

Dong, Y., Jin, P., Zhang, H., Hu, J., Lamour, K. and Yang, Z. (2022) Distribution and prevalence of plant-parasitic nematodes of turfgrass at golf courses in China. *Biology* 2022, 11, 1322. DOI: 10.3390/biology11091322

Eisenback, J.D. (2002) *Identification Guides for the Most Common Genera of Plant-Parasitic Nematodes*. Mactode Publications, Blacksburg, Virginia.

Entwistle, K., Fleming, T., Kerr, R., Maule, A., Martin, T., *et al.* (2014) Biosecurity and emerging plant health problems in turf production and maintenance. *European Journal of Horticultural Science* 79, 108–115.

Fleming, C., Craig, D., Hainon-McDowell, M. and Entwistle, K. (2008) Plant parasitic nematodes: a new turf war. *Biologist* 55, 76–82.

Fleming, C., Kerr, R., Moreland, B.P., Maule, A.G. and Entwistle, K. (2010) Plant parasitic nematodes in European turfgrass. In: *Proceedings of the 2nd European Turfgrass Society Conference, Angers, France, 11–14 April 2010*. European Turfgrass Society, Livorno, Italy, pp. 75–77.

Fleming, T.R., Maule, A.G., Martin, T., Hainon-McDowell, M., Entwistle, K., et al. (2015) A first report of *Anguina pacificae* in Ireland. *Journal of Nematology* 47, 97–104.

Fushtey, S.G. and McElroy, F.D. (1977) Plant parasitic nematodes in turfgrass in southern British Columbia. *Canadian Plant Disease Survey* 57, 54–56.

Giblin-Davis, R.M., Cisar, J.L., Bilz, F.G. and Williams, K.E. (1992) Host status of different bermudagrass (*Cynodon* spp.) for sting nematode, *Belonolaimus longicaudatus*. *Journal of Nematology* 24, 749–756.

Giblin-Davis, R.M., Busey, P. and Center, B.J. (1995) Parasitism of *Hoplolaimus galeatus* on diploid and polyploid St Augustinegrass. *Journal of Nematology* 27, 472–477.

Groover, W., Held, D., Lawrence, K. and Carson, K. (2020) Plant growth-promoting rhizobacteria: a novel management strategy for *Meloidogyne incognita* on turfgrass. *Pest Management Science* 76, 3127–3138. DOI: 10.1002/ps.5867

Handiseni, M., Cromwell, W., Zidek, M., Zhou, X.-C. and Young, K.-J. (2017) Use of Brassicaceae seed meal extracts from managing root-knot nematode in bermudagrass. *Nematropica* 47, 55–62.

Hannan, B. (2018) Controlling dagger nematode. *New Zealand Turf Management Journal* 34, 24–27.

Haydock, P.P.J., Woods, S.R., Grove, I.G. and Hare, M.C. (2006) Chemical control of nematodes. In: Perry, R.N. and Moens, M. (eds) *Plant Nematology*, 2nd edn. CAB International, Wallingford, UK, pp. 392–459.

Henn, R.A. and Dunn, R.A. (1989) Reproduction of *Hoplolaimus galeatus* and growth of seven St Augustinegrass (*Stenotaphrum secundatum*) cultivars. *Nematropica* 19, 81–87.

Hixson, A.C., Crow, W.T., McSorley, R. and Trenholm, I.E. (2004) Host status of 'Sealsle 1' seashore paspalum (*Paspalum vaginatum*) to *Belonolaimus longicaudatus* and *Hoplolaimus galeatus*. *Journal of Nematology* 36, 493–498.

Howard, D. and Ormsby, D. (2017) 40 years of nematodes. *New Zealand Turf Management Journal* 34, 24–27.

Inguagiato, J.C. and Martin, S.B. (2015) Diseases of cool- and warm-season putting greens. *USGA Green Section Record* 53, 1–19.

Johnson, A.W. (1970) Pathogenicity and interaction on three nematode species on six bermudagrasses. *Journal of Nematology* 2, 36–41.

Jones, W.B., Kruse, J.K., Enloe, A. and Crow, W.T. (2020) Effects of pre-planting incorporation or post-planting top-dressing of organic amendments on bermudagrass for tolerance to *Belonolaimus longicaudatus*. *Nematropica* 50, 59–66.

Jordan, K.S. and Mitkowski, N.A. (2006) Population dynamics of plant-parasitic nematodes in golf course greens in southern New England. *Plant Disease* 90, 501–505.

Kaapro, J. (2020) Nematodes population dynamics and soil testing. *New South Wales Golf Course Superintendents' Association Newsletter* (Summer), 12–15.

Knight, K.W., Barber, C.J. and Page, G.D. (1997) Plant-parasitic nematodes of New Zealand recorded by host association. *Journal of Nematology* 29(4S), 640–656.

Luc, J.E., Crow, W.T., Pang, W., McSorley, R. and Giblin-Davis, R.M. (2011) Effects of irrigation, thatch, and wetting agent on movement of *Pasteuria* sp. endospores in turf. *Nematropica* 41, 185–190.

McCarty, L.B. (2010) *Best Golf Course Management Practices: Construction, Watering, Fertilizing, Cultural Practices and Pest Management Strategies to Maintain Golf Course Turf with Minimal Environmental Impact*, 3rd edn. Prentice-Hall, Upper Saddle River, New Jersey.

McClure, P.C., Schmidt, M.A. and McCullough, M.D. (2008) Distribution, biology and pathology of *Anguina pacificae*. *Journal of Nematology* 40, 226–239.

McSorley, R. (2011) Overview of organic amendments for management of plant-parasitic nematodes, with case studies from Florida. *Journal of Nematology* 43, 69–81.

Migunova, V.D. and Sasanelli, N. (2021) Bacteria as biocontrol tool against phytoparasitic nematodes. *Plants* 10, 389. DOI: 10.3390/plants10020389

Mitkowski, N.A. and Jackson, N. (2003) *Subanguina radiciola*, the root-gall nematode infesting *Poa annua* in New Brunswick, Canada. *Plant Disease* 87, 1263.

Mwamula, A.O. and Lee, D.W. (2021) Occurrence of plant-parasitic nematodes of turfgrass in Korea. *Plant Pathology Journal* 37, 446–454. DOI: 10.5423/PPJ.OA.04.2021.0059

Neylan, J. (1996) *Nematode Types and Levels in Bowling Greens*. HRDC Final Report No. TU201. Horticultural Research & Development Corporation, Gordon, Australia.

Nisa, R.U., Tantray, A.Y. and Shah, A.A. (2022) Shift from morphological to recent advanced molecular approaches for the identification of nematodes. *Genomics* 114, 110295. DOI: 10.1016/j.ygeno.2022.110295

Noling, J.W. (2019) *Movement and Toxicity of Nematodes in the Plant Rootzone*. Leaflet No. ENY-041. University of Florida IFAS Extension, Gainesville, Florida.

Ntalli, N.G. and Carboni, P. (2012) Botanical nematicides: a review. *Journal of Agricultural and Food Chemistry* 60, 9929–9940. DOI: 10.1021/jf303107j

Oka, Y., Karssen, G. and Mor, M. (2004) First report of root-knot nematode *Meloidogyne marylandi* on turfgrass in Israel. *Plant Disease* 88, 309.

Onoda, M., Hirano, K., Amemiya, Y., Yazawa, Y. and Yaegashi, T. (1998) Root-knot nematode parasitising on bermudagrass from Japan (abstract in English). *Japanese Journal of Turfgrass Science* 26, 131–142.

Pang, W., Luc, J.E., Crow, W.T., Kennelly, K.E., Giblin-Davis, R.M., et al. (2011) Field responses of bermudagrass and seashore paspalum cultivars to sting and spiral nematodes. *Journal of Nematology* 43, 201–208.

Park, C.G., Son, J.-K., Lee, B.-H., Cho, J.H. and Ren, Y. (2014) Comparison of ethanedinitrile (C_2N_2) and metam sodium for control of *Bursaphelenchus xylophilus* (Nematoda: Aphelenchidae) and *Monochamus alternaus* (Coleoptera: Cerambycidae) in naturally infested logs at low temperature. *Journal of Economic Entomology* 107, 2055–2060. DOI: 10.1603/EC14009

Perry, V.G., Smart, G.C. and Horn, G.C. (1970) Nematode problems of turfgrasses in Florida and their control. *Proceedings of the Florida State Horticultural Society* 83, 489–492.

Poveda, J., Abril-Urias, P. and Escobar, C. (2020) Biological control of plant-parasitic nematodes by filamentous fungi, inducers of resistance: *Trichoderma*, mycorrhizal and endophytic fungi. *Frontiers in Microbiology* 11, 992. DOI: 10.3389/fmicb.2020.00992

PPDB (2021) Pesticide Properties Database. A to Z list of pesticide active ingredients. University of Hertfordshire. Hatfield, UK (accessed 12 March 2023).

Robbins, R.T. and Barker, K.R. (1974) The effects of soil type, particle size, temperature, and moisture on reproduction of *Belonolaimus longicaudatus*. *Journal of Nematology* 6, 1–6.

Rosskopf, E., Di Gioia, F., Hong, J.C., Pisani, C. and Kokalis-Burelle, N. (2020) Organic amendments for pathogen and nematode control. *Annual Review of Phytopathology* 58, 277–311.

Rungrassamee, W., Wick, R.L. and Dicklow, B. (2003) Relationship of *Pasteuria* to root-knot and stunt nematodes (abstract). *Phytopathology* 93, S135.

Ruscoe, P.E. and Stirling, G.R. (2020) Southern sting nematode (*Ibipora lolii*), a serious pest of turf grasses in Australia: a review of what can be learnt from research on *Belonolaimus longicaudatus*, a closely related pest of turfgrass and many crops in the United States. *Australasian Plant Pathology* 49, 493–504. DOI: 10.1007/s13313-020-00729-1

Sasanelli, N., Konrad, A., Migunova, V., Todera, I., Iurca-Straistaru, E., et al. (2021) Review on control methods against plant parasitic nematodes applied in southern member states (C Zone) of the European Union. *Agriculture* 11, 602. DOI: 10.3390/agriculture11070602

Schwartz, B.M., Kenworthy, K.E., Crow, W.T., Ferrell, J.A., Miller, G.M. and Quesenberry, K.H. (2010) Variable responses of zoysiagrass genotypes to the sting nematode. *Crop Science* 50, 723–729. DOI: 10.2135/cropsci2008.11.0672

Settle, D.M., Fry, J.D., Millikens, G.A., Tisserat, N.A. and Todd, T.C. (2007) Quantifying the effects of lance nematode parasitism in creeping bentgrass. *Plant Disease* 69, 939–943.

Simard, L., Belair, G., Powers, T., Tremblay, N. and Dionne, J. (2008) Incidence and population density of plant-parasitic nematodes on golf courses in Ontario and Quebec, Canada. *Journal of Nematology* 40, 241–251.

Siviour, T.R. and McLeod, R.W. (1979) Redescription of *Ibipora lolli* (Siviour 1978) comb. n. (Nematoda: Belonolaiminae) with observations on its host range and pathogenicity. *Nematologica* 25, 487–493.

Stirling, G.R. (2008) Managing nematode pests on turfgrass. *Australian Turfgrass Management Journal* 10, 44–48.

Stirling, G.R. (2014) *Biological Control of Plant-Parasitic Nematodes: Soil Ecosystem Management in Sustainable Agriculture*, 2nd edn. CAB International, Wallingford, UK.

Stirling, G.R., Stirling, A.M., Giblin-Davis, R.M., Ye, W., Porazinska, D.L., et al. (2013) Distribution of southern sting nematode, *Ibipora lolli* (Nematoda: Belonolaimidae), on turfgrass in Australia and its taxonomic relationship to other belonolaimids. *Nematology* 15, 401–415.

Stirling, G.R., Stirling, A.M. and Eden, L. (2021) Plant parasitic nematodes on turfgrass in Queensland, Australia, and biosecurity issues associated with interstate transfer and eradication of southern sting nematode (*Ibipora lolii*). *Australasian Plant Pathology* DOI: 10.1007/s13313-021-00820-1

Stynes, B. (1968) Parasitic nematode attacks Newcastle couch roots. *Bowls in New South Wales* (June), 23, 28.

Sun, H., Schmidt, R.E. and Eisenback, J.D. (1997) The effect of seaweed concentrate on the growth of nematode-infected bent grown under low soil moisture. *International Turfgrass Society Research Journal* 8, 1336–1343.

Swart, A., Marais, M. and Schoeman, A.S. (2000) Plant nematodes in South Africa 2. Golf-course putting greens. *African Plant Protection* 6, 35–39.

Tredway, L.P., Tomaso-Peterson, M., Kerns, J.P. and Clarke, B.B. (2023) *Compendium of Turfgrass Diseases*, 4th edn. The American Phytopathological Society, St. Paul, Minnesota.

Vandenbossche, B., Viaene, N., Sutter, N., De Maes, M., Karsen, G. and Bert, W. (2011) Diversity and incidence of plant-parasitic nematodes in Belgium turf grass. *Nematology* 13, 245–256.

Viaene, N., Coyne, D.C. and Davies, K.G. (2013) Biological and cultural management. In: Perry, R.N. and Moens, M. (eds) *Plant Nematology*, 2nd edn. CAB International, Wallingford, UK, pp. 383–410.

Walker, N.R. and Martin, D.L. (2001) Effects of *Tylenchorhynchus claytonia* on creeping bentgrass (abstract). *Phytopathology* 91, S93.

Walker, N.R. and Martin D.L. (2002) Effects of sand particle size on population of the ring nematode (abstract). *Phytopathology* 92, S84.

Wilson, M. and Jackson, T.A. (2019) Progress in the commercialisation of bionematicides. *Biocontrol* 58, 715–722.

Yeates, G.W. (1980) *Nematodes: are they a problem? Sports Turf Review* 130, 126–128. Turf Culture Institute, Inc., Palmerston North, New Zealand.

Yu, Q., Potter, J.W. and Gilby, G. (1998) Plant-parasitic nematodes associated with turfgrass in golf courses in southern Ontario. *Canadian Journal of Plant Pathology* 20, 304–307.

Zeng, Y., Ye, W., Martin, B., Martin, M. and Tredway, L. (2012) Diversity and occurrence of plant parasitic nematodes associated with golf course turfgrasses in North and South Carolina. *Journal of Nematology* 44, 337–347.

11

Understanding Turfgrass Protectant Pesticides

Abstract
This chapter focuses on the widening range of synthetic fungicide and novel biofungicide formulations available for turfgrass application. A confident understanding of how they work against the pathogens targeted and what happens once on the plant treated is required to know how best they are applied to maximize their efficacy. Adoption of tank mixtures using fungicide(s) or biofungicide(s) combined with one or multiple chemical products requires careful consideration to avoid potential unwanted effects.

An increasing range of turfgrass protectants formulated as fungicides and biofungicides continue to contribute significantly to the presentation of high-quality turfgrass worldwide. Newer-generation fungicides and novel biofungicides offer unique modes of action with greater pathogen selectivity and fewer environmental risks.

Fungicide resistance among several fungal pathogens has been responsible partly for the development of new chemistries. Biofungicides provide alternatives in countries where conventional fungicides are unavailable. More precise application of turfgrass protectants is possible because of advances in application technologies. Correct selection, placement and timing is fundamental to maximize fungicide and biofungicide efficacy.

Disclaimer

The authors have undertaken all reasonable efforts to provide accurate details of the contents of this chapter. Information contained in this chapter must only be used as a guide and does not endorse any individual product's performance.

No responsibility is taken by the authors for any damage caused by actual or implied suggestions.

Registration of Turfgrass Protectants

Registration of specific fungicides and biofungicides for turfgrass application is subject to change over time within all countries. Readers are strongly advised to consult current product labels to obtain correct application information and ensure the product is registered for a specific situation. Turfgrass fungicide and biofungicide labels are legally binding documents in most countries and states.

Turfgrass protectant label nomenclature

A fungicide or biofungicide label states the name of the manufacturer, chemical or active ingredient and common or trade name together with the concentration(s) of the active ingredient(s) either per volume (i.e. chemical) or population (i.e. microbial) basis.

Turfgrass Protectant Formulations

Turfgrass protectants are formulated as either a dry granular product or a sprayable solution depending on the active ingredient. Granular products comprise an active ingredient(s) coated on an inert solid material and applied in solid form. Sprayable solutions are formulated to be applied in a water solution (Box 11.1). Each formulation type has strengths and limitations.

Emulsifiable concentrates (EC) and emulsions in water (EW) are solvent-based liquids. Microemulsions (ME) and microemulsion concentrates (MEC) are liquids similar to emulsifiable concentrates but possess smaller-sized particles preventing separation and settling. Suspoemulsions (SE) comprise suspended solids and emulsion droplets. Soluble liquids (SL) consist of water-soluble active ingredients dissolved in water. Soluble liquids are true solutions requiring no agitation.

Flowables (F) or water-dispersible granules (WDG) and suspension concentrates (SC) have very fine particles suspended in water. These formulations are prone to settling, requiring constant agitation once in water. Wettable powders (WP) are finely ground particles similar to flowables and are solids suspended in water. Wettable powders tend to be dusty and prone to settling after dilution. Water-soluble packet (WSP) formulations are enclosed in a special plastic that reduces the chance of dust inhalation. The plastic is water-soluble releasing the active ingredient(s).

Formulation components

Fungicides and biofungicides are formulated with multiple components (Box 11.2) depending on the specific product and manufacturer.

Pesticide formulations contain one or more active ingredients and various adjuvants. The active ingredient(s) is the pesticidally active component(s). Adjuvant describes substances that alter a physical and/or chemical property to enhance product efficacy. Key adjuvants are surfactants, petroleum- or plant-derived oils, activators, and acidifying, buffering and compatibility agents. Certain adjuvants may assist in the retention and absorption of fungicides or assist to disguise the plant surface against colonization by foliar fungi (Walters, 2006).

Surfactants (surface-acting agents) improve the suspension of water-insoluble, active ingredients and are broadly categorized as emulsifiers, stickers or wetting agents. Refined petroleum- or plant-derived oils assist to reduce spray droplet evaporation and improve retention on treated foliage. Acidifying and buffering agents are either organic or inorganic acids that modify or maintain the tank solution pH. Compatibility agents prevent chemical and physical separation of the active ingredient(s) when in solution. The multitude of propriety adjuvants has practical implications when considering fungicide- or biofungicide-based tank mixtures. Readers are strongly advised to consult product labels.

How Fungicides and Biofungicides Work

Fungicides and biofungicides are categorized in various ways (Box 11.3). The combination of

---

Box 11.1. Sprayable plant protectant formulations.

- Emulsifiable concentrate (EC) and emulsion in water (EW)
- Microemulsion (ME) and microemulsion concentrate (MEC)
- Suspoemulsion (SE) and soluble liquid (SL)
- Flowable (F), water-dispersible granules (WDG) and suspension concentrate (SC)
- Wettable powder (WP) and water-soluble packet (WSP)

Box 11.2. Formulation components.

- Active ingredient
- Diluent
- Solvent
- Synergist
- Surfactant
- Antifreeze
- Antimicrobials
- Suspension agents

Box 11.3. Categorization of plant protectants.

Origin – synthetic or biologically derived.
Mode of action – how they affect plant pathogens.
Phytomobility – absorption and movement in turfgrass.

active ingredient(s), formulation and effect on plant physiological processes influences their efficacy. Understanding the biochemical effects on plant pathogens by fungicides and biofungicides and their degree of plant absorption is fundamental for effective turfgrass disease management.

Biochemical mode of action

Fungicides and biofungicides are categorized in accordance with their biochemical mode of action (MoA) as defined by their FRAC (Fungicide Resistance Action Committee) code. The biochemical mode of action is well known for most fungicides. FRAC classification has practical significance in minimizing fungicide resistance.

Fungicide mode of action describes how an active ingredient interferes with fungal growth and physiology at the molecular, cellular and whole-microbe level. Active ingredients of fungicides in the same mode of action category have a common target site in pathogens. Active ingredients categorized with the same number–letter combination target pathogens in the same way (Table 11.1). Mode of action categories having the same number but different letters have closely related target sites in the pathogen.

Fungicides are further categorized into two broad groups (Table 11.2). Multi-site fungicides or protectants affect many different biochemical processes in fungal cells. Single-site fungicides or curatives only affect a specific biochemical process in the target fungi which varies depending on the active ingredient.

Table 11.1. Fungicide group name and FRAC categories of turfgrass fungicides.

FRAC code	Fungicide group name	Relative risk of resistance	Common names
1	Methyl benzimidazole carbamates (MBCs)	High	Thiabendazole, thiophanate-methyl
2	Dicarboximides	Medium–high	Iprodione, procymidone, vinclozolin
3	Demethylation inhibitors (DMIs)	Medium	Difenoconazole, fenarimol, mefentrifluconazole, metconazole, myclobutanil, prochloraz, propiconazole, tebuconazole, triadimefon, triadimenol, triticonazole
4	Phenylamides	High	Mefenoxam, metalaxyl
7	Succinate dehydrogenase inhibitors (SDHIs)	Medium–high	Benzovindiflupyr, boscalid, fluopyram, flutolanil, fluxapyroxad, isofetamid, penthiopyrad, pydiflumetofen
11	Quinone outside inhibitors (QoIs)	High	Azoxystrobin, fluroxastrobin, mandestrobin, pyraclostrobin, trifloxystrobin
12	Phenylpyrroles	Low–medium	Fludioxonil
14	Aromatic hydrocarbons (AHs)	Low–medium	Chloroneb, etridiazole, quintozene, tolclofos-methyl
19	Polyoxins	Medium	Polyoxin D zinc salt
20	Phenylureas	Not known	Pencycuron
21	Quinone inside inhibitors (QiIs)	Medium–high	Cyazofamid
28	Carbamates	Low–medium	Propamocarb
29	Dinitroanilines	Low	Fluazinam
43	Benzamides	Medium	Fluopicolide
M03	Dithiocarbamates	Low	Mancozeb, thiram, zineb
M04	Phthalimides	Low	Captan
M05	Chloronitriles	Low	Chlorothalonil
P01	Benzothiadiazole (BTH)	Not significant	Acibenzolar-S-methyl
P07	Phosphanates	Low	Fosetyl-Al
BM02	Microbial	Not known	*Bacillus* spp., *Pseudomonas* spp., *Trichoderma* spp.

Table 11.2. Multi-site and site-specific turfgrass fungicides.

Category	Fungicide	Reference
Multi-site	Chlorothalonil, mancozeb, thiram	Latin (2021)
Single-site	Azoxystrobin, boscalid, chloroneb, cyazofamid, etridiazole, fenarimol, fludioxonil, fluopicolide, fluroxastrobin, flutolanil, fosetyl-Al, mefenoxam, metconazole, myclobutanil, phosphonic acid, polyoxin D, propamocarb, propiconazole, pyraclostrobin, quintozene, tebuconazole, thiophanate-methyl, triadimefon, trifloxystrobin, triticonazole, vinclozolin	

Two related terms describe how fungicides function. 'Fungistatic' products only inhibit fungal growth and are typically site-specific. 'Fungicidal' products kill the pathogens. For the purpose of this book, the term fungicide is adopted to cover both types of activity.

Box 11.4. Phytomobility terminology.

Contact – remains on a treated plant surface unless dislodged by water or mowing.
Acropetal penetrant – absorbed into a treated surface and translocated upwards in xylem.
Local penetrant – moves between adjacent cells to opposite side of the leaf.
Systemic – moves through cells from regions of high photosynthate concentration (mature leaves) towards regions of low concentration (roots and immature leaves).

Fungicide phytomobility

The degree to which fungicides penetrate and translocate in a treated plant is termed phytomobility (Box 11.4). Generally, fungicides in the same chemical group are likely to have the same phytomobility. Many older-generation fungicides (e.g. mancozeb) remain on treated foliage without epidermal penetration. Most newer-generation turfgrass fungicides are acropetal or local penetrants.

For a fungicide to have efficacy the targeted fungus must be actively growing. Contact fungicides have efficacy against germinating fungal spores and mycelium but have no impact once fungi are inside plant tissues. Penetrants offer an advantage in their ability to move into plant tissues to inhibit the targeted fungi provided adequate concentration of the active ingredient remains inside plant tissues.

A small percentage of certain acropetal penetrant fungicides may have some basipetal (downwards) movement in phloem tissues. Evidence of fungicide distribution inside treated turfgrass is limited. Azoxystrobin, propiconazole, pyraclostrobin and thiophanate-methyl applied to creeping bentgrass showed experimentally to have low concentrations in the sampled roots but pyraclostrobin was nearly undetectable (Hockemeyer and Latin, 2015). Few fungicides are truly systemic penetrants (Table 11.3) but do possess dual movement depending on plant species–active ingredient combination.

Preventive versus curative strategy

The two terms preventive (prophylactic) and curative (eradicant) describe two opposing disease management strategies. Lower rates and/or longer intervals between applications are stated on product labels when disease outbreaks have not occurred (i.e. pre-symptoms) but are expected. Higher rates and/or shorter intervals are stated when disease outbreaks have occurred (i.e. symptoms). In practice, fungicide applications particularly on highly managed turfgrass (e.g. bowling and golf greens) range somewhere between both strategies.

Product Application and Placement

Correct application and placement of turfgrass protectants is critical. Key decisions are how to apply (i.e. dry, solution or gas) and where to target (i.e. foliage or roots). Each placement poses technical and practical challenges.

Granular products can be applied using fertilizer-type spreading equipment. Sprayable solutions are applied using either small-capacity, compressed-air knapsacks or medium- to large-size, powered boomspray units (Fig. 11.1). Fumigants generally can only be applied by licensed operators. Injection-based technologies have been used experimentally for subsurface placement of specific products.

Foliar application and rootzone targeting

Turfgrass protectants may be applied directly on to the foliage or targeted to the rootzone. Foliar application and rootzone targeting of applied turfgrass protectants are both governed initially by wind speed. Minimization of spray drift can be managed by early-morning application

Table 11.3. Phytomobility of turfgrass fungicides.

Phytomobility	Fungicide	References
Contact	Anilazine, captan, chloroneb, chlorothalonil, etridiazole, fluazinam, mancozeb, quintozene, thiram	Clarke and Kenna (2000); Latin (2021)
Acropetal penetrant	Azoxystrobin, boscalid, fenarimol, fluopicolide, flutolanil, fluxapyroxad, fluroxastrobin, isofetamid, mefenoxam, metconazole, myclobutanil, penthiopyrad, propamocarb, propiconazole, tebuconazole, thiophanate-methyl, triadimefon, triticonazole	
Local penetrant	Cyazofamid, fludioxonil, iprodione, polyoxin D, pyraclostrobin, trifloxystrobin, vinclozolin	
Systemic penetrant	Fosetyl-Al, phosphonic acid	

Fig. 11.1. A turfgrass boomspray unit. (Courtesy of Gary W. Beehag.)

as wind speed is often low or use of enclosed boomspray technologies (Fig. 11.2). Correct travel speed and boom height are imperative. Information about maximum wind speeds during application may be stated on product labels.

Managing foliar application

Successful foliage application is associated with numerous issues (Box 11.5). One or more factors may play a greater role which may require decisions depending on circumstances. Attainment of uniform coverage on individual leaves is imperative and for this reason, smaller-sized water droplets provide greater uniformity and effectiveness.

Atmospheric influences

Phytotoxicity can be potentially problematic for certain fungicides applied during hot, drying conditions on sensitive turfgrasses (e.g. bentgrass). Biofungicides based on living microbes are highly susceptible to degradation in sunlight, requiring application early in the morning or during cool and cloudy conditions.

Extended periods of dew formation raise the question whether to remove dew formation or whether to apply to wet foliage (Fig. 11.3). Dew formation is often associated with early mornings without wind. Applications of chlorothalonil and propiconazole alone and in tank mixes against dollar spot on creeping bentgrass showed no significant differences when fungicides were sprayed in the morning in the presence and absence of dew (McDonald *et al.*, 2006).

Box 11.5. Foliar application factors.

- Wind speed and direction
- Ambient temperature, humidity and light intensity
- Presence of dew and leaf orientation
- Threat of rainfall and post-application drying time
- Mowing and clippings

Fig. 11.2. An enclosed boomspray system. (Courtesy of Gary W. Beehag.)

Fig. 11.3. Early-morning spraying in the presence of dew formation on a golf green. (Courtesy of Gary W. Beehag.)

The impact of dew formation at the time of fungicide and biofungicide applications remains unclear. Laterally orientated foliage, as opposed to more vertically orientated foliage, is more likely to retain water droplets containing active ingredients.

Fungicides or biofungicides may demand application under threat of or during periods of light rainfall. A minimum drying time is commonly stated on most fungicide labels; however modern formulations are designed to be rain-fast in less than that time. The time period for fungicide-treated foliage to become dry is variable subject to formulation, droplet size and water volume together with ambient temperature and light intensity; thus difficult to precisely determine. This scenario is particularly relevant for contact-type fungicides and whether a second application is warranted.

Several investigations have shown simulated rain applied 30–60 min following fungicide application resulted in significant but variable reductions in the amount of each fungicide retained on foliage (e.g. Carroll *et al.*, 2001; Pigati *et al.*, 2010). Overall, these findings strongly indicate that fungicides can be adversely affected by rainfall and the inclusion of an effective 'sticking' agent will assist in disease reduction.

Influence of mowing and rolling

Timing of mowing and rolling with allowance of drying time for foliage applications are key factors, particularly on bowling and golf greens. Application of a foliar-targeted product on to turfgrass not immediately mown means greater leaf surface, thus increased potential for foliar absorption. Turfgrass mown without catchers immediately prior to treatment presents a greater physical barrier for vertical movement of applied products, even granules.

Managing rootzone placement

Once the protectant product has been applied additional issues influencing the vertical movement of rootzone-targeted products come into play (Box 11.6). Direct placement by subsurface injection technology of certain plant protectant products to the rootzone overcomes deposition factors associated with surface-applied products. Light rainfall events subsequent to product application are beneficial, aiding the vertical movement of applied products. Programming of certain aerification operations (e.g. narrow-diameter solid or hollow tines or high-pressure air or water injection technology) prior to product application potentially aids in greater movement of rootzone-targeted fungicides and biofungicides on bowling and golf greens.

Sprayable solutions with increased water volumes in combination with soil-wetting agents and post-application irrigation have long been practised to move fungicides into the rootzone. Recommendations of water volumes are commonly stated on fungicide labels. However, scientific-based evidence supporting the view that increased water volumes, soil-wetting agents and post-application water significantly move fungicides away from the region of initial application and reduce disease has not been universally demonstrated.

Larger-sized spray droplets are more appropriate against root-infecting pathogens. Several studies have investigated experimentally the influence of post-application water and soil-wetting agents on the vertical distribution of selected fungicides (Hockemeyer and Latin, 2015; Ou and Latin, 2018). Generally, while slight differences were reported between certain fungicide, water and wetting agent combinations, most authors concluded most fungicides were retained in the upper thatch regions. A minimum volume and uniformly applied water is required to dislodge applied fungicides and biofungicides from treated foliage.

In contrast, the results of another study investigating the effect of thatch-mat accumulation, irrigation and soil-wetting agent application on the vertical movement of *Pasteuria* sp. bacteria indicated under the test conditions that endospores readily moved downwards under the influence of applied water but were not hindered by the presence of thatch (Luc *et al.*, 2011). Always consult the product label for water volume recommendations.

Maximizing fumigant efficacy

Fumigants are broard-spectrum, volatile biocides used against certain soil-borne pathogens (Fermanian *et al.*, 2003). Fumigant application requires special precautions to optimize their efficacy. A limited number of fumigants are available as restricted use for turfgrass application in certain countries.

The chemical 1,3-dichloropropene is applied in liquid form then vaporizes to move in the pore space once in the rootzone. Metam-sodium (metham-sodium) belongs to the dithiocarbamate chemical class and is applied as an aqueous solution that undergoes hydrolysis to the primary degradation product methyl isothiocyanate (MITC). The granular product dazomet also generates methyl isothiocyanate once applied. Methyl isothiocyanate can also be biologically derived from *Brassica* plants (Stirling, 2014). Methyl isothiocyanate moves passively aided by rootzone moisture and is not a true fumigant.

Optimal fumigant efficacy can be obtained by applying it to a well-aerated rootzone during warm conditions (generally 10–27°C) and a moist (field capacity) but not saturated rootzone. Following application, the treated rootzone needs to be covered or irrigated to assist sealing the surface. A non-entry period is required (3–6 weeks generally to ensure product degradation) depending on the product.

Fungicide Tank Mixtures, Compatibility and Synergism

An expanding range of fungicides, biofungicides and other water-soluble chemistries are possible tank mix combinations (Box 11.7). Combining different fungicides with other chemistries is legal unless otherwise stated on the product labels. Pesticide manufacturers understand the advantages

---

Box 11.6. Rootzone targeting factors.

- Incidence of rainfall
- Verdure factors
- Thatch-mat thickness and density
- Rootzone conditions
- Depth of functioning roots
- Use of a soil-wetting agent
- Pre- and post-application irrigation

of combination products by having label statements about mixing of products and further offering combination products based on their own formulations.

Combinations of two fungicides with different modes of action and phytomobilities offer to widen the range of activity against pathogens and increase overall efficacy. None the less, there are associated risks and application issues and label statements about fungicide and biofungicide mixtures may caution against certain mixtures.

Tank mixtures of certain formulations (e.g. synthetic fungicides with biofungicides) are not recommended unless the pesticide label or formal research states the individual products are compatible. However, in many cases definitive information about fungicide–chemical interactions may be unclear or unknown. Readers are strongly advised to consult all product labels.

Chemistry compatibility testing

Spray tank mixtures will be either compatible or incompatible (Box 11.8). Compatibility simply refers to the mixing in solution of two or more components in the absence of any reduction in efficacy or change in physical, chemical or placement properties of the products (McCarty, 2010).

Physical incompatibility between different products can be readily assessed by progressively mixing each product in a clear container partly filled with water, shaking, then observing for the presence of a non-uniform mixture (Fig. 11.4). Signs of any non-uniformity indicate physical incompatibility; thus the intended mixture cannot be considered.

Chemical incompatibility may cause increased chemical degradation and phytotoxicity. Placement incompatibility refers to mixing two or more products having different stated target sites and water volumes (e.g. mixing a foliar-targeted fungicide mixed with a soil-drench product). The effects of one product will be of reduced benefit, depending on application placement.

Fungicide–fungicide synergism

Synergism of fungicide–fungicide mixtures is based on the premise that the level of disease control provided by the mixture exceeds the sum of control afforded by its individual components

---

Box 11.7. Possible fungicide tank mix combinations.

- Fungicide–fungicide
- Fungicide–insecticide or miticide
- Fungicide–liquid fertilizer
- Fungicide–plant-growth regulator

Box 11.8. Categorization of compatibility.

Physical – unstable mixture causing foaming, precipitation and/or clogged spray nozzles.
Chemical – reduction or loss of efficacy, possible formation of new insoluble and phytotoxic chemistries.
Placement – inappropriate mixing of different products requiring application to either foliage or roots but not both.

Fig. 11.4. Clear water (left), compatible (centre) and incompatible (right) solutions. (Courtesy of Gary W. Beehag.)

(Latin, 2021). The concept of fungicide synergism is problematic to prove and outcomes may vary anywhere between antagonism and synergism (Burpee and Latin, 2008; Chang et al., 2012). If making one's own fungicide mixtures, fungicide interactions must be reasonably predicted in order to make the initial decision to mix and apply a fungicide mixture advantageous in time, effort and outcome.

Fungicide–plant-growth regulator mixtures

Interactive effects of tank mixes of plant-growth regulators (flurprimidol, paclobutrazol and trinexapac-ethyl) and fungicides from several chemical groups (carboxamide; chloronitrile; demethylation inhibitor or DMI; quinone outside inhibitor or QoI) on creeping bentgrass against dollar spot have shown variable results (e.g. Fidanza et al., 2006; Stewart et al., 2008).

The underlying reasons accounting for the reported differences are speculative. Certain triazole fungicides (e.g. propiconazole) have growth-regulatory properties. Flurprimidol and paclobutrazol are closely related to DMI fungicides in their chemical structure and known to have fungistatic properties. Caution is strongly advised when applying fungicide–plant-growth regulators during periods of severe disease incidence and/or plant stress.

Fungicide–fertilizer mixtures

Soluble fertilizers may be either acidic (e.g. ammonium sulfate, ammonium polyphosphate) or alkaline (e.g. urea). Addition of other fertilizers (e.g. iron sulfate and zinc sulfate) because of their high ionic concentration may result in certain fungicide formulations (e.g. emulsifiable concentrates and flowables) becoming chemically incompatible.

Fungicide–nematicide mixtures

Soluble fungicide–nematicide mixtures are uncommon. A mixture of azoxystrobin and the nematicide abamectin applied to nematode-infected Japanese zoysiagrass in pot trials demonstrated experimentally a positive effect (Shaver et al., 2013).

Generalized tank-mixing guidelines

The first step when considering fungicide-based tank mixtures is to consult the labels of all intended products. The following information is provided as an overall guide when considering tank mixing fungicides with other products (i.e. insecticides, miticides and plant-growth regulators) or water-soluble fertilizers (Fermanian et al., 2003):

- Only mix products having the same targeted site whether foliage, thatch-mat or rootzone.
- Carefully pour a pre-mixed slurry or solution of each pesticide slowly into the water-filled spray tank with agitation to prevent settling.
- Mix different pesticide formulations carefully in the following order: wettable powders first, then flowables, other powdered formulations, adjuvants and lastly emulsifiable concentrates.
- Apply caution when mixing wettable powders with emulsifiable concentrates or a soluble fertilizer. Soluble fertilizers or biostimulants containing zinc or iron sulfate and chelated compounds in a tank mix can degrade the emulsion or suspension of other pesticides.
- Avoid mixing soluble alkaline (e.g. lime and urea) with strongly acidic (e.g. sulfur, zinc sulfate, iron sulfate, ammonium sulfate) materials for incompatibility reasons.
- Avoid mixing synthetic fungicides with biofungicides or liquid organic-based products.
- Always test an experimental mix with new combinations on an unimportant turfgrass area in case of phytotoxicity problems. Apply immediately after mixing.

Caution must be taken when mixing different companies' products as manufacturers are reluctant to advise on other products. Pesticide formulations may change over time with respect to formulation (i.e. adjuvants, solvents and other inerts).

Continuous agitation is essential for tank mixes comprising certain formulations (e.g. flowables or water-dispersible and wettable powders) to avoid settling. Excessive agitation caused by the presence of entrapped air bubbles in the tank–pipe system may lead to foaming. Suspension and antifoaming agents assist to reduce settling and foaming.

Spray tank water considerations

The significance of key physical and chemical quality criteria of spray water (Box 11.9) is often reflected in statements on many product labels.

Water quality may require correction within any limitations of statements or recommendations on fungicide and biofungicide labels. Extremes of water temperature may further impact fungicide performance.

Spray water quality

Turbidity (clarity) can be an issue in storage dams, if considering to use the water, following significant rainfall events. Turbidity can qualitatively be assessed by observation of a sample in a clear glass jar. The level of acidity/alkalinity of water can vary widely between water sources. Most fungicides generally have acceptable efficacy under slightly acidic conditions, but several appear to be lesser affected by pH of the water solution (Latin and Stacey, 2020). Caution is required as certain fungicides may undergo unacceptable hydrolysis in highly alkaline water.

Water may be described as 'hard' (i.e. high calcium and/or magnesium concentration) or 'soft' (i.e. relatively high concentrations of sodium and potassium). The chemistry of the fungicide solution can be altered following combination with positively charged ions (e.g. calcium, magnesium) in the spray water. Salinity is the total concentration of soluble salts and can be significant in sewage-treated water. The impact of highly biologically active water sources (e.g. sewage water) on fungicide performance is unknown.

Spray water volume

There is not a single water spray volume for all fungicide and biofungicide applications. Generally speaking, most fungicides are optimized for application at 7.57 litres (2 gallons) to 15.1 litres (4 gallons) per 93 sq. metres (1000 sq. feet). Always consult the product label.

Box 11.9. Key spray water quality criteria.

Turbidity – influenced by suspended organic and clay-sized impurities.
pH – level of acidity/alkalinity.
Hardness – concentration of calcium and magnesium ions in solution.
Salinity – concentration of total amount of soluble salts.

Spray nozzle selection

An extensive range of spray nozzle designs are available (Fermanian *et al.*, 2003). Spray nozzles are the last part of the spray system through which solutions move prior to contacting the turfgrass. Advances in spray nozzle technology have allowed for a wide selection of spray patterns (e.g. flat fan, hollow cone, air-induction and flooding) and water droplet sizes that can influence applied water volumes. Spray nozzles are colour-coded (Fig. 11.5) in accordance with ISO (International Organization for Standardization). Operators are strongly advised to consult product labels in order to make an informed decision about nozzle selection and required pressure, as nozzle type can impact pesticide efficacy.

Greater efficacy of foliar-applied fungicide applications is generally achieved with more uniform surface coverage (Shepard *et al.*, 2006; Kennelly and Wolf, 2009). Uniformity of surface coverage can be visually assessed using water-sensitive test strips having a coloured, coated surface which changes colour following contact with water (Fig. 11.6). Water-sensitive, coloured test strips are highly useful when targeting foliage-borne pathogens. Placement of water-sensitive strips prior to application provides a guide to the correct spray–nozzle combination, aiding in the assessment of droplet uniformity and size.

Spray system cleaning

Thorough cleaning and rinsing of a spray system is required immediately after use to avoid

Fig. 11.5. Colour-coded, multiple spray nozzles. (Courtesy of Gary W. Beehag.)

Fig. 11.6. Yellow-coloured, water-sensitive strips changing blue following a light (left), medium (centre) and high (right) spray water application. (Courtesy of Gary W. Beehag.)

contaminating future spray solutions. Consultation with the spray technology manuals and product labels may provide specific instructions. Rinsing of spray equipment must ensure all rinse water is treated and recycled accordingly or applied to turfgrass, depending on label instructions.

Environmental Behaviour and Fates of Turfgrass Protectants

The environmental behaviour and fate of turfgrass protectants is determined largely by their specific biological and physicochemical characteristics. Investigation of pesticide movement in groundwater beneath golf courses in Cape Cod, Massachusetts, USA (Cohen et al., 1990) was a significant driving force behind the understanding of the environmental behaviour and fate of turfgrass pesticides.

Information about the environmental behaviour and fates of turfgrass pesticides is continually under review (e.g. Clark and Kenna, 2000; Sigler et al., 2000; Magri and Haith, 2009) particularly on golf courses. The environmental behaviour and fate of plant protectants applied to turfgrass is complex (Fig. 11.7) and not fully understood. A basic understanding of their behaviours and fates assists in making informed decisions about product choice and application to minimize environmental risks.

Aboveground behaviour and fates

Sprayable solutions are subject to the effects of wind drift, volatilization, photodecomposition, foliage absorption and surface runoff to varying degrees. Granular formulations are generally less impacted by the same atmospheric factors.

Wind drift

While spray drift is not technically a pesticide fate process, chemical drift loss constitutes a financial and possibly environmental or legal issue, depending on outcome. Increased nozzle pressure is associated with decreased droplet size and spray volume resulting in greater potential for drift.

Volatilization

Volatilization (i.e. transformation from a liquid into a gaseous phase) is more likely during periods of elevated temperature, low humidity and

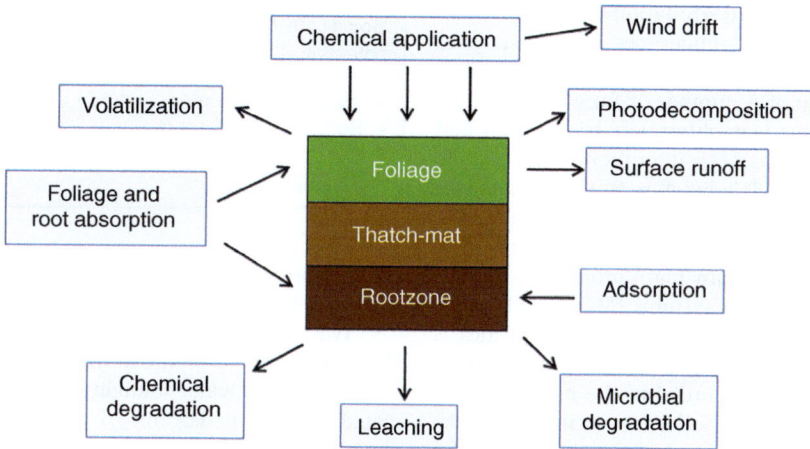

Fig. 11.7. Diagrammatic illustration of environmental fates of turfgrass protectants.

low moisture, resulting in a significant reduction of the amount and concentration of an active ingredient.

Photodecomposition

Photodecomposition (i.e. pesticide degradation in the presence of ultraviolet radiation) is significantly reduced, even eliminated, once the active ingredient reaches the rootzone. Photodecomposition has been investigated for azoxystrobin, fludioxonil and iprodione. Biofungicides generally, as opposed to synthetic formulations, are more susceptible to photodecomposition.

Foliar absorption

Actively growing foliage, as opposed to dormant foliage, absorbs and possibly translocates fungicides, depending on the turfgrass–fungicide combination.

Surface runoff

Physical dislodgement of applied fungicides occurs following prolonged irrigation and unexpected rainfall events. The highest concentration of fungicide generally occurs following the first runoff event and will decrease over time. Liquid as opposed to granular formulations may be more susceptible to runoff, depending on circumstances (e.g. turfgrass height, stand density, product formulation, etc.).

Belowground behaviour and fates

Sprayable solutions and granular fungicides once in the thatch-mat or rootzone region are subject to the combined effects of plant absorption, soil adsorption/absorption, chemical and microbial degradation, and leaching, depending on circumstances.

Stem and root absorption

Absorption of applied fungicides by turfgrass stems and roots is not well documented; thus the extent of translocation and internal fate remain unclear.

Sorption

Sorption describes two associated processes: adsorption and absorption. Adsorption refers to bonding of a pesticide to the outer surface of soil particles while absorption relates to their penetration into soil particles. Irrespective, sorption is a dynamic process, depending on the amount of clay and organic matter in the rootzone. High-sand-content media, as opposed to high-clay-content media, possess minimal sorptive surfaces for fungicide adsorption and absorption.

Chemical and microbial degradation

Chemical and microbial degradation of plant protectants reduces their concentration and activity. Chemical degradation occurs when the active ingredient undergoes hydrolysis and transformation into its breakdown metabolites. Degradation of a number of fungicides in turfgrass systems (e.g. chloroneb, triadimefon and vinclozolin) has been documented (Frederick et al., 1994). Triadimefon is the breakdown product of triadimenol (Schumann et al., 2000).

Microbial biodegradation is a primary means by which many plant protectants are degraded in turfgrass systems (Magri and Haith, 2009). Microbial degradation of fungicides is mediated by certain species of bacteria, actinomycetes and fungi utilizing the products as a nutritional source. Soil bacteria in the genera *Flavobacterium* and *Pseudomonas* and fungi in the genus *Trichoderma* have been associated with biodegradation of pesticides (Aislabie and Lloyd-Jones, 1995).

The rate and circumstances under which chemical and microbial degradation proceeds differ widely between fungicides, depending on their chemistry, formulation and interacting conditions. Chemical and microbial degradation generally proceeds more rapidly under conditions of higher temperatures, balanced oxygen–moisture level, near-neutral pH, high microbial populations, and presence of wetting agents and other chemicals (Koch and Kerns, 2015; Latin and Ou, 2018).

Minimizing fungicide leaching

Leaching, when the active ingredients have downward movement through turfgrass profiles via water and into groundwater, is influenced by their chemistry and complex environmental factors. Factors that influence leaching are temperature, soil texture (sandy versus clay), soil microbial activity and past applications of the same chemistry.

Leaching of numerous fungicides of varying concentrations has been demonstrated experimentally under controlled conditions and varying concentrations of the same fungicides have been reported in different studies (e.g. Stromqvist and Jarvis, 2005; Larsbo et al., 2008; Aamlid et al., 2021). The potential of downward-moving fungicide leachate is dynamic, being difficult to accurately and consistently predict, and numerous indices are adopted (Box 11.10). The value of each factor varies widely between fungicides

---

Box 11.10. Key pesticide properties of leaching potential.

- Water solubility (ppm at 20°C)
- Sorption coefficient (Koc)
- Soil half-life (days)
- Soil texture

(Balogh and Anderson, 1992; Ferruzzi and Gan, 2004).

Water solubility is the amount of pesticide that will dissolve or move in a known water volume. Higher water solubility values are associated with greater leaching. The sorption coefficient (Koc) or soil adsorption coefficient refers to the affinity of a pesticide to bind to soil particles and organic matter. Theoretically, the larger the Koc value, the more strongly the pesticide undergoes sorption and persistence and the less likely it is to migrate to groundwater. Soil half-life is the measurement of the persistence of a pesticide in soil. The longer the half-life, the more persistent the pesticide and the potential it will eventually migrate into groundwater.

Values of water solubility, sorption coefficient and half-life of turfgrass fungicides vary widely (Table 11.4). However, the indices are either calculated based on model predictions or measured under laboratory-controlled conditions and should only be used as a guide for interpretative purposes.

A pesticide is said to be slightly soluble between 1 and 100 ppm, soluble between 100 and 10,000 ppm and very soluble in excess of 10,000 ppm. Pesticide persistence ratings range from low (< 30 days), moderate (30–100 days) to high (>100 days) (Kamrin, 1997).

Leaching of fungicides is highly probable in newly established turfgrass sites constructed of near-pure-sand media. By contrast, potential leaching is reduced in well-established turfgrass sites given accumulation and density of thatch-mat, depending on pesticide. Leaching potential is increased in regions and seasons subject to severe and/or consistent rainfall events. Potential leaching of pesticides is indirectly associated with the degradation rate (Table 11.5).

Several chemical and physical indices have been developed that indicate the potential for surface water and groundwater contamination (Box 11.11). Newer-generation fungicides as opposed to older-generation products have

Table 11.4. Water solubility, organic carbon coefficient and half-life of nominated fungicides.

Fungicide	Solubility in water (ppm)	Organic carbon coefficient (Koc)	Soil half-life (days)	Reference
Chloroneb	8	1,159–1,653	90–180	Balogh and Anderson (1992)
Chlorothalonil	0.6	1,380–5,800	14–90	
Etridiazole	50–200	1,000–4,400	20	
Fenarimol	14	600–1,030	20	
Fosetyl-Al	1.2×10^5	20	1	
Iprodione	13–14	500–1,300	7–30	
Metalaxyl	7100–8400	29–287	7–160	
Quintozene	0.03–0.44	350–10,000	21–434	
Propiconazole	100–110	387–1,147	109–123	
Thiophanate-methyl	3.5	1,830	10	
Triadimefon	70	73	16–28	

Table 11.5. Fungicide persistence classification and leaching potential of nominated fungicides.

Fungicide	Persistence classification	Leaching potential	References
Chlorothalonil	Moderate to short-lived	Small	Balogh and Anderson (1992); Ferruzzi and Gan (2004)
Fosetyl-Al	Very short-lived	Very small	
Iprodione	Moderate–very short-lived	Small	
Mancozeb	High–moderate	Small	
Metalaxyl	High to short-lived	Large	
Quintozene	High–moderately short-lived	Small	
Triadimefon	Moderately short–short-lived	Medium	

> **Box 11.11.** Leaching interpretative values of pesticides (Kenna, 1995).
>
> - Water solubility: greater than 30 ppm
> - Sorption coefficient: less than 300–500
> - Soil half-life: greater than 21 days

much improved environmental properties (rapid degradation), thus posing lesser risk.

Fungicide leaching can be minimized by first consulting product labels for relevant statements, avoiding application during periods of high risk from high or extended rainfall events and to saturated rootzones.

Minimizing fungicide non-target impacts

Side effects or non-target impacts describe unscheduled or unpredictable impacts by pesticides directly on organisms or indirectly on their biological functioning. Non-target impacts of several fungicides in turfgrass systems have been published (Smiley, 1981; Dernoeden, 1992; Harman et al., 2006). Documented non-target impacts by fungicides on turfgrasses and their diseases range from beneficial to deleterious (Box 11.12) depending on underlying circumstances still not fully understood.

> **Box 11.12.** Non-target impacts by turfgrass fungicides.
>
> - Changed turfgrass morphology and foliage colour
> - Inhibition of beneficial organism populations
> - Stimulation of non-target pathogens
> - Enhanced disease resurgence
> - Microbial-enhanced biodegradation
> - New-encounter turfgrass diseases
> - Fungicide resistance

Phytotoxicity

Phytotoxicity or temporary change of normal foliage colour, root growth and delayed rate of immature leaf senescence (hormonal) has been documented among certain triazole and benzimidazole-based fungicides (Table 11.6). The results of most studies investigating thatch-mat accumulation generally indicated increased rates were due to a decrease of microbial activity of the thatch rather than from a stimulated rate of root and rhizome production (Smiley et al., 1985).

Inhibition of beneficial organisms

Published information of inhibitory or suppressive impacts on beneficial bacterial and fungal populations has been documented for several fungicides (e.g. chlorothalonil, fenarimol and iprodione). Generally, the degree of inhibition on endophyte and mycorrhizal species under test has shown mixed and variable results (Dernoeden et al., 1990; Bary et al., 2005; Harman et al., 2006).

Fumigation generally causes temporary changes to bacterial populations and the associated processes of ammonification and nitrification (Stevens et al., 2020). The results of one fumigation study conducted over 2 years investigating the effects of fumigation (i.e. methyl bromide, dazomet and metam-sodium) on bacterial populations in newly constructed, experimental golf greens constructed of a sand–peat mix reported the size of certain bacterial populations (e.g. Pseudomonas and Stenotrophomonas) was either greater than or similar to the size of the populations prior to fumigation (Elliott and Des Jardin, 2001).

Stimulation of non-target pathogens

Unexpected appearance of non-target pathogens by specific older-generation fungicides was reported during the 1970s. Benomyl- and thiabendazole-treated Kentucky bluegrass stimulated development of leaf spot caused by *Drechslera poae* in Rhode Island, USA (Jackson, 1970) and benomyl-treated bentgrass stimulated an undescribed basidiomycete fungus in Australia (Smith et al., 1970). Stimulation of several fungal diseases has been further documented (Table 11.7). Microbial imbalance, whereby applied fungicides inhibit growth of microbial antagonists and/or stimulate the fungal pathogen, is attributed to stimulation of non-target diseases (Smiley et al., 2005).

Enhanced disease resurgence

Enhanced disease resurgence or the increase in severity of targeted diseases following application of older-generation fungicides has been reported for damping-off, dollar spot, red thread, Rhizoctonia blight and summer patch (Baldwin, 1989; Dernoeden et al., 1985; Couch and Smith, 1991). Evidence-based occurrence and explanation of enhanced disease resurgence generally among newer-generation turfgrass fungicides remain unclear.

Microbial-enhanced biodegradation

Enhanced microbial biodegradation (i.e. a pesticide or its metabolites being transformed by soil-borne microbes at an accelerated rate resulting

Table 11.6. Morphological effects on turfgrasses by fungicides.

Effect	Fungicide active ingredients	References
Darker leaf colour	Mancozeb, myclobutanil, chlorothalonil, propiconazole, triadimefon	Smiley et al. (1985); Couch (1995); Reicher and Throssell (1997)
Increased leaf weight	Chlorothalonil, propiconazole	
Increased root growth	Chlorothalonil, propiconazole, triadimefon	
Increased thatch	Benzimidazole fungicides, iprodione	

Table 11.7. Documented cases of stimulation of non-target pathogens by fungicides.

Disease	Fungicide active ingredients	References
Dollar spot	Thiophanates	Smiley (1981); Couch and Smith (1991); Dernoeden and McIntosh (1991); Tredway et al. (2023)
Leaf spots	Triadimefon	
Rhizoctonia blight	Cyclohexamide	
Summer patch	Chlorothalonil, thiram	
Yellow spot	DMIs	

in loss of pesticidal activity) has been demonstrated in agricultural crops (Felsot and Shelton, 1993; Arbeli and Fuentes, 2007). Occurrence of enhanced microbial biodegradation of pesticides in turfgrass systems is unclear. Enhanced biodegradation of organophosphate pesticides has been documented in the United States (Niemcyzk and Chapman, 1989) and Australia (Beehag, 1995). All turfgrass pesticides are potentially subject to more rapid microbial biodegradation. Minimization even avoidance of microbial biodegradation can be achieved by applying fungicides in accordance with label statements.

Fungicide Resistance in Turfgrass Systems

Fungicide resistance occurs when a fungicide fails to inhibit growth or activity of the targeted fungal pathogen. Most fungicides are potentially at risk of the development of resistance over time with repeated applications. The risk of fungicide resistance in highly managed turfgrass systems (e.g. bowling and golf greens) is ever present because of the trend away from multi-site fungicides and greater reliance on site-specific fungicides in managing diseases.

Fungicide resistance is a genetic phenomenon that can occur through successive fungal generations over time (Latin, 2021). The biological mechanisms of fungicide resistance are unpredictable and not fully understood. Delaying and minimizing potential risk of fungicide resistance is supported by information stated on fungicide labels.

Determination of fungicide resistance

Documentation of true fungicide resistance, as opposed to reduced fungicide sensitivity implying resistance, can only be demonstrated under laboratory-controlled conditions. The relative degree of fungicide sensitivity is assessed by measuring the growth of the suspected resistant isolate against a standard sensitive isolate of the same fungal species or strain to a series of increasing concentrations of the active ingredient in question. Theoretically, biofungicides are not susceptible to development of fungicide-resistant pathogens.

Fungicide resistance risk components

Development of fungicide resistance is governed by continual interaction of several factors (Fig. 11.8). Any one factor may attribute significantly more to the development of resistance.

Pathogen biology

Obligate and facultative saprophytic fungi pose the greatest risk for the development of fungicide resistance. However, not all fungi have an equal risk of becoming fungicide resistant. Pathogen biology factors contributing to increased risk of fungicide resistance are short generation times and high frequency of reproduction (i.e. polycyclic) and long periods favourable for growth.

Fungicide mode of action

Collectively, single-site fungicides, as opposed to multi-site ones, interrupt only one physiological process; hence fungi are more likely to develop fungicide resistance to these active ingredients.

Fungicide application regime

Fungicide resistance is more commonly associated with foliage-borne fungi on highly managed cool-season turfgrasses in subtropical or transitional zones. Turfgrass cultivars displaying high levels of susceptibility to one or more fungal diseases, as opposed to lesser-susceptible cultivars, demand greater fungicide use; hence increasing risk for the development of fungicide resistance (Table 11.8). Every fungicide application is a potential fungicide-resistance selection event.

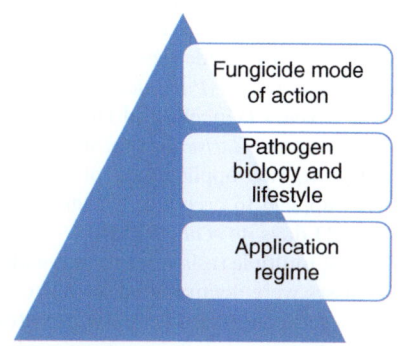

Fig. 11.8. Fungicide resistance risk factors. (Source: Latin, 2021.)

Table 11.8. Relative risk ratings of fungicides and respective diseases.

Relative risk	Fungicide group name	Relative risk	Disease	Reference
High	Benzimidazoles, phenylamides, QoIs, strobilurins, thiophanates	High	Anthracnose, dollar spot, grey leaf spot, microdochium patch, Pythium blight	Latin (2021)
High/medium	Dicarboximides, SDHIs	Moderate	Brown patch, Pythium root rot, southern blight, yellow patch	
Medium	Benzamides, DMIs	Lowest	Spring dead spot, summer patch, take-all patch	
Medium/low	AHs, carbamates, phenylpyrroles			
Low	Dithiocarbamates, phosphonates, phthalimides, thiazines			

Cross-resistance and multiple resistance

Fungicide resistance can manifest in one of two forms (Box 11.13). Cases of cross-resistance to fungicides were reported for strains of the pathogens causing anthracnose, dollar spot, grey leaf spot, microdochium patch and Pythium blight from the mid-1900s for benzimidazole, dicarboximide and QoI class fungicides (Hartill, 1986; Pennucci et al., 1990; Kerns and Tredway, 2013; Latin, 2021) primarily in the United States.

Reports of cross-resistance to fungicides continue to be documented for several classes of fungicides in numerous countries (Table 11.9). The global significance of cross-resistance in turfgrass systems is unknown.

Development of fungicide resistance, if and when it occurs, is not uniform in time nor space. Initially a very small population of a fungicide-resistant strain arises by selection and then the population of the resistant strain increases from continual exposure to the fungicide. Resistance to the QoI fungicide azoxystrobin by the grey leaf spot pathogen (*Pyricularia grisea*) developed after 3 years (Vincelli and Dixon, 2002). Fungicide-resistant *Clarireedia* spp. strains developed following two applications of thiophanate-methyl applied to creeping bentgrass at an interval of 21 days (Jo et al., 2008).

Cases of multiple resistance to fungicides in several classes were documented from the early 2000s in several countries for pathogen strains causing anthracnose, dollar spot, microdochium patch and Pythium blight for combinations of several fungicide classes (Mann, 2002; Latin, 2021). Documented cases of multiple resistance are relatively uncommon in turfgrass systems. None the less, instances of multiple resistance have been confirmed in more recent decades among foliar-borne pathogens for several fungicide class combinations (Table 11.10).

Fungicide resistance is only an issue for pathogen strains being susceptible to the fungicide in the first instance. The level of fungicide application against soil-borne diseases is generally lower compared to foliage-borne diseases. Soil-borne fungal diseases (e.g. spring dead spot and take-all patch) have relatively low rates of reproduction, growth, and generally have lower populations.

Box 11.13. Cross-resistance and multiple resistance.

Cross-resistance – a pathogen strain resistant to one active ingredient develops resistance to another active ingredient having the same biochemical mode of action.
Multiple resistance – a pathogen strain resistant to two or more active ingredients with different modes of action.

Turfgrass fungicide resistance case studies

Cases of true fungicide resistance in turfgrass systems are fortunately limited to a few pathogen–fungicide combinations (Box 11.14). The following case reviews of fungicide resistance provide an insight into their occurrence.

Table 11.9. Selected examples of fungicide cross-resistance.

Pathogen name	Fungicide class	Country	References
Colletotrichum cereale	DMI	USA	Young et al. (2010)
Sclerotinia homoeocarpa	Carboxamide	USA	Gonzalez and Kennelly (2012)
	DMI	Korea	Lee et al. (2017)
Clarireedia spp.	SDHI	Japan/USA	Jung and Jo (2008)
Monographella nivalis	Dicarboximide	Canada	Gourlie and Hsiang (2017)

Table 11.10. Selected examples of fungicide multiple resistance.

Pathogen name	Fungicide class	Country	References
Colletotrichum cereale	Benzimidazole, QoI	USA	Wong et al. (2008)
Sclerotinia homoeocarpa	Benzimidazole, dicarboximide, DMI	USA	Bishop et al. (2008)
Clarireedia spp.	Carboxamide, dicarboximide, DMI	China	Zhang et al. (2021)
Monographella nivalis	Benzimidazole, dicarboximide	UK	Mann (2002)
Pythium aphanidermatum	Phenylamide, QoI	USA	Olaya et al. (2003)

Box 11.14. Cases of fungicide resistance in turfgrass systems.

- Anthracnose (*Colletotrichum cereale*)
- Dollar spot (*Clarireedia* spp.)
- Grey leaf spot (*Pyricularia grisea*)
- Microdochium patch (*Monographella nivalis*)
- Pythium blight (*Pythium* spp.)

Anthracnose

Anthracnose is a high-risk disease for its pathogens to develop fungicide resistance in cool-season grass golf greens, as highlighted in the 1980s in the United States to the older-generation benzimidazole fungicide benomyl (Shane and Danneberger, 1989).

Fungicide resistance among *Colletotrichum cereale* populations has been more recently documented against other benzimidazole (thiophanate-methyl), DMI (myclobutanil, propiconazole, tebuconazole and triadimefon) and QoI (azoxystrobin, pyraclostrobin and trifloxystrobin) fungicides. Resistance to QoI fungicides in some instances occurred very rapidly; thus highlighting the high-risk category of anthracnose. Resistance to benzimidazole and DMI fungicides among *C. cereale* populations, as opposed to QoI fungicides, appears to be relatively stable (Wong and Midland, 2004).

Dollar spot

Dollar spot is another high-risk disease for its pathogens to develop fungicide resistance in cool-season grass golf greens. The high risk of fungicide resistance developing among dollar spot pathogen populations was first demonstrated in the United States in the 1960s to inorganic fungicides and subsequently to the benzimidazole fungicides benomyl and thiophanate-methyl (Goldberg and Cole, 1973).

Fungicide resistance among *Clarireedia* spp. populations to benzimidazole, carboximide, dicarboximide, DMI, phenylurea and QoI fungicides has more recently been confirmed throughout the world. The results of one study investigating numerous plant-growth regulator (flurprimidol, paclobutrazol and trinexapac-ethyl) and DMI fungicide interactions on resistance to the dollar spot pathogen led the authors to conclude the interaction between plant-growth regulators and DMI fungicides may have contributed to resistance to the DMI fungicides (Ok et al., 2011).

Grey leaf spot

Grey leaf spot has been identified as a high-risk disease to fungicide resistance to the QoI fungicides, i.e. azoxystrobin and trifloxystrobin (Vincelli and Dixon, 2002). The emergence of grey leaf spot resistance to QoI fungicides was rapid given their registration in 1998 in the United States and resistance appearance shortly thereafter.

Microdochium patch

Microdochium patch is a relatively low-risk disease for the pathogen to develop fungicide resistance. Fungicide resistance among *Monographella nivalis* populations was initially reported to early-generation benzimidazole (benomyl) and dicarboximide (iprodione) fungicides in New Zealand (Hartill, 1986; Pennucci et al., 1990), Western Europe (Huth and Schlosser, 1980) and the United States (Chastagner and Vassey, 1982).

Fungicide resistance among *M. nivalis* populations to aromatic hydrocarbon (quintozene), benzonitrile (chlorothalonil), benzimidazole (carbendazim) and dicarboximide (iprodione, procymidone, vinclozolin) classes has been confirmed more recently in several countries.

Pythium blight

Confirmed development of fungicide resistance of Pythium blight is widespread. Fungicide resistance among *Pythium* spp. populations was confirmed in the United States in the 1980s to metalaxyl (Sanders, 1984). Since 2000, cases of reduced sensitivity and true resistance have been reported among carbamate, ethyl phosphate, phenylamide and QoI fungicides (Olaya et al., 2003; Titone et al., 2009).

Fungicide resistance management strategies

The key aim of a fungicide resistance management strategy is to limit or delay any opportunities for the evolution of fungicide-resistant strains. However, absolute adherence to resistance management principles does not guarantee fungicide resistance will not occur. First and foremost, adherence to an integrated disease management strategy incorporating disease-resistant varieties, in combination with sound cultural and biological practices, is fundamental for the purpose of reducing fungicide applications and hence minimizing the selection of a fungicide-resistant strain.

A multifaceted fungicide resistance management strategy is required (Box 11.15). Integration of appropriate chemical and proven non-chemical strategies, as opposed to reliance on cultural practices alone, generally will have a far greater impact on delaying fungicide resistance. The FRAC code documents fungicides' and biofungicides' biochemical mode of action as the basis of their selection in a fungicide resistance management strategy.

Employing rotation (i.e. alternate) or tank mixing (i.e. two or more) single-site fungicides in a fungicide resistance management strategy remains controversial (Vargas, 2005; Latin, 2021). Rotating or tank mixing fungicides is based on the premise that alternating or mixing different biochemical modes of action will slow evolution of fungicide resistance. None the less, general guidelines based on published FRAC codes should be followed.

Box 11.15. General fungicide resistance management strategy.

- Implementation of proven cultural practices
- Application of appropriately registered biofungicides
- Adoption of a preventive approach to fungicide use
- Application of multi-site fungicides
- Limited adoption of site-specific fungicides
- Mixing a multi-site and a site-specific fungicide
- Rotate or tank mix site-specific fungicides

References

Aamlid, T.S., Almvik, M., Petterson, T. and Bolli, R. (2021) Leaching and surface runoff after fall application of fungicides on putting greens. *Agronomy Journal* 113, 3743–3763. DOI: 10.1002/agj2.20549

Aislabie, J. and Lloyd-Jones, G. (1995) A review of bacterial-degradation of pesticides. *Australian Journal of Soil Research* 33, 925–942. DOI: 10.1071/SR9950925

Arbeli, Z. and Fuentes, C.L. (2007) Accelerated biodegradation of pesticides: an overview of the phenomenon, its basis and possible solutions; and a discussion on the tropical dimension. *Crop Protection* 26, 1733–1746.

Baldwin, N.A. (1989) Increases in post-emergence damping-off of *Agrostis castellana* and *Poa pratensis* following treatment with benomyl. *Journal of the Sports Turf Research Institute* 65, 151–154.

Balogh, J.C. and Anderson, J.L. (1992) Environmental impacts of turfgrass pesticides. In: Balogh, J.C. and Walker, W.J. (eds) *Golf Course Management and Construction: Environmental Issues*. CRC Press, Boca Raton, Florida, pp. 221–354.

Bary, F., Gange, A.C., Crane, M. and Hagley, J.K. (2005) Fungicide levels and arbuscular mycorrhizal fungi in golf putting greens. *Journal of Applied Ecology* 42, 171–180.

Beehag, G.W. (1995) A review of enhanced biodegradation of fenamiphos in turf in Australia. In: *Proceedings of the 2nd Turf Research Conference, Sydney, Australia*. Australian Turfgrass Research Institute, Sydney, Australia, pp. 61–65.

Bishop, P., Sorochan, J., Ownley, B.H., Samples, T.J., Windham, A.S., et al. (2008) Resistance in *Sclerotinia homoeocarpa* to iprodione, propiconazole, and thiophanate-methyl in Tennessee and Mississippi. *Crop Science* 48, 1615–1620.

Burpee, L.L. and Latin, R. (2008) Reassessment of fungicide synergism for control of dollar spot. *Plant Disease* 92, 601–606. DOI: 10.1094/PDIS-92-4-0601

Carroll, M.J., Hill, R.L., Pfeil, E. and Krouse, J.M. (2001) Effect of residence on washoff chlorothalonil from turf foliage. *International Turfgrass Society Research Journal* 9, 1–5.

Chang, S.-W., Jung, S.-W., Kim, S.-H., Park, J.-H. and Lee, J.-Y. (2012) Synergistic interaction of fungicides in mixtures under different conditions of dollar spot disease caused by *Sclerotinia homoeocarpa*. *Asian Journal of Turfgrass Science* 26, 96–101.

Chastagner, G.A. and Vassey, W.E. (1982) Occurrence of iprodione-tolerant *Fusarium nivale* under field conditions. *Plant Disease* 66, 112–114.

Clark, I.M. and Kenna, M.P. (eds) (2000) *Fate and Management of Turfgrass Chemicals*. American Chemical Society, Washington, DC.

Cohen, S.Z., Nickerson, S., Maxey, R., Dupuy, A. and Senita, J.A. (1990) A groundwater monitoring study for pesticides and nitrates associated with golf courses on Cape Cod. *Ground Water Monitoring and Remediation* 10, 160–173. DOI: 10.1111/j.1745-6592.1990.tb00333.x

Couch, H.B. (1995) *Diseases of Turfgrasses*, 3rd edn. Krieger Publishing Company, Malabar, Florida.

Couch, H.B. and Smith, B.D. (1991) Synergistic and antagonistic interactions of fungicides against *Pythium aphanidermatum* on perennial ryegrass. *Crop Protection* 10, 386–390.

Dernoeden, P.H. (1992) The side effects of fungicides. *Golf Course Management* 50, 88–110.

Dernoeden, P.H. and McIntosh, M.S. (1991) Disease enhancement and Kentucky bluegrass quality as influenced by fungicides. *Agronomy Journal* 83, 322–326.

Dernoeden, P.H., Murray, J.J. and O'Neill, N.R. (1985) Non-target effects of fungicides on turfgrass growth and enhancement of red thread. In: Lemaire, F. (ed.) *Proceedings of the Fifth International Turfgrass Research Conference, Avignon, France, 1–5 July 1985*. INRA, Paris, pp. 579–594.

Dernoeden, P.H., Krusberg, L.R. and Sardanelli, S. (1990) Fungicides effects on *Acremonium* endophyte, plant parasitic nematodes, and thatch in Kentucky bluegrass. *Plant Disease* 74, 879–881.

Elliott, M.l. and Des Jardin, E.A. (2001) Fumigation effects on bacterial populations in new golf course bermudagrass putting greens. *Soil Biology and Biochemistry* 33, 1841–1849. DOI: 10.1016/S0038-0717(01)00112-2

Felsot, A.S. and Shelton, D.R. (1993) Enhanced biodegradation of soil pesticides: interactions between physiochemical processes and microbial ecology. In: Linn, D.M., Carski, F.H., Brusseau, M.L. and Chang, T.-H. (eds) *Sorption and Degradation of Pesticides and Organic Chemicals in Soils*. SSA Special Publication No. 32. Soil Science Society of America, Inc. and American Society of Agronomy, Inc., Madison, Wisconsin, pp. 227–251.

Fermanian, T.W., Shurtleff, M.C., Randell, R., Wilkinson, H.T. and Nixon, P.L. (2003) *Controlling Turfgrass Pests*, 3rd edn. Prentice-Hall, Upper Saddle River. New Jersey.

Ferruzzi, G. and Gan, J. (2004) *Pesticide Selection to Reduce Impacts on Water Quality*. Publication No. 8119. University of California, Division of Agriculture and Natural Resources, Davis, California.

Fidanza, M.A., Wetzel, H.C., Agnew, M.L. and Kaminski, J.E. (2006) Evaluation of fungicide and plant growth regulator tank-mix programs on dollar spot severity of creeping bentgrass. *Crop Protection* 25, 1032–1038.

Frederick, E.K., Bischoff, M., Throssell, C.S. and Turco, R.F. (1994) Degradation of chloroneb, triadimefon, and vinclozolin in soil, thatch, and grass clippings. *Bulletin of Environmental Contamination and Toxicology* 53, 536–542.

Goldberg, C.W. and Cole, H. (1973) *In vitro* study of benomyl tolerance exhibited by *Sclerotinia homoeocarpa* (abstract). *Phytopathology* 63, 201.

Gonzalez, B.R. and Kennelly, M.M. (2012) Sensitivity of Kansas isolates of *Sclerotinia homoeocarpa* to boscalid. *Online Plant Health Progress*. DOI: 10.1094/PHP-2012-0402-01-RS

Gourlie, R. and Hsiang, T. (2017) Resistance to dicarboximide fungicides in a Canadian population of *Microdochium nivale*. *International Turfgrass Society Research Journal* 13, 133–138.

Harman, G.E., Nelson, E.B. and Ondik, K.L. (2006) Fungicide application effects on non-target microbial populations on putting greens. *USGA Turfgrass and Environmental Research Online* 5, 1–6.

Hartill, W.F.T. (1986) Resistance of plant pathogens to fungicides in New Zealand. *New Zealand Journal of Agricultural Research* 14, 239–245.

Hockemeyer, K.R. and Latin, R. (2015) Spatial and temporal distribution of fungicides applied to creeping bentgrass. *Journal of Environmental Quality* 44, 841–848. DOI: 10.2134/jeq2014.08.0335

Huth, G. and Schlosser, E. (1980) Tolerance of *Fusarium nivale* to benzimidazole fungicides in golf greens. *Zeitschrift für Vegetationstechnik im Landschafts- und Sportstättenbau* 4, 161–164.

Jackson, N. (1970) Evaluation of some chemicals for control of stripe smut in Kentucky bluegrass. *Plant Disease Reporter* 54, 168–170.

Jo, Y.-K., Chang, S.W., Boehm, M. and Jung, G. (2008) Rapid development of fungicide resistance by *Sclerotinia homoeocarpa* on turfgrass. *Phytopathology* 98, 1297–1304.

Jung, G. and Jo, Y.-K. (2008) New challenges to an old foe, dollar spot fungicide resistance. *Golf Course Management* 76, 117–121.

Kamrin, M.A. (1997) *Pesticide Profiles: Toxicity, Environmental Impact, and Fate*. CRC Press, Boca Raton, Florida.

Kennelly, M.M. and Wolf, R.E. (2009) Effect of nozzle type and water volume on dollar spot control in greens-height creeping bentgrass. *Applied Turfgrass Science* 6, 1–8. DOI: 10.1094/ATS-2009-0921-01-RS

Kerns, J.P. and Tredway, L.P. (2013) Advances in turfgrass pathology since 1990. In: Stier, J.C., Horgan, B.P. and Bonos, S.A. (eds) *Turfgrass: Biology, Use, and Management*. American Society of Agronomy, Inc., Crop Science Society of America, Inc. and Soil Science Society of America, Inc., Madison, Wisconsin, pp. 733–776.

Koch, P.L. and Kerns, J.P. (2015) Temperature influences persistence of chlorothalonil and iprodione on creeping bentgrass foliage. *Plant Health Progress* 16, 107–112.

Larsbo, M., Aamlid, T.S., Persson, L. and Jarvis, N. (2008) Fungicide leaching from golf greens: effects of root zone composition and surfactant use. *Journal of Environmental Quality* 37, 1527–1535.

Latin, R. (2021) *A Practical Guide to Turfgrass Fungicides*, 2nd edn. The American Phytopathological Society, St. Paul, Minnesota.

Latin, R. and Ou, L. (2018) Influence of irrigation and wetting agent on fungicide residues in creeping bentgrass. *Plant Disease* 102, 2352–2360. DOI: 10.1094/PDIS-11-17-1844-RE

Latin, R. and Stacey, T. (2020) Influence of water carrier pH on efficacy of fungicides for dollar spot control. *Golf Course Management* 88, 59–63.

Lee, J.W., Choi, J. and Kim, J.-W. (2017) Triazole fungicides resistance of *Sclerotinia homoeocarpa* in Korean golf courses. *Plant Pathology Journal* 33, 589–596.

Luc, J.E., Crow, W.T., Pang, W., McSorley, R. and Giblin-Davis, R.M. (2011) Effects of irrigation, thatch, and wetting agent on movement of *Pasteuria* sp. endospores in turf. *Nematropica* 41, 185–190.

McCarty, L.B. (2010) *Best Golf Course Management Practices: Construction, Watering, Fertilizing, Cultural Practices and Pest Management Strategies to Maintain Golf Course Turf with Minimal Environmental Impact*, 3rd edn. Prentice-Hall, Upper Saddle River, New Jersey.

McDonald, S.J., Dernoeden, P.H. and Bigelow, C.A. (2006) Dollar spot control in creeping bentgrass as influenced by fungicide spray volume and application timing. *Applied Turfgrass Science* 3, 1–11. DOI: 10.1094/ATS-2006-0531-01-RS

Magri, A. and Haith, D.A. (2009) Pesticide decay in turf: a review of processes and experimental data. *Journal of Environmental Quality* 38, 4–12.

Mann, R.L. (2002) In vitro fungicide sensitivity of *Microdochium nivale* isolates from the UK. *Journal of Turfgrass and Sports Science* 78, 25–30.

Niemcyzk, H.D. and Chapman, R.A. (1989) Evidence of enhanced degradation of isophenfos in turfgrass thatch and soil. *Journal of Economic Entomology* 80, 880–882. DOI: 10.1093/jee/80.4.880

Ok, C.-H., Popko, J.T., Campbell-Nelson, K. and Jung, G. (2011) In vitro assessment of *Sclerotinia homoeocarpa* resistance to fungicides and plant growth regulators. *Plant Disease* 95, 51–56.

Olaya, G., Cleere, S., Stanger, C., Burbidge, J., Hall, A. and Windass, J. (2003) A novel potential target site QoI fungicide resistance mechanism in *Pythium aphanidermatum* (abstract). *Phytopathology* 93, S67.

Ou, L. and Latin, R. (2018) Influence of management practices on distribution of fungicides in golf course turf. *Agronomy Journal* 110, 2523–2533. DOI 10.2134/agronj2018.02.0115

Pennucci, A., Beever, R.E. and Laracy, E.P. (1990) Dicarboximide-resistant strains of *Michrochium nivale* in New Zealand. *Australasian Plant Pathology* 19, 38–41.

Pigati, R.L., Dernoeden, P.H., Grybauskas, A.P. and Momen, B. (2010) Simulated rainfall and mowing impact fungicide performance when targeting dollar spot in creeping bentgrass. *Plant Disease* 94, 596–603.

Reicher, A. and Throssell, C. (1997) Fungicides do more than control disease. *Golf Course Management* 65, 53–58.

Sanders, P.L. (1984) Failure of metalaxyl to control Pythium blight on turfgrass in Pennsylvania. *Plant Disease* 68, 776–777.

Schumann, G.L., Clark, J.M., Doherty, J.J. and Clarke, B.B. (2000) Application of DMI fungicides to turfgrass with three delivery systems. In: Clark, J.M. and Kenna, M.P. (eds) *Fate and Management of Turfgrass Chemicals*. American Chemical Society, Washington, DC, pp. 150–163.

Shane, W.W. and Danneberger, K. (1989) First report of field resistance of *Colletotrichum graminicola* on turf to benzimidazole fungicides in the United States. *Plant Disease* 73, 755.

Shaver, J.B., Agudelo, P. and Martin, S.B. (2013) Azoxystrobin and abamectin improve root health of zoysiagrass infected with *Trichodorus obtusus* (abstract). *Journal of Nematology* 45, 304.

Shepard, D., Agnew, M., Fidanza, M., Kaminski, J. and Dant, L. (2006) Selecting nozzles for fungicide spray applications. *Golf Course Management* 74, 83–88.

Sigler, W.V., Taylor, C.P., Throssell, C.S., Bischoff, M. and Turea, R.F. (2000) Environmental fates of fungicides in the turfgrass environment: a minireview. In: Clark, J.M. and Kenna, M.P. (eds) *Fate and Management of Turfgrass Chemicals*. American Chemical Society, Washington, DC, pp. 127–149.

Smiley, R.W. (1981) Nontarget effects of pesticides on turfgrasses. *Plant Disease* 65, 17–23.

Smiley, R.W., Craven-Fowler, M., Kane, R.T., Petrovic, A.M. and White, R.A. (1985) Fungicide effects on thatch depth, thatch decomposition rate and growth of Kentucky bluegrass. *Agronomy Journal* 77, 597–602.

Smiley, R.W., Dernoeden, P.H. and Clarke, B.B. (2005) *Compendium of Turfgrass Diseases*. 3rd edn. The American Phytopathological Society, St Paul, Minnesota.

Smith, A.M., Stynes, B.A. and Moore, K.J. (1970) Benomyl stimulates the growth of a basidiomycete on turf. *Plant Disease Reporter* 54, 774–775.

Stewart, J.M., Latin, R., Reicher, Z. and Hallet, S.G. (2008) Influence of trinexapac-ethyl on the efficacy of chlorothanil and propiconazole for control of dollar spot on creeping bentgrass. *Applied Turfgrass Science* 5, 1–18. DOI: 10.1094/ATS-2008-0319-01-RS

Stevens, M.C., Yang, R. and Freeman, J.H. (2020) Deposition and transformation of nitrogen after soil fumigation with ethanedinitrile. *HortScience* 55, 2023–2027. DOI: 10.21273/HORTSCI15397-20

Stirling, G.R. (2014) *Biological Control of Plant-Parasitic Nematodes*, 2nd edn. CAB International, Wallingford, UK.

Stromqvist, J. and Jarvis, N. (2005) Sorption, degradation and leaching of the fungicide iprodione in a golf green under Scandinavian conditions: measurements and modelling. *Pest Management Science* 61, 1168–1178. DOI: 10.1002/ps.1101

Titone, P., Mocioni, M., Garibaldi, A. and Gullino, M.L. (2009) Fungicide failure to control Pythium blight on turf grass in Italy. *Journal of Plant Diseases and Protection* 116, 55–59.

Tredway, L.P., Tomasco-Peterson, M., Kerns, J.P. and Clarke, B.B. (2023) *Compendium of Turfgrass Diseases*. 4th edn. The American Phytopathological Society, St Paul, Minnesota.

Vargas, J.M. (2005) *Management of Turfgrass Diseases*, 3rd edn. Wiley, Hoboken, New Jersey.

Vincelli, P. and Dixon, E. (2002) Resistance to QoI (strobilurin-like) fungicides in isolates of *Pyricularia grisea* from perennial ryegrass. *Plant Disease* 86, 235–240. DOI:10.1094/PDIS.2002.86.3.235

Vincelli, P. and Dixon, E. (2007) Does spray coverage influence fungicide efficacy against dollar spot? *Applied Turfgrass Science* 4, 1–5. DOI: 10.1094/ATS-2007-1218-01-RS

Walters, D. (2006) Disguising the leaf surface: the use of leaf coatings for disease control. *European Journal of Plant Pathology* 114, 255–260. DOI: 10.1007/s10658-005-5463-7

Wong, F.P. and Midland, S. (2004) Fungicide-resistant anthracnose – bad news for greens management. *Golf Course Management* 87, 1426–1432.

Wong, F.P., de la Cerda, K.A., Hernandez-Martinez, R. and Midland, S.L. (2008) Detection and characterisation of benzimidazole resistance in California populations of *Colletotrichum cereale*. *Plant Disease* 92, 239–246.

Young, J.R., Tomasco-Peterson, M., de la Cerda, K. and Wong, F.P. (2010) Two mutations in B-tubulin 2 gene associated with thiophanate-methyl resistance in *Colletotrichum cereale* isolates from creping bentgrass in Mississippi and Alabama. *Plant Disease* 94, 207–212.

Zhang, H., Jiang, S., Zhao, Z., Guan, J., Dong, Y., et al. (2021) Fungicide sensitivity of *Clarireedia* spp. isolates from golf courses in China. *Crop Protection* 149, 105785. DOI: 10.1016/j.cropro.2021.105785

Glossary

abiotic	A non-living environmental factor.
acervulus	A fruiting body containing conidia.
acropetal	Upward movement of fluids or chemicals inside a plant.
active ingredient	The active component of a pesticide formulation.
adjuvant	A chemical which increases the efficacy of a pesticide.
aerobic	Relating to the presence of oxygen.
aetiology	The study of causes of disease.
agar	A jelly-like material used to culture microbes for identification purposes.
ammonification	Process by which microbes mineralize organic compounds into ammonia.
anaerobic	The absence of oxygen.
anamorph	The asexual form of a fungus.
anastomosis	Fusion between the mycelium of the same fungal species or strains.
antagonist	An organism that inhibits or supresses the growth and development of others.
antibiotic	A compound that inhibits or suppresses bacteria.
ascomycetes	Group of fungi characterized by producing sexual spores within an ascus.
ascus	Sexual spore-bearing cell produced by ascomycete fungi.
avirulent	Non-pathogenic and unable to cause disease.
bacterium	A prokaryotic, single-celled organism that multiplies by cell division.
basidiomycetes	Group of fungi characterized by producing basidiospores.
basidiospores	Sexual spores produced by special structures called basidia in basidiomycetes.
basipetal	Downward movement of fluid or chemicals in a plant.
biotic	A living factor.
biological control	Antagonism, competition, suppression or parasitism of organisms.

biopesticide	A chemical compound derived from an organism or its metabolites.
biotroph	An organism that gains its nutrition from living host cells rather than dead cells.
binucleate	Having two nuclei per cell.
causal agent	A disease-causing agent.
chlorosis	Change from normal green to yellowish-green caused by lack of chlorophyll.
coalesce	Joining of leaf lesions or advancing disease symptoms.
conidium	An asexual, non-motile fungal spore.
cultivar	A cultivated variety of a turfgrass species with defined characteristics.
cyst	A sac containing eggs, usually enclosed within a dead female nematode body.
disease complex	A disease caused by several pathogens on the same host.
disease cycle	Progression of disease from arrival of inoculum, infection through disease expression and survival in the absence of a host or during environmental conditions non-conducive to disease.
DNA	Deoxyribonucleic acid containing the genetic code of an organism.
ectoparasite	A parasite that lives and feeds from the exterior surface of the host.
ectotrophic	Soil-borne fungi gaining nutrition from plant roots, rhizomes, stolons or crowns.
edaphic	Pertaining to the rootzone.
endoparasite	A parasite that lives and feeds from inside the host.
endophyte	An organism that resides inside a non-symptomatic host.
enzyme	A protein that catalyses specific biochemical processes.
epidemiology	The study of all factors associated with development and dissemination of disease in organisms.
epidemic	A severe outbreak of disease in an organism population.
epiphytotic	Widespread outbreak of disease in a plant population.
eradicant	A chemical compound applied to eliminate a pathogen.
etiolation	Abnormal elongation of plant growth as result of reduced light or chemicals.
eukaryote	An organism having cells with a membrane-bound nucleus and other organelles.
exudate	Secretion of liquid biochemical compounds from organism tissues.
facultative parasite	An organism primarily saprophytic but capable of being pathogenic.
forb	A dicotyledonous plant.
fruiting body	A fungal spore-bearing structure.
fumigant	A volatile, lethal compound often in a gaseous state that diffuses through soil.
fungistat	A chemical compound capable only of inhibiting fungi.
fungicide	An active ingredient in a compound capable of acting against fungi.
gene	A unit of DNA controlling heritable characteristics.
genus	A taxonomic category of closely related species.

graminicolous	Pertaining to microbes living on grass.
guttation fluid	Exudation of liquids from plant hydathodes.
hormone	A signalling compound that regulates the functioning of organisms.
host plant	A living plant from which an organism survives and gains nutrition.
humus	Dark-coloured, organic compounds formed from decomposition of organic matter.
hydathode	Leaf structure that exudes water and plant-produced sugars.
hypha (*pl*. hyphae)	Fungal filament or mycelium of a fungus.
hyphopodium	Specialized fungal structure, usually shaped like a foot, that is used for host attachment and infection.
infectious disease	Spread of disease between plants.
inoculum	Various structures of a pathogen capable of causing disease.
isolate	A culture or subpopulation of microbes.
latent period	The length of time between infection and appearance of symptoms.
lesion	A localized area exhibiting disease or symptoms.
metabolite	A chemical involved in metabolism.
mosaic	A disease symptom characterized by a patchwork of normal green and abnormal chlorotic symptoms.
microorganism	A small-sized organism only observable with the aid of magnification.
mycelium	Massed hyphae of a fungus.
mycoplasma	A bacterium that does not possess a cell wall around its cell membrane.
mycorrhiza	A symbiotic relationship between a non-pathogenic fungus and plant roots.
mycoparasite	A fungus that parasitizes another fungus.
necrosis	Death of cells.
necrotroph	A pathogen that kills host cells to gain nutrition from dead tissues.
nematode	A parasitic or free-living non-segmented roundworm.
nematicide	A chemical compound capable of killing nematodes.
nemastatic	A chemical compound capable only of inhibiting nematodes.
nitrification	Process by which microbes oxidize ammonia to nitrite and then nitrate.
non-target effect	Any non-intentional, usually negative, impact to an organism or the environment by a pesticide.
obligate parasite	An organism that is only capable of survival in association with a living host.
obligate saprophyte	An organism that is only capable of survival on dead tissues.
oomycete	A fungus-like microbe.
oospore	Thick-walled spore formed by oomycetes.
PCR (polymerase chain reaction)	A laboratory technique involving the production of copies of specific portions of DNA used often for microbial identification.
parasite	An organism that colonizes and gains nutrition from another organism.

pathogen	An organism that colonizes and gains nutrition from another, thereby causing disease.
pathogenicity	The relative ability to cause disease.
pathology	The study of diseases.
pathovar	A subdivision of a plant pathogen defined by host range.
pesticide	A chemical used against pests.
phloem	The carbohydrate-conducting tissues of plants.
phytoplasma	A plant-parasitic prokaryote often found in phloem tissues.
phytotoxicity	Growth inhibition of or adverse effects to plants caused by certain chemicals.
plant pathology (phytopathology)	The study of plant disease.
preventive (protectant)	A chemical compound applied prior to infection by a pathogen.
prokaryote	An organism lacking internal membrane-bound organelles and a distinct nucleus.
protein	Nitrogen-containing organic compounds composed of amino acids.
pycnidium	Specialized asexual, flask-shaped fungal fruiting body producing conidia.
resistant	Possessing genetic properties that prevent or impede disease development.
rhizosphere	The microenvironment immediately surrounding plant roots.
saprophyte	An organism that obtains nutrients from non-living organic matter.
sclerotium	A resting structure of a fungus capable of being viable for an extended time.
seed-borne	Carried on or in seeds.
senescence	The decline of plant organs post-maturity or from stress.
septate	Having one or more septa or cellular cross walls.
Seta (*pl.* setae)	A hair-like structure formed by certain fungi.
sp. (*pl.* spp.)	Abbreviation of species – the singular form refers to a species; the plural form refers to two or more species.
spore	A reproductive structure of a fungus.
sterile fungus	A fungus that is not known to produce spores.
strain	A distinct biological form of an organism.
susceptible	Prone to disease when infected by a pathogen.
symbiosis	A mutually beneficial association between two different types of organisms.
symptom	The visual expression of a disease.
symptomatology	The set of symptoms that characterizes a specific disease.
symptomless host	An infected plant that produces no obvious symptoms.
syndrome	The combined effects of a plant disease.
synergism	An interaction where the effects of two or more pathogens in combination is greater than that caused by each pathogen alone.
systemic	Pertaining to the internal movement of chemicals throughout a plant.
taxonomy	The science of naming and classifying organisms.
teleomorph	The sexual stage of a fungus.
uninucleate	Having a single nucleus per cell.
vector	A living organism capable of transmitting or moving a pathogen.

virulent	Describing a pathogenic microorganism capable of causing disease.
virus	A submicroscopic, obligate pathogen.
volatilization	Process by which a portion of an applied product is lost to the atmosphere.
wetting agent	A surface-active chemical used to lower the surface tension of water.
xylem	The moisture-conducting tissue of plants.

Index

Note: The page numbers in italics and bold represent figures and tables respectively.

acidifying/alkaline fertilisers 105
Acidovorax avenae (bacterial decline) 230, *231*
Adelaide patch (*Wongia garrettii*) *132*, **137**
    see also root decline of warm-season grasses 130–135
adjuvants 272
Agaricus 97
    see also fairy ring formations 93–91
Agrostis (bentgrass) **3**
alkali grass (*Puccinella distans*) 4
Alternaria leaf spot 218
anaerobic rootzone 35
annual bluegrass (*Poa annua*) **3**
annual meadow grass **3**
    see also annual bluegrass
Anthracnose (*Colletotrichum*-incited diseases) 159–164, **159**, *160*, **160**, *161*, *162*
    see also fungicide resistance 289, **289**
Arachis (creeping groundnut) **5**
arsenical herbicides 190
Ascochyta leaf blight **148**
ascospores 133
Athelia rolfsii (southern blight) 204–208, *205*, *206*
Australian pennywort (*Hydrocotyle tripartita*) **5**
Axonopus (carpetgrass) **3**

bacterial and phytoplasma diseases 227–231
    causal agents **228**
    bacterial wilt 228–230, *230*
    bacterial etiolation 229
    bacterial decline 230
    bermudagrass white leaf 230–231
    cyanobacteria and yellow spot 231–234, *232*, *233*

bahiagrass (*Paspalum notatum*) **3**
Bambusoideae **4**
basal leaf blight (*Waitea prodiga*) 201–202
    see also warm-weather *Rhizoctonia* and *Waitea*-incited diseases 195–204
basidiomycete fungal mycelium 95
beneficial microbes 30, 35, 70, *148*, 254, **254**, **262**, 263
beneficial nematodes 245–261
    see also nematodes
bentgrass (creeping bentgrass) **3**
bermudagrass (*Cynodon*) **3**
    see also couchgrass **3**
bermudagrass decline 134, *135*
    see also root decline of warm-season grasses
biofumigation 72, 73
biofungicides 272
biological disease management 70–73
    biofumigation 72, 73
    microbial inoculants 70–72
    non-pathogenic bacterial and fungal genera 70, **71**
    organic amendments/suppressive composts 72
Bipolaris 174–180, *176*, **176**, *178*, **178**
    see also Helminthosporium group diseases 174–180
biostimulants 70
biotroph 30, **31**
black layer 95
blue grama (*Bouteloua gracilis*) 4
blue-green algae (cyanobacteria) 29, 231
Blumeria graminis (Powdery mildew) 82–83, *83*
boomspray *275*, *276*
botanical and animal extracts 263, **263**

301

Bothriochloa (pitted beard grass and red leg grass) **4**
Bouteloua 3
   see also buffalograss 3
brown blight (*Pyrenophora siccans*) **178**
   see also Helminthosporium group diseases 175–176, *175*, **175**
brown patch (*Rhizoctonia solani*) 125, 195–199, *195*, *196*, **195**, *197*, *198*, *199*, *221*
   global distribution 196, **195**
   see also warm-weather *Rhizoctonia* and *Waitea*-incited diseases 195–202
brown ring patch (*Waitea circinata*) 199–201, **199**, *200*, *201*
   see also warm-weather *Rhizoctonia* and *Waitea*-incited diseases 195–202
brown stripe (*Bipolaris heveae*) **178**
   see also Helminthosporium group diseases 175–176, *175*, **175**
Brunswick grass (*Paspalum nicorae*) **4**
Budhanggurabania cynodonticola **132**, *137*, *138*
   see also root decline of warm-season grasses 130–135
   see also Deniliquin patch
buffalograss (*Bouteloua dactyloides* and also St Augustine grass) **3**

Calvin-Benson photosynthetic/C3 pathway 5
Canada bluegrass (*Poa compressa*) **4**
Candidacolonium cynodontis **132**
   see also root decline of warm-season grasses
carpetgrass (*Axonopus*) **3**
Cenchrus clandestinus syn. *Pennisetum clandestinum* (kikuyu) **3**
centipede grass (*Eremochloa ophiuroides*) **3**
Cercospora leaf spot (*Cercospora*) 218
climate change 26
copper spot (*Microdochium sorghi*) 164–166, *165*
chamomile (*Chamaemelum*) **215**
chemicals
   biodegradation 266, 283–284
   compatibility and synergism 278
   fungicide-fungicide 280
   fungicide-plant growth regulator 281
   fungicide-fertiliser 281
   fungicide-nematicide 281
Chloridoideae 3
Chrysochloa orientalis (gingerspike) **4**
clovers **215**
   see also Trifolium
Colletotrichum-incited diseases 159–164
   see also Anthracnose 159–164, **159**, *160*, **160**, *161*, *162*, **289**
cool-season (C3) turfgrasses
   geographic distribution 5, 6, **6**
   species **3**, 4
cool-weather fungal diseases 78–143
cool-weather *Pythium* diseases 83–87
   damping-off 86, *86*
   Pythium crown and root rot 87
   Pythium spring dead spot 87
cool-weather *Rhizoctonia* diseases 127–129
coin spot 166–171
   see also dollar spot
Curvularia-incited diseases 176–177
   Curvularia leaf blight 177
   Ink spot 177
Clarireedia 166–171, **167**
   see also coin spot, dollar spot
clippings removal 61
Colobanthus 216
composts and organic amendments 72, 262, **262**
consumable chemical products 68, **68**
Coprinus snow mould (*Coprinus psychromorbidus*) **144**
   see also snow moulds 143–148
cotula 5, *216*
   see also Leptinella
cottony blight 86
   see also *Pythium*-incited diseases
couchgrass 3
   see also bermudagrass and also Cynodon
cricket wicket soil profile *20*
cream leaf blight (*Limonomyces roseipellis*) 80–82
creeping groundnut (*Arachis prostrata*) **6**
creeping speedwell (*Veronica filiformis*) **5**
crested dogs tail (*Cynosurus cristatus*) **3**
cultural management intensity 19, *21*
cultural management practices
   disease-resistant cultivars 59–60, *60*
   primary
      fertilisation regime 63–65, **64**
      irrigation scheduling 62–63, *63*
      mowing regime 61, 62, *62*, **64**
      rolling scheduling 65, *65*
   sanitation 61
   secondary **65**
      dew removal 65, *277*
      fraze mowing 67
      growth cloths/covers 66
      oscillating fans 66
      portable lighting 66
      sand topdressing 67
      surface-aerification practices 66, **67**
      surface burning 68
      syringing 65
Curvularia leaf blight (*Curvularia*) **178**
cut leaf end *62*, *231*
cyanobacteria 29, 231–234, *232*, **232**
cyanotoxins 233
Cynodon 3
   see also couchgrass, bermudagrass
Cynosurus cristatus (crested dogs tail) **3**

Dactylis glomerata (orchard grass) **3**
damping-off (*Pythium*) 86–87, *86*
    see also *Pythium*-incited diseases 83–87
Deniliquin patch (*Budhanggurabania cynodonticola*) **132**, *138*
    see also root decline of warm-season grasses 130–135
dichondra 5, **5**, *217*
Digitaria **3**, 4
    see also Queensland blue couch, Pangola grass and Richmond grass **3**, 4
disease diagnosis 50–55
    complexes 38
    cycles 36
    economic significance 10–11, *11*
    epidemics 38
    Kochs postulates 55
    management versus control 12
    predictive disease models 43
    pressure 23
    sample collection 51–53, *51*, *52*
    succession 38
    symptomology 45–49, *46*, *47*
Distichlis **3**
dog print 178
    see also Curvularia
dollar spot (*Clarireedia*) 166–171, *166*, *168*, *169*, *170*, **167**, **171**
    fungicide resistance 289, **289**
downy mildew (yellow tuft) 79, *79*
Drechslera (*Pyrenophora*) 176, **178**
    see also Helminthosporium group diseases 175–176
Dropseed (*Sporobulus*) **4**
Durban grass (*Dactyloctenium*) 4, *4*

ecological groups of microorganism
    microflora 29
    microfauna 29
ectotrophic root-infecting fungi (ERI) 132, **132**, *133*
elephant footprint 125, 195
    see also *Rhizoctonia* and *Waitea*-incited diseases
endophyte 30, **31**, 263
environment
    canopy moisture **33**
    changes in crop production 26
    climate change 26
    leaf wetness **34**
    light duration and quality **36**, *37*
    microorganism temperature categorization 32, **32**
    phyllosphere 16
    rhizosphere 17–18
    subalpine *17*
    subtropical *18*
    suppressive soils 35
    thatch-mat region 17

enzyme-linked immunosorbent assay (ELISA) 50
Epichloe 30
epidemiology factors 17, *17*
epidemics/epiphytics 38
Eremochloa (centipede grass) **3**
Erysiphe (*Blumeria*) 82, *83*
eriophyid mites 231
Enterobacter 70, **71**
Experimental grasses 3
Exserohilum 176, **176**
    see also Helminthosporium group diseases 174–176

facultative parasites, saprophytes (hemibiotrophs) 30
fairway patch (*Phialocephala bamuru*) 134–135, *136*
    see also root decline of warm-season grasses 130–135
fairy ring formations 93–101
    fungal genera 93, **94**
    signs and symptoms 93, **95**
    edaphic 95–97, *96*, *100*
    lectophilic 97–98, **97**, *98*
    thatch collapse 98, *99*
    biology of intersecting rings 99, **100**
fescue (*Festuca*) **3**, 4
Festuca **3**, 4
    see also tall fescue, meadow fescue and sheep fescue
forbs, mown **5**, 215–216
    Australian pennywort **5**
    Chamomile **5**
    clover **5**
    cotula **5**
    creeping groundnut **5**
    creeping speedwell **5**
    Dichondra **5**
    fungal diseases 215–225
    Lippia **5**
    Pinto peanut **5**
    starweed **5**
frog-eye *112*
frost 34
fumigants 278
fungi 29–30
    cardinal temperatures 32, **33**
    foliage-infecting species 78–91, 158–182
    stem- and root-infecting species 92–143, 184–208
fungicides and biofungicides
    application and placement 274–278, *276*, *277*
    biochemical mode of action 273, **273**
    colour sensitive test strips 282, *282*
    environmental behaviour and fate 282–287, *283*, *284*, **285**
    formulations 272
    group name/FRAC categories 273

fungicides and biofungicides (*continued*)
    minimising non-target impacts 285–287, **285**, **286**
    mixing and compatibility 278–280, *279*
    multisite/site-specific 273, **274**
    nozzle selection 281
    phytomobility 274, **275**
    preventive vs curative strategy 274
    resistance/case studies 287–290, **287**, **288**, **289**, *290*
    resistance management 290
    spray water quality and volume 281
Fusarium 107

Gaeumannomyces- incited diseases
    take-all patch of cool-season grasses 101–107, *102*, **102**, *103*, *104*
    see also root decline of warm-season grasses 130–135
gingerspike (*Chrysochloa orientalis*) 4
gold bracelet disease (*Rhizoctonia* sp.) 218–220, *219*, *220*
Globisporangium 83–84, **84**
    see also Pythium-incited diseases
grasses 2–5, **3**, **4**
grey leaf spot (*Pyricularia oryzae*) 171–174, **172**, *173*
    fungicide resistance 289–290
guttation fluid 34, *34*

hard fescue (*Festuca longifolia*) **4**
Hatch-Slack photosynthetic/C3 pathway 5
Helminthosporium-group diseases 175–179, *175*, **175**
    Bipolaris 176
    Curvularia 176, *177*
    Exserolium 176
    Pyrenophora (*Drechslera*) 176
hemibiotroph 30, **31**
Hilaria 3
host defence mechanisms 21–22, **21**
hydrophobicity (water repellence) 35, *36*
hypersensitive reaction (HR) 23

induced systemic reaction (ISR) 23
ink spot 177
integrated management systems
    international committee on taxonomy of viruses (ICTV) 238
    plant-parasitic nematodes 259–266
    Rhizoctonia and *Waitea*-incited diseases 204
    root decline of warm-season grasses **142**
    turfgrass disease management plan 73
irrigation scheduling 62–63, *63*, **64**

Java lawn grass (*Polytrias indica*) **4**

Kentucky bluegrass (*Poa pratensis*) 3
kikuyu grass (*Cenchrus clandestinus*) 3
kikuyu yellows (*Verrucalvus flavofaciens*) 184–185, *184*, 185
Koeleria 3

Labyrinthula (rapid blight) 235–236
Laetisaria fuciformis (red thread) 88–90, *89*
large patch (*Rhizoctonia solani* AG-2-2LP) 127–129, *126*, *127*, *129*
    see also cool-weather *Rhizoctonia* and *Waitea*-incited diseases 125–130
leaf blight (*Pyrenophora catenaria*) **178**
    see also Helminthosporium-group diseases 175–179
leaf blotch (*Pyrenophora cynodontis*) **178**
leaf and sheath rot (*Waitea oryzae*, W. *prodiga* and W. *zeae*) 202
    see also warm weather *Rhizoctonia* and *Waitea*-incited diseases 195–204
leaf spot (*Bipolaris sorokiniana*) **178**
    see also Helminthosporium-group diseases 175–179
leaf wetness 34
Leptosphaerulina leaf blight (*Leptosphaerulina*) 180–181, *180*, **181**
lesser-known, cool-weather fungal diseases 148–149, **148**
lesser-known, warm-weather fungal diseases **183**
Limonomyces roseipellis (pink patch) 80–82, *81*
Lippia (*Phyla nodifolia*) 5
Lolium (perennial ryegrass) 3

Magnaporthiopsis-incited diseases
    summer patch 187–191, **188**, **189**, *189*
    see also root decline of warm-season grasses 130–135
marine net slime mould 235
    see also rapid blight
meadow fescue (*Festuca pratensis*) **4**
melting-out 175–176
    see also Pyrenophora
microbiome 6
microbial antagonists 71, **71**, *143*, **148**, **254**, **263**
Microdochium patch (*Monographella nivalis*) 107–114, *107*, *108*, **109**, *109*, *110*, *111*, *112*, *113*
    fungicide resistance **289**, 290
    see also Microdochium bolleyi, *M. paspali* and *M. poae* 108
Microlaena stipoides (weeping grass) **4**
moisture 62, **64**

mycorrhizae 30, 263
monoculture 7, **7**
Monographella nivalis (Microdochium patch) 107–114
mowing regime 61–63, *62*
mycelium *62, 83, 88, 109,* 110, *112, 169, 182*

necrotic ring spot (*Ophiosphaerella korrae*) 121–122, **121**, *121, 122*
    see also Ophiosphaerella-incited diseases 114–124
necrotroph 30, **31**
new encounter pathogens and diseases 9–10, *9*, **10**
nematicides and bionematicides 263–266, **263**, *264*, **265**, **266**
nematodes 245–261
    significance 245–246
    feeding mechanisms 246–247, *247*
    genera and feeding groups 247–252, **248**
    economically important 248–252, **249**, *249, 250, 251*
    basic biology 252–254, *252*
    population dynamics 252–254
    damage thresholds and relative pathogenicity 254–256, *255*, **256**
    microbial antagonists 254, **254**
    relative pathogenicity 255–256, *255, 256*
    signs and damage symptoms 256–258, *246, 253, 257, 258, 259, 264*
    confirmation 258–259
    systems management approach 259–266, *261*, **262**, **265**, **266**
net blotch (*Pyrenophora dictyoides*) **178**
    see also Helminthosporium-group diseases 175–179
Nigrospora leaf blight (*Nigrospora sphaerica*) 182–183, *182*
nutrition 35, 63–65, **64**, **65**, 66

obligate parasites (biotrophs) 30
oomycetes 83
Ophiobolus patch 101–106
    see also take-all of cool-season grasses
Ophiosphaerella-incited diseases
    dead spot 123–125, **123**
    necrotic ring spot 121–122, **121**, *121, 122*
    spring dead spot 114–121, **114**, *115*, **116**, *117, 118, 119*
orchard grass (*Dactylis glomerata*) **3**

Pangola grass (*Digitaria decumbens*) **4**
Panicoideae **3**
Panicum repens (Torpedo grass) **4**
parasite 30

Paspalum **3**, 4
    see also bahia grass, seashore paspalum, sour grass and Brunswick grass
pathogen 30
    pathogen attack mechanisms 23–25
    host disruption 25–26
pathosystem 6–7
pembagrass (*Stenotaphrum dimidiatum*) **4**
Penicillifer martini 134–135, 136
    see also fairway patch
    see also root decline of warm-season grasses **132**, *140*
perennial ryegrass (*Lolium perenne*) **3**
pesticide registration and labelling 271
Phialocephala bamuru (fairway patch) 134–135, 136
    see also root decline of warm-season grasses **132**, *140*
Phleum pratense (Timothy grass) **4**
phytoplasmas 228
Physopella 90–91
    see also rust and smut diseases
pitted beard grass (*Bothriochloa pertusa*) **4**
plant growth-promoting rhizobacteria 31, 263, **263**
plant-growth regulators (PGR's) 68, **69**
plant-parasitic nematodes 245–261
    see also nematodes
plant and animal residues **262**, **263**
Poaceae 1
polyculture 7, **7**
Polytrias indica (Java grass) **4**
Pooideae **3**
Powdery mildew (*Blumeria graminis*) 82–83, *83*
Pyrenophora 175–176, **178**
    see also Helminthosporium group diseases
Pyricularia oryzae (grey leaf spot) 171–174
Pythium-incited diseases
    Cool-weather *Pythium* diseases 86–87, *85, 255*
    damping-off 86, *86*
    fungicide resistance 289, **290**
    Pythium blight 191–194, *191*, 193
    Pythium crown and root rot 87
    Pythium root dysfunction 194
    Pythium spring dead spot 87
    Warm-weather *Pythium* diseases 191–194
Phytophthora root rot (*Phytophthora cryptogea*) 220–221
primitive foliar and root-infecting diseases 234–237
    Polymyxa root rot (*Polymyxa graminis*) 235, *235*
    rapid blight (*Labyrinthula terrestris*) 235
    slime moulds 237, *238*
Puccinella (alkali grass) **4**
Puccinia graminis (rust) **90**, *91*

Queensland blue couch (*Digitaria didactyla*) **3**

red leaf spot (*Pyrenophora gigantea*) **178**
    see also Helminthosporium-group diseases 175–179
red leg grass (*Bothriochloa macra*) **4**
red thread (*Laetisaria fuciformis*) 88–90, *89*
redtop bentgrass (*Agrostis alba*) **4**
Rhizoctonia and *Waitea*-incited diseases 125–129, 195–202
    basal leaf blight 201–202
    brown patch 125, 195–199, *195*, *196*, **195**, *197*, *198*, 199, *221*
    brown ring patch 199–201, **199**, *200*, *201*
    large patch 127–129, *126*, *127*, *129*
    leaf and sheath rot 202
    yellow patch 129–130, *129*, **129**
Richmond grass (*Digitaria diversinervis*) **4**
Rolfs disease 204–208, *205*, *206*
    see also southern blight
rolling scheduling 65, *65*
root decline of warm-season grasses 130–135
    bermudagrass decline 134, *135*
    causal agents characterisation 132–133, **132**
    fairway patch 134–135, *136*
    global distribution 133–134
    lesser known root decline diseases of warm-season grasses 135
    summer decline 135, *137*
    symptomology 138–139, *139*
root-feeding nematodes 245–261
    see also nematodes
root system architecture 135, *139*, 260, 253, 255, 256, *258*, *258*, *259*, *261*
rusts (*Puccinia, Physopella, Uromyces*) 90–91, *91*

sampling 51, *52*, *53*, *261*
sand topdressing 66–67
saprophytes 30
Sclerophthora macrospora (Downy mildew) 78, *79*
sclerotia *146*, *222*
Sclerotinia minor (Sclerotinia patch) 222–224, *223*, *224*
Sclerotinia patch (*Sclerotinia minor*) 222–224, *223*, *224*
seashore paspalum (*Paspalum vaginatum*) **3**
senectopathic 163
sheep fescue (*Festuca ovina*) **4**
slime moulds (*Physarium*) 237, *238*
smoke ring 191, 197
smuts (*Entyloma, Tilletia, Urocystis, Ustilago*) 91–92, *92*
snow molds 143–148
    coprinus snow mould *144*
    global distribution 144–145, **145**
    grey snow mould *144*
    pink snow mold *144*, *145*, *146*
    Sclerotinia snow mould *144*
    snow root rot *144*
    Typhula blight *144*, 145

soil wetting agents 68
sour grass (*Paspalum conjugatum*) **4**
southern blight 204–208, *205*, *206*
    see also Rolfs disease
St. Augustinegrass **3**
starweed (*Plantago triandra*) **5**, 216
Stenotaphrum (St. Augustinegrass and Pembagrass) **3**, **4**
summer decline (*Wongia griffinii*) 135, *137*, **137**
    see also root decline of warm-season grasses 130–135
summer patch (*Magnaporthiopsis poae*) 187–191, **188**, **189**, *190*
surfactants 272
symptomatology 45–49, *46*, *47*, **179**, **204**, 256–257
systemic acquired resistance (SAR) 23

take-all patch of cool-season grasses (*Gaeumannomyces avenae*, *G. tritici*) 101–106, **101**, *101*, *103*, *104*
take-all root rot 130–135
    see also root decline of warm-season grasses
tall fescue (*Festuca arundinacea*) **3**
tank mixtures 278–280, *279*
thatch-mat management 66–68, *67*, *68*
timothy grass (*Phleum pratense*) **4**
torpedo grass (*Panicum repens*) **4**
turfgrasses (cool-season and warm-season) **3**, **4**
    hosts 18–21
    plant disease resistance 22, 59, *60*
    resistant varieties 83, 90, 92, 106, 114, 121, 125, 129, 147, 171, 174, 180, 191, 199, 237, 261
    stress 20

Uromyces (rust and smut diseases) 89–92, **90**
Ustilago (rust and smut diseases) 89–92, **90**

vectors 227
velvet bentgrass (*Agrostis canina*) **4**
Verrucalvus flavofaciens (kikuyu yellows) 184–185
Viruses and viral diseases 237–241, *238*, *239*, **239**, *240*
    centipede grass mosaic disease 240
    mosaic disease of Queensland blue couch 240
    mosaic disease of St. Augustinegrass 240
    St. Augustine decline 239
    Zoysia dwarf virus/mosaic virus 241

Waitea circinata (brown ring patch) 199–201, **199**, *200*, *201*

Waitea patch (brown ring patch) 199–201, **199**, *200, 201*
Waitea prodiga (basal leaf blight) 201–202
    see also warm-weather *Rhizoctonia* and *Waitea* diseases 195–202
warm-season (C4) turfgrasses 3–4
    distribution 4
    species **3, 4**
warm-weather *Rhizoctonia* and *Waitea* diseases 195–202
water repellence (hydrophobicity) 35, *36*
weeping grass (*Microlaena stipoides*) **4**
white patch disease 225
winter Pythium patch 224, *225*
Wongoonoo patch (*Gaeumannomyces wongoonoo*) **132, 137**, *139*
    see also root decline of warm-season grasses 130–135

Wongia garrettii and W. *griffini* **132**, *133*, 137
    see also root decline of warm-season grasses 130–135

Xanthomonas (bacterial wilt) 228

yellow patch (*Rhizoctonia cerealis*) 129–130, *129*, **129**
    see also cool-weather *Rhizoctonia* diseases 125–127
yellow spot (cyanobacteria) 231–234
yellow tuft (*Sclerophthora macrospora*) 79, *79*

zonate leaf spot **178**
zoysiagrass (*Zoysia* spp.) **3**

CABI – who we are and what we do

This book is published by **CABI**, an international not-for-profit organisation that improves people's lives worldwide by providing information and applying scientific expertise to solve problems in agriculture and the environment.

CABI is also a global publisher producing key scientific publications, including world renowned databases, as well as compendia, books, ebooks and full text electronic resources. We publish content in a wide range of subject areas including: agriculture and crop science / animal and veterinary sciences / ecology and conservation / environmental science / horticulture and plant sciences / human health, food science and nutrition / international development / leisure and tourism.

The profits from CABI's publishing activities enable us to work with farming communities around the world, supporting them as they battle with poor soil, invasive species and pests and diseases, to improve their livelihoods and help provide food for an ever growing population.

CABI is an international intergovernmental organisation, and we gratefully acknowledge the core financial support from our member countries (and lead agencies) including:

Discover more

To read more about CABI's work, please visit: **www.cabi.org**

Browse our books or explore our online products at
https://www.cabidigitallibrary.org

Interested in writing for CABI? Find our author guidelines here:
www.cabi.org/publishing-products/information-for-authors/

Printed and bound by CPI Group (UK) Ltd, Croydon, CR0 4YY
08/10/2024

14570392-0001